Representations of Algebras, Geometry and Physics

CONTEMPORARY MATHEMATICS

769

Representations of Algebras, Geometry and Physics

Maurice Auslander Distinguished Lectures
and International Conference
April 25–30, 2018
Woods Hole Oceanographic Institute
Woods Hole, MA

Kiyoshi Igusa
Alex Martsinkovsky
Gordana Todorov
Editors

AMS AMERICAN MATHEMATICAL SOCIETY

2020 *Mathematics Subject Classification.* Primary 13C15, 14A20, 16G20, 18D50, 18E10, 18E15, 18E30, 18F99, 18G55, 20F55.

For additional information and updates on this book, visit
www.ams.org/bookpages/conm-769

Library of Congress Cataloging-in-Publication Data

Names: Maurice Auslander Distinguished Lectures and International Conference (2018 : Woods Hole, Mass.), issuing body. | Martsinkovsky, A. (Alex), editor. | Igusa, Kiyoshi, 1949- editor. | Todorov, G. (Gordana), editor.

Title: Representations of algebras, geometry and physics : Maurice Auslander Distinguished Lectures and International Conference, April 25 - 30, 2018, Woods Hole Oceanographic Institute, Woods Hole, MA / Alex Martsinkovsky, Kiyoshi Igusa, Gordana Todorov, editors.

Description: Providence, Rhode Island : American Mathematical Society, [2021] | Series: Contemporary mathematics, 0271-4132 ; volume 769 | Includes bibliographical references.

Identifiers: LCCN 2020043192 | ISBN 9781470452308 (paperback) | 9781470464257 (ebook)

Subjects: LCSH: Representations of algebras–Congresses. | Geometry, Algebraic–Congresses. | Commutative algebra–Congresses. | AMS: Commutative algebra – Theory of modules and ideals – Dimension theory, depth, related rings (catenary, etc.). | Algebraic geometry – Foundations – Generalizations (algebraic spaces, stacks). | Associative rings and algebras – Representation theory of rings and algebras – Representations of quivers and partially ordered sets. | Category theory; homological algebra | Group theory and generalizations – Special aspects of infinite or finite groups – Reflection and Coxeter groups.

Classification: LCC QA176 .M385 2018 | DDC 512/.22–dc23

LC record available at https://lccn.loc.gov/2020043192

∞ The paper used in this book is acid-free and falls within the guidelines
established to ensure permanence and durability.
Visit the AMS home page at `https://www.ams.org/`

10 9 8 7 6 5 4 3 2 1 26 25 24 23 22 21

Contents

Preface

This volume of expository lectures is an outgrowth of the annual Maurice Auslander Distinguished Lectures and International Conference that took place April 25 - 30, 2018 at the Woods Hole Oceanographic Institute in Woods Hole, MA.

The continued – and gratefully acknowledged – support from the National Science Foundation and Bernice Auslander results in a combination of established experts and a large number of graduate students and young researchers attending the conference. Their interest and unabating enthusiasm is *raison d'être* for this series of expository works (for the previous collections, see Contemporary Mathematics, vols. 607, 673, and 716).

The selected expository lectures in this volume cover a rather wide area of research. Karin Baur and Charlie Beil introduce a new class of quiver algebras called geodesic ghor algebras which, unlike dimer algebras, exhibit good behavior on surfaces of higher genus. Ralph Kaufmann's detailed paper deals with algebraic aspects of Feynman categories, a subject that has found applications in algebra, category theory, geometry, number theory and physics. An intriguing connection between representations of quivers and Coxeter groups is discussed in Mark Kleiner's contribution. Amnon Neeman surveys the relatively new concept of approximable triangulated categories that leads to new powerful results. The question of when the heart of a t-structure is a Grothendieck category is dealt with by Carlos E. Parra and Manuel Saorín. Sebastian Posur introduces the reader to methods of constructive category theory by describing two computational examples. One deals with natural transformations between finitely presented functors and the other – with the differentials on the pages of a spectral sequence.

As we mentioned before, the primary audience for this volume are graduate students and young researchers working in and around representation theory. We believe, however, that this multifaceted volume may be of interest to a much wider audience.

The Editors

Contemporary Mathematics
Volume **769**, 2021
https://doi.org/10.1090/conm/769/15414

Examples of geodesic ghor algebras on hyperbolic surfaces

Karin Baur and Charlie Beil

Abstract. Cancellative dimer algebras on a torus have many nice algebraic and homological properties. However, these nice properties disappear for dimer algebras on higher genus surfaces. We consider a new class of quiver algebras on surfaces, called 'geodesic ghor algebras', that reduce to cancellative dimer algebras on a torus, yet continue to have nice properties on higher genus surfaces. These algebras exhibit a rich interplay between their central geometry and the topology of the surface. We show that (nontrivial) geodesic ghor algebras do in fact exist, and give explicit descriptions of their central geometry. This article serves a companion to the article 'A generalization of cancellative dimer aglebras to hyperbolic surfaces', where the main statement is proven.

1. Introduction

Cancellative dimer algebras on a torus form a prominent class of noncommutative crepant resolutions and Calabi-Yau algebras, e.g., [**Br**, **D**, **B2**]. In particular, they are homologically homogeneous endomorphism rings of modules over their centers, and their centers are 3-dimensional toric Gorenstein coordinate rings. These interesting properties vanish, however, once dimer algebras are placed on higher genus surfaces. For example, the center of a dimer algebra on a surface of genus $g \geq 2$ is simply the polynomial ring in one variable, and so there are no interesting interactions between the topology of the surface and the algebras central geometry.

In this article we consider special quotients of dimer algebras, called 'ghor algebras', whose central geometries are, in contrast, closely related to the topology of the surface. A ghor algebra is a quiver algebra whose quiver embeds in a surface, and with relations determined by the perfect matchings of its quiver (the precise definition is given in Section 2). Ghor algebras were introduced in [**B1**, **B4**] to study nonnoetherian dimer algebras on a torus.[1] In Section 2 we introduce a special property that certain ghor algebras possess, called 'geodesic'. On a torus, a

2020 *Mathematics Subject Classification.* Primary 13C15, 14A20.

Key words and phrases. Dimer algebra, hyperbolic surface, non-noetherian ring, noncommutative algebraic geometry.

The authors were supported by the Austrian Science Fund (FWF) grant P 30549-N26. The first author was also supported by the Austrian Science Fund (FWF) grant W 1230 and by a Royal Society Wolfson Fellowship.

[1] Ghor algebras were originally called 'homotopy algebras' (e.g., in [**B4**]) because their relations are homotopy relations on the paths in the quiver when the surface is a torus. However, in the higher genus case their relations also identify homologous cycles, and therefore the name 'homotopy' is less suitable. The word 'ghor' is Klingon for surface.

ghor algebra is geodesic if and only if it is a cancellative dimer algebra (if and only if it is a noncommutative crepant resolution, if and only if it is noetherian [**D**, **B2**]). On higher genus surfaces, geodesic ghor algebras remain endomorphism rings of modules over their centers, but new features arise. The purpose of this article is to present explicit examples of ghor algebras that exhibit these new features.

In each example, we will describe the 'cycle algebra', which is a commutative algebra formed from the cycles in the quiver. The cycle algebra and center coincide for geodesic ghor algebras on a torus, but may differ on higher genus surfaces. However, the two commutative algebras remain closely related: if the center is nonnoetherian, then the cycle algebra is a depiction of the center. The cycle algebra thus plays a fundamental role in describing the geometry of the center when the center is nonnoetherian.

We will also show that polynomial rings in at least three variables are geodesic ghor algebras. The results stated in Section 2 are proven in the companion article [**BB**].

2. Ghor algebras: Background and main results

NOTATION 2.1. Throughout, k is an uncountable algebraically closed field. We denote by $\operatorname{Spec} S$ and $\operatorname{Max} S$ the prime ideal spectrum (or scheme) and maximal ideal spectrum (or affine variety) of S, respectively. Given a quiver Q, we denote by kQ the path algebra of Q; by Q_ℓ the paths of length ℓ; by $\mathrm{t}, \mathrm{h} : Q_1 \to Q_0$ the tail and head maps; and by e_i the idempotent at vertex $i \in Q_0$. By *cyclic subpath* of a path p, we mean a subpath of p that is a nontrivial cycle.

In this article will consider surfaces Σ that are obtained from a regular $2N$-gon P, $N \geq 2$, by identifying the opposite sides of P. This class of surfaces includes all smooth orientable compact closed connected genus $g \geq 1$ surfaces. Specifically,

- if P is a $4g$-gon, then Σ is a smooth genus g surface; and
- if P is a $2(2g+1)$-gon, then Σ is a genus g surface with a pinched point.

The polygon P is then a fundamental polygon for Σ.

If $N = 2$, then Σ is a torus, and the covering space of Σ is the plane $\mathbb{R}^2$. For $N \geq 3$, the covering space of Σ is the hyperbolic plane $\mathbb{H}^2$. The hyperbolic plane may be represented by the interior of the unit disc in $\mathbb{R}^2$, where straight lines in $\mathbb{H}^2$ are segments of circles that meet the boundary of the disc orthogonally. In the covering, the hyperbolic plane is tiled with regular $2N$-gons, with $2N$ such polygons meeting at each vertex. In this case, Σ is said to be a hyperbolic surface.

DEFINITION 2.2.

- A *dimer quiver* on Σ is a quiver Q whose underlying graph $\overline{Q}$ embeds in Σ, such that each connected component of $\Sigma \setminus \overline{Q}$ is simply connected and bounded by an oriented cycle, called a *unit cycle*.
  - A *perfect matching* of a dimer quiver Q is a set of arrows $x \subset Q_1$ such that each unit cycle contains precisely one arrow in x. Throughout, we assume that each arrow is contained in at least one perfect matching.
  - A perfect matching x is called *simple* if $Q \setminus x$ contains a cycle that passes through each vertex of Q (equivalently, $Q \setminus x$ supports a simple kQ-module of dimension vector $(1, 1, \ldots, 1)$).

We denote by $\mathcal{P}$ and $\mathcal{S}$ the set of perfect and simple matchings of Q, respectively. We will consider the polynomial rings generated by these matchings, $k[\mathcal{P}]$ and $k[\mathcal{S}]$.

• Denote by $e_{ij} \in M_n(k)$ the $n \times n$ matrix with a 1 in the ij-th slot and zeros elsewhere. Consider the two algebra homomorphisms

$$\eta : kQ \to M_{|Q_0|}(k[\mathcal{P}]) \qquad \text{and} \qquad \tau : kQ \to M_{|Q_0|}(k[\mathcal{S}])$$

defined on the vertices $i \in Q_0$ and arrows $a \in Q_1$ by

$$\eta(e_i) = e_{ii}, \qquad \eta(a) = e_{\mathrm{h}(a),\mathrm{t}(a)} \prod_{\substack{x\in\mathcal{P}:\\ x\ni a}} x,$$
$$\tau(e_i) = e_{ii}, \qquad \tau(a) = e_{\mathrm{h}(a),\mathrm{t}(a)} \prod_{\substack{x\in\mathcal{S}:\\ x\ni a}} x,$$

and extended multiplicatively and k-linearly to kQ. We call the quotient

$$A := kQ/\ker\eta$$

the *ghor algebra* of Q.

• The *dimer algebra* of Q is the quotient of kQ by the ideal

$$I = \langle p - q \mid \exists a \in Q_1 \text{ such that } pa, qa \text{ are unit cycles}\rangle \subset kQ,$$

where p, q are paths.

A ghor algebra $A = kQ/\ker\eta$ is the quotient of the dimer algebra kQ/I since $I \subseteq \ker\eta$: if pa, qa are unit cycles with $a \in Q_1$, then

$$\eta(p) = e_{\mathrm{h}(p),\mathrm{t}(p)} \prod_{\substack{x\in\mathcal{P}:\\ x\not\ni a}} x = \eta(q).$$

Dimer algebras on non-torus surfaces have been considered in the context of, for example, cluster categories [**BKM, K**], Belyi maps [**BGH**], and gauge theories [**FGU, FH**].

NOTATION 2.3. Let $\pi : \Sigma^+ \to \Sigma$ be the projection from the covering space Σ^+ (here, $\mathbb{R}^2$ or $\mathbb{H}^2$) to the surface Σ. Denote by $Q^+ := \pi^{-1}(Q) \subset \Sigma^+$ the (infinite) covering quiver of Q.

We introduce the following special class of ghor algebras that generalizes cancellative dimer algebras on a torus.

DEFINITION 2.4. Let $A = kQ/\ker\eta$ be a ghor algebra and P a fundamental polygon of Σ. A cycle $p \in A$ is *geodesic* if the lift to Q^+ of each cyclic permutation of each representative of p does not have a cyclic subpath. The algebra A is *geodesic* if for each $i \in Q_0$ and side j of the polygon P, there is a geodesic cycle at i that intersects side j transversely before intersecting any other sides.

PROPOSITION 2.5. [**BB**, Corollary 3.14] *If a ghor algebra A is geodesic, then*

$$A := kQ/\ker\eta \cong kQ/\ker\tau.$$

In particular, it suffices to only consider the simple matchings of Q to determine the relations of A.

The algebra homomorphisms η and τ on kQ induce algebra homomorphisms on A,

$$\eta : A \to M_{|Q_0|}(k[\mathcal{P}]) \qquad \text{and} \qquad \tau : A \to M_{|Q_0|}(k[\mathcal{S}]).$$

For $i, j \in Q_0$, consider the k-linear maps

$$\bar{\eta} : e_jAe_i \to k[\mathcal{P}] \qquad \text{and} \qquad \bar{\tau} : e_jAe_i \to k[\mathcal{S}]$$

defined by sending $p \in e_jAe_i$ to the single nonzero matrix entry of $\eta(p)$ and $\tau(p)$ respectively; that is,

$$\eta(p) = \bar{\eta}(p)e_{ji} \qquad \text{and} \qquad \tau(p) = \bar{\tau}(p)e_{ji}.$$

Then

$$\bar{\eta}(p)|_{\substack{x=1:\\ x \notin \mathcal{S}}} = \bar{\tau}(p).$$

These maps are multiplicative on the paths of Q. Moreover, for $a \in Q_1$, we have $a \in x$ if and only if $x \mid \bar{\eta}(a)$.

By the definition of A, if two paths $p, q \in e_jAe_i$ satisfy $\bar{\eta}(p) = \bar{\eta}(q)$, then $p = q$. Furthermore, if A is geodesic, then $\bar{\tau}(p) = \bar{\tau}(q)$ implies $p = q$, by Proposition 2.5. An important monomial is the $\bar{\tau}$-image of each unit cycle in Q, namely

$$\sigma := \prod_{x \in \mathcal{S}} x.$$

The following lemma will be useful in the next section.

LEMMA 2.6. *If $p \in A$ is a cycle satisfying $\sigma \nmid \bar{\tau}(p)$, then p is a geodesic cycle.*

PROOF. If p is not a geodesic cycle, then there is a cyclic permutation of a lift of a representative of p with a cyclic subpath q in Q^+. Since q is a cycle in Q^+, we have $\bar{\tau}(q) = \sigma^\ell$ for some $\ell \geq 1$ by [**BB**, Lemma 3.1.i]. □

REMARK 2.7. Let A be a geodesic ghor algebra on a surface Σ, and fix $i \in Q_0$. If Σ is a torus, then the geodesic assumption implies that for each $j, k \in \pi^{-1}(i)$, there is a path p^+ from j to k in Q^+ such that $\sigma \nmid \bar{\tau}(p)$ [**B1**, Proposition 4.20.iii]. However, if Σ is hyperbolic, then this implication no longer holds.

Indeed, suppose Σ is hyperbolic, that is, $N \geq 3$. Fix unit vectors $u_1, u_2 \in H_1(\Sigma, \mathbb{Z}) = \mathbb{Z}^N$ for which $u_1 \neq \pm u_2$. Let p_1, p_2, q_1, q_2 be paths in kQ^+e_j that are lifts of cycles with homology classes

$$[\pi(p_1)] = -[\pi(p_2)] = u_1 \qquad \text{and} \qquad [\pi(q_1)] = -[\pi(q_2)] = u_2.$$

Consider the paths

$$s := q_2p_2q_1p_1 \qquad \text{and} \qquad t := q_2q_1p_2p_1$$

in Q^+. Then $\bar{\tau}(s) = \bar{\tau}(t)$, and so $\pi(s) = \pi(t)$ in A. Observe that t is a cycle in Q^+, whereas s is not. Furthermore, the $\bar{\tau}$-image of any cycle in Q^+ is a power of σ [**BB**, Lemma 3.1.i]. Thus, although s is not a cycle, we have $\bar{\tau}(s) = \bar{\tau}(t) = \sigma^\ell$ for some $\ell \geq 1$. It therefore follows that the monomial of each path in Q^+ from $\mathrm{t}(s)$ to $\mathrm{h}(s)$ is a power of σ [**BB**, Lemma 3.1.ii], a feature that never occurs if Σ is flat.

If Σ is a torus, then a ghor algebra A is geodesic if and only if it is noetherian, if and only if its center R is noetherian, if and only if A is a finitely generated R-module [**B2**, Theorem 1.1]. If Σ is hyperbolic, then only one direction of the implication survives:

PROPOSITION 2.8. [**BB**, Section 4.1] *If the center R of a ghor algebra A is noetherian, then*

(1) *A is geodesic;*
(2) *A is noetherian; and*
(3) *A is a finitely generated R-module.*

In contrast to the torus case, the centers of geodesic ghor algebras on hyperbolic surfaces are almost always nonnoetherian.[2] We can nevertheless view such a center as the coordinate ring on a geometric space, using the framework of nonnoetherian geometry introduced in [**B3**] (see also [**B5**]). In short, the geometry of a nonnoetherian coordinate ring of finite Krull dimension looks just like a finite type algebraic variety, except that it has some positive dimensional closed points.

DEFINITION 2.9. A *depiction* of an integral domain k-algebra R is a finitely generated overring S such that the morphism

$$\operatorname{Spec} S \to \operatorname{Spec} R, \quad \mathfrak{q} \mapsto \mathfrak{q} \cap R,$$

is surjective, and

$$U_{S/R} := \{\mathfrak{n} \in \operatorname{Max} S \mid R_{\mathfrak{n}\cap R} = S_{\mathfrak{n}}\} = \{\mathfrak{n} \in \operatorname{Max} S \mid R_{\mathfrak{n}\cap R} \text{ is noetherian}\} \neq \emptyset.$$

For example, the algebra $S = k[x, y]$ is a depiction of its nonnoetherian subalgebra $R = k + xS$. We thus view $\operatorname{Max} R$ as the variety $\operatorname{Max} S = \mathbb{A}^2_k$, except that the line $\{x = 0\}$ is identified as a 1-dimensional (closed) point of $\operatorname{Max} R$. In particular, the complement $\{x \neq 0\} \subset \mathbb{A}^2_k$ is the 'noetherian locus' $U_{S/(k+xS)}$ [**B3**, Proposition 2.8].

The following is the main theorem of the companion article [**BB**].

THEOREM 2.10. *Suppose $A = kQ/\ker \eta$ is a geodesic ghor algebra on a surface Σ obtained from a regular $2N$-gon P by identifying the opposite sides of P. Set*

$$R = k\left[\cap_{i\in Q_0}\bar{\tau}(e_iAe_i)\right] \qquad \text{and} \qquad S = k\left[\cup_{i\in Q_0}\bar{\tau}(e_iAe_i)\right];$$

then R is isomorphic to the center of A. Furthermore, the following holds.

(1) *If there is a cycle p such that $\bar{p}^n \notin R$ for each $n \geq 1$, then A and R are nonnoetherian. In this case, R is depicted by the cycle algebra S.*
(2) *The center R and cycle algebra S have Krull dimension*

$$\dim R = \dim S = N + 1.$$

In particular, if Σ is a smooth genus $g \geq 0$ surface, then

$$\dim R = \operatorname{rank} H_1(\Sigma) + 1 = 2g + 1.$$

(3) *A is an endomorphism ring of a module over its center: for each $i \in Q_0$, we have*

$$A \cong \operatorname{End}_R(Ae_i).$$

3. Examples

In the following, we consider explicit examples of geodesic ghor algebras. Recall that Σ is a closed surface obtained from a regular $2N$-gon P by identifying the opposite sides of P (and thus necessarily identifying all the vertices of P). Set $[m] := \{1, \ldots, m\}$. For a path p, set $\bar{p} := \bar{\tau}(p)$.

[2]It is possible that there is only a finite number of geodesic ghor algebras that are noetherian for each genus $g \geq 2$.

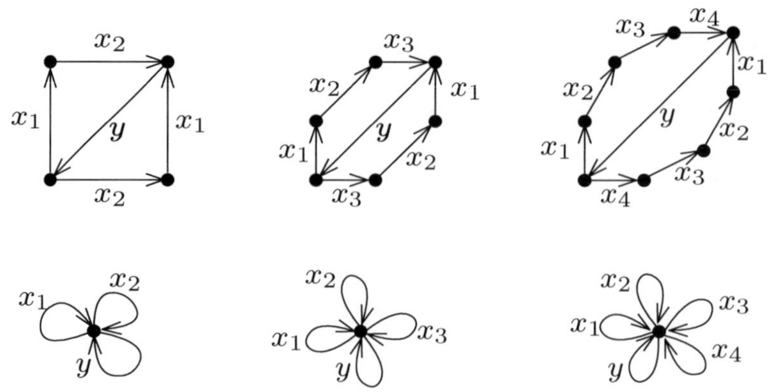

FIGURE 1. The polynomial ghor algebras $k[x_1, x_2, y]$, $k[x_1, x_2, x_3, y]$, and $k[x_1, x_2, x_3, x_4, y]$.

3.1. Polynomial rings are geodesic ghor algebras. The simplest possible geodesic ghor algebra on a $2N$-gon P has one arrow along each side, and a single diagonal arrow in the interior of P. The cases $N = 2, 3, 4$ are shown in Figure 1.

In such a quiver, each arrow is contained in a unique simple matching. Thus there is one simple matching for each arrow. Since each arrow is a cycle, the center, cycle algebra, and ghor algebra all coincide. Consequently, the ghor algebra is the polynomial ring in $N+1$ variables. It follows that every polynomial ring in at least 3 variables arises as a ghor algebra.

3.2. A geodesic ghor algebra on a genus 2 surface. Consider the ghor algebra $A = kQ/\ker\eta$ on a smooth genus 2 surface with quiver Q given in Figure 2.i (the identifications of the sides of the polygon P are indicated by color). We will determine the center R and cycle algebra S of A explicitly, and show that R is nonnoetherian of Krull dimension 5.

Q has twelve simple matchings, shown in Figure 3. The nontrivial simple matchings y_j and z_j may be determined using special partitions of Q_1 called subdivisions, which are described in [**BB**, Section 3]. The polynomial ring $k[\mathcal{S}]$ is thus

$$k[\mathcal{S}] = k[x_1, \ldots, x_4, y_1, \ldots, y_4, z_1, \ldots, z_4].$$

Note that x_{j+1} is the rotation of x_j by $\frac{\pi}{2}$ in the counter-clockwise direction, and similarly for the simple matchings y_j and z_j. Using the four x_j simple matchings and Lemma 2.6, it is straightforward to verify that A is geodesic.

Consider the eight cycles $\alpha_1, \ldots, \alpha_8$ that run along the sides of the fundamental polygon P, as shown in Figure 2.ii. These cycles have $\bar{\tau}$-images

$$\bar{\alpha}_j = \begin{cases} x_{j+1}x_{j+2}^2x_{j+3}y_jy_{j+2}y_{j+3}^2z_jz_{j+1}^2z_{j+2} & \text{if } j \text{ is odd} \\ (x_{j+2}x_{j+3}y_{j+2}y_{j+3}z_{j+2}z_{j+3})^2 & \text{if } j \text{ is even} \end{cases}$$

Observe that A has cycle algebra

$$S = k[\sigma, \bar{\alpha}_j \mid j \in [8]] \subset k[\mathcal{S}],$$

and center

$$R = k[\sigma] + (\bar{\alpha}_j\bar{\alpha}_{j+1}\bar{\alpha}_{j+2}, \sigma^2 \mid j \in [8])S,$$

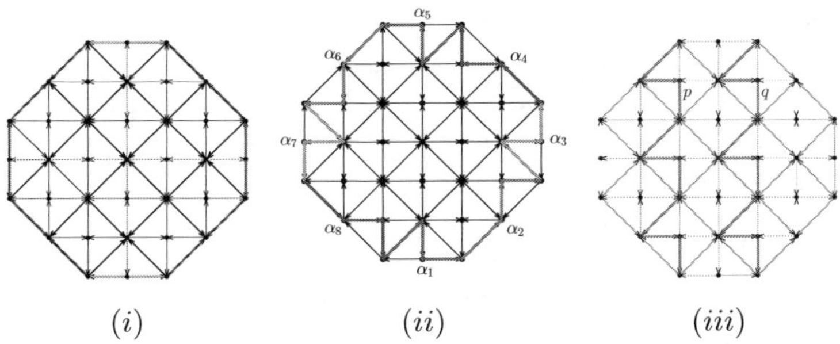

FIGURE 2. (i) The dimer quiver Q of Section 3.2. (ii) The cycles α_j run along the sides of the fundamental polygon P. (iii) The homologous cycles p, q are equal in the ghor algebra A, but not in the dimer algebra of Q.

with indices taken modulo 8.

The center R is nonnoetherian since it contains the infinite ascending chain of ideals

$$\bar{\alpha}_1\sigma^2 R \subset (\bar{\alpha}_1, \bar{\alpha}_1^2)\sigma^2 R \subset (\bar{\alpha}_1, \bar{\alpha}_1^2, \bar{\alpha}_1^3)\sigma^2 R \subset \cdots .$$

Nevertheless, we claim that R satisfies the ascending chain condition on prime ideals, and in particular has Krull dimension $5 = 2g + 1$.

If $\mathfrak{n} \in \operatorname{Max} S$ is a point for which none of the monomial generators of S vanish (for example, if each monomial is set equal to 1), then $R_{\mathfrak{n}\cap R} = S_{\mathfrak{n}}$. Consequently, the noetherian locus $U_{S/R}$ is nonempty. Therefore $\dim R = \dim S$, by [**B3**, Theorem 2.5.4]. It thus suffices to show that S has Krull dimension 5.

For $j \in [5]$ set

$$\mathfrak{p}_j := (\sigma, \bar{\alpha}_1, \ldots, \bar{\alpha}_{3+j})S,$$

and consider the chain of ideals of S

$$0 \subseteq \mathfrak{p}_1 \subseteq \mathfrak{p}_2 \subseteq \cdots \subseteq \mathfrak{p}_5. \tag{1}$$

By [**BB**, Theorem 3.11], two cycles p and q are homologous if and only if $\bar{p} = \bar{q}\sigma^\ell$ for some $\ell \in \mathbb{Z}$. Furthermore, suppose $\mathfrak{q}$ is a prime ideal of S. Then $\bar{\alpha}_i \in \mathfrak{q}$ implies $\sigma \in \mathfrak{q}$ since $\bar{\alpha}_i\bar{\alpha}_{i+4} = \sigma^2$. But this then implies $\bar{\alpha}_k$ or $\bar{\alpha}_{k+4}$ is in $\mathfrak{q}$ for each $k \in [4]$. Therefore each $\mathfrak{p}_j$ is prime, and $\mathfrak{p}_1$ is a height one prime. Consequently, $\dim S \leq 5$. But the inclusions in (1) are strict, again since any two cycles p and q are homologous if and only if $\bar{p} = \bar{q}\sigma^\ell$ for some $\ell \in \mathbb{Z}$. Whence $\dim S \geq 5$. It follows that $\dim R = \dim S = 5$.

3.3. Flower of life: A geodesic ghor algebra on a pinched torus. Finally, consider the ghor algebra $A = kQ/\ker\eta$ on a pinched torus with quiver Q given in Figure 4.i (the identifications of the sides of the polygon P are indicated by color).[3] We will determine the center R and cycle algebra S of A explicitly, and show that R is nonnoetherian of Krull dimension 4.

Q has ten simple matchings, three of which are shown in Figure 5. Specifically, Q has one simple matching consisting of concentric circles, three simple matchings

[3]The quiver in this example fits into a pattern of overlapping circles called the 'flower of life' in the New Age literature.

FIGURE 3. The simple matchings for the geodesic ghor algebra of Section 3.2.

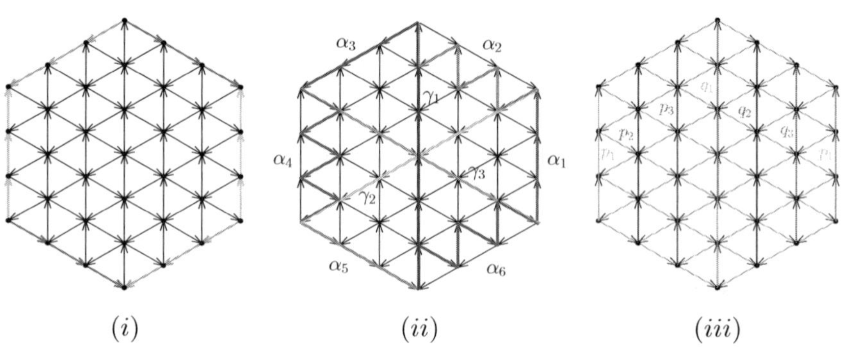

FIGURE 4. (i) The dimer quiver Q of Section 3.3. (ii) The cycles α_j run along the sides of the fundamental polygon P. (iii) The cycles q_jp_j each have $\bar{\tau}$-image $\overline{\gamma_1\alpha_1}$ (note that $p_1 = \alpha_1$ and $q_1 = \gamma_1$).

for the three straight directions of Q (Q is symmetric upon rotations by $\frac{2\pi}{3}$ and $\frac{4\pi}{3}$), and six simple matchings that are 'wiggly' (and identical up to rotation by a multiple of $\frac{\pi}{3}$). Using these simple matchings and Lemma 2.6, it is straightforward to verify that A is geodesic.

As shown in Figure 4.ii, denote by α_j, $j \in [6]$, the cycles that run along the sides of the fundamental polygon P; by γ_k, $k \in [3]$, the 'straight' cycles that pass through the center of P; and by δ_k, $k \in [3]$, the cycles that run opposite to γ_k and

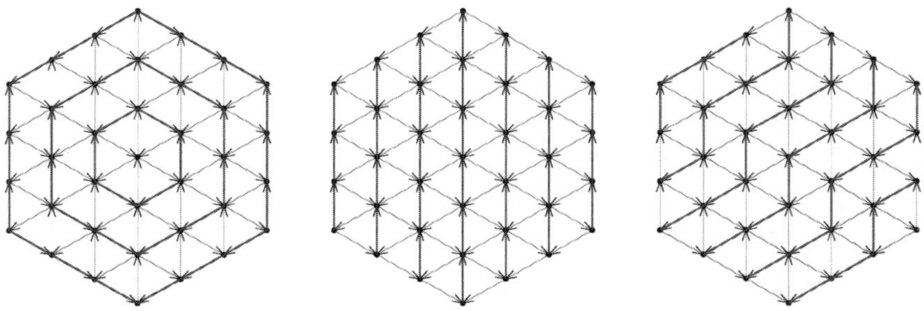

FIGURE 5. The three types of the ten simple matchings for the geodesic ghor algebra of Section 3.3.

also pass through the center of P. We claim that A has cycle algebra

$$S = k[\sigma, \bar{\alpha}_j, \bar{\gamma}_k \mid j \in [6],\ k \in [3]] \subset k[\mathcal{S}]. \tag{2}$$

Indeed, consider the set of all vertical paths p_j, q_j, $j \in [3]$, shown in Figure 4.iii. The concatenations $q_j p_j$ are cycles in Q. These cycles satisfy

$$\overline{q_j p_j} = \overline{\gamma_1 \alpha_1}, \quad j \in [3].$$

Furthermore, the cycle δ_1 satisfies

$$\bar{\delta}_1 = \overline{\alpha_3 \alpha_4 \alpha_5}.$$

Similar equalities hold for the southeasterly and southwesterly directions. Thus the ten monomials given in (2) generate S.

Observe that A has center

$$R = k[\sigma] + (\bar{\alpha}_{2j}(\bar{\alpha}_{2j+1}, \bar{\alpha}_{2j+2}, \bar{\alpha}_{2j+3}, \sigma) \mid j \in [3])S,$$

with indices taken modulo 6. R is nonnoetherian since it contains the infinite ascending chain of ideals

$$\bar{\alpha}_1 \sigma^3 R \subset (\bar{\alpha}_1, \bar{\alpha}_1^2)\sigma^3 R \subset (\bar{\alpha}_1, \bar{\alpha}_1^2, \bar{\alpha}_1^3)\sigma^3 R \subset \cdots.$$

To show that R has Krull dimension 4, it suffices to show that S has Krull dimension 4, as in Section 3.2. For $j \in [6]$ and $k \in [3]$, we have the relations

$$\bar{\alpha}_j \bar{\alpha}_{j+3} = \sigma^3 \quad \text{and} \quad \bar{\gamma}_k \bar{\delta}_k = \sigma^6.$$

We also have the three relations

$$\bar{\gamma}_1 \bar{\alpha}_4 = \bar{\alpha}_6 \bar{\alpha}_2, \qquad \bar{\gamma}_2 \bar{\alpha}_6 = \bar{\alpha}_2 \bar{\alpha}_4, \qquad \bar{\gamma}_3 \bar{\alpha}_3 = \bar{\alpha}_4 \bar{\alpha}_6,$$

and the six homotopy relations

$$\bar{\gamma}_1 \bar{\alpha}_5 = \bar{\alpha}_6 \bar{\alpha}_1, \qquad \bar{\gamma}_1 \bar{\alpha}_3 = \bar{\alpha}_2 \bar{\alpha}_1, \qquad \ldots, \qquad \bar{\gamma}_3 \bar{\alpha}_3 = \bar{\alpha}_4 \bar{\alpha}_5, \qquad \bar{\gamma}_3 \bar{\alpha}_1 = \bar{\alpha}_6 \bar{\alpha}_5.$$

Thus, following the arguments of Section 3.2, the chain of ideals of S

$$\begin{aligned} 0 \ &\subset\ (\sigma, \bar{\alpha}_1, \bar{\alpha}_2, \bar{\alpha}_3, \bar{\alpha}_4, \bar{\alpha}_5, \bar{\gamma}_2)S =: \mathfrak{p}_1 \\ &\subset\ (\mathfrak{p}_1, \bar{\alpha}_6)S \\ &\subset\ (\mathfrak{p}_1, \bar{\alpha}_6, \bar{\gamma}_1)S \\ &\subset\ (\mathfrak{p}_1, \bar{\alpha}_6, \bar{\gamma}_1, \bar{\gamma}_3)S \end{aligned}$$

is a maximal chain of primes. Therefore $\dim R = \dim S = 4$.

References

[BB] K. Baur, C. Beil, *A generalization of cancellative dimer algebras to hyperbolic surfaces*, in preparation.

[BKM] K. Baur, A. D. King, and B. R. Marsh, *Dimer models and cluster categories of Grassmannians*, Proc. Lond. Math. Soc. (3) **113** (2016), no. 2, 213–260, DOI 10.1112/plms/pdw029. MR3534972

[B1] C. Beil, *Ghor algebras, dimer algebras, and cyclic contractions*, arXiv:1711.09771.

[B2] C. Beil, *Noetherian criteria for dimer algebras*, arXiv:1805.08047.

[B3] C. Beil, *Nonnoetherian geometry*, J. Algebra Appl. **15** (2016).

[B4] C. Beil, *Nonnoetherian homotopy dimer algebras and noncommutative crepant resolutions*, Glasg. Math. J. **60** (2018), no. 2, 447–479, DOI 10.1017/S0017089517000209. MR3784058

[B5] C. Beil, *On the central geometry of nonnoetherian dimer algebras*, J. Pure Appl. Algebra **225** (2020).

[B6] C. Beil, *On the noncommutative geometry of square superpotential algebras*, J. Algebra **371** (2012), 207–249, DOI 10.1016/j.jalgebra.2012.07.051. MR2975394

[BGH] S. Bose, J. Gundry, Y. He, *Gauge Theories and Dessins d'Enfants: Beyond the Torus*, J. High Energy Phys. (2015) no. 1, 135.

[Br] N. Broomhead, *Dimer models and Calabi-Yau algebras*, Mem. Amer. Math. Soc. **215** (2012), no. 1011, viii+86, DOI 10.1090/S0065-9266-2011-00617-9. MR2908565

[D] B. Davison, *Consistency conditions for brane tilings*, J. Algebra **338** (2011), 1–23, DOI 10.1016/j.jalgebra.2011.05.005. MR2805177

[FGU] S. Franco, E. García-Valdecasas, and A. M. Uranga, *Bipartite field theories and D-brane instantons*, J. High Energy Phys. **11** (2018), 098, front matter+42, DOI 10.1007/jhep11(2018)098. MR3908478

[FH] S. Franco and A. Hasan, *Graded quivers, generalized dimer models and toric geometry*, J. High Energy Phys. **11** (2019), 104, 46, DOI 10.1007/jhep11(2019)104. MR4058841

[K] M. C. Kulkarni, *Dimer models on cylinders over Dynkin diagrams and cluster algebras*, Proc. Amer. Math. Soc. **147** (2019), no. 3, 921–932, DOI 10.1090/proc/14344. MR3896043

School of Mathematics, University of Leeds, Leeds, LS2 9JT, United Kingdom; On leave from the University of Graz

Email address: pmtkb@leeds.ac.uk

Institut für Mathematik und Wissenschaftliches Rechnen, Universität Graz, Heinrichstrasse 36, 8010 Graz, Austria

Email address: charles.beil@uni-graz.at

Contemporary Mathematics
Volume **769**, 2021
https://doi.org/10.1090/conm/769/15419

Feynman categories and representation theory

Ralph M. Kaufmann

ABSTRACT. We give a presentation of Feynman categories from a representation–theoretical viewpoint.

Feynman categories are a special type of monoidal categories and their representations are monoidal functors. They can be viewed as a far reaching generalization of groups, algebras and modules. Taking a new algebraic approach, we provide more examples and more details for several key constructions. This leads to new applications and results.

The text is intended to be a self–contained basis for a crossover of more elevated constructions and results in the fields of representation theory and Feynman categories, whose applications so far include number theory, geometry, topology and physics.

CONTENTS

Introduction

This paper concentrates on the algebraic aspects of Feynman categories. Feynman categories where introduced to have an enveloping theory for several types

The author gratefully acknowledges support from the Simons Foundation, the Institut des Hautes Etudes Scientifiques, and the Max–Planck–Institut for Mathematics in Bonn for their hospitality and their support.

of generalizations of algebras [**KW17**]. Going beyond this and the original intended application [**KWZ12**] the theory has found applications outside its initial intention in algebra, category theory, geometry, number theory and physics, [**KL16, BK17, GCKT20a, GCKT20b, War19**]. Further applications to representation theory are present, but an extension of results and approaches is desirable and anticipated. The present treatment is designed to aid this development.

Although treated within the general theory, the algebraic aspects have not been presented in full detail. There are many constructions and results that are subtle when going beyond set based categories, which is necessary to study algebra representations, that is modules. The current paper bridges this gap providing an algebraically motivated, example based introduction to the theory while at the same time providing new level of detail for these constructions. This clarifies previous results and constructions, while providing new results and concrete examples along the way.

The basic idea underlying the formalism of Feynman categories is to separate objects and their structures. This is in a similar spirit as Galois' insight to separate the group from its representations, or, in modern terminology, the category of its representations. Taking this approach leads to a hierarchy of abstraction, and allows one to operate on a higher level.

Continuing with the group analogy, many things about groups and their representations just follow from the axioms of a group, and are hence true in general for all groups and their representations —for instance restriction and Frobenius reciprocity. Other results depend only on the group and hence work in all representation of that group. Finally there are results about particular representations. In keeping with this theme, the Feynman categories are the analogues of the groups, and their representations are given by functors; that is monoidal functors to be precise.

There is a natural categorical transition from groups to groupoids or quivers, which is discussed in the first section. In this version, the groups are indeed an example of Feynman categories and restriction, induction and Frobenius reciprocity are generalized to a pair of adjoint functors, see §1 and §2. More generally, Feynman categories can be understood as having two constituents, a groupoid providing basic objects and isomorphisms, and a set of morphism encoding operations and their relations. Up to isomorphism the morphisms further decompose into tensor products of a basic morphisms, those whose target are the basic objects. The morphisms can be thought of as "proto–operations" on "proto–objects" that get realized to operations on objects if a representation functor is applied. A presentation of a Feynman category, will be a set of basic generating morphism and relations among them.

A good example for the presentation of proto–operations, or morphisms, and their relations is given by considering commutative algebras. This example also illustrates the hierarchy of abstraction. An algebra is a linear object A together with a multiplication $\mu : A \otimes A \to A$ which is bi–linear and associative. The structure so far is an object in a linear category and a 2-variable morphism with a relation —associativity. There are two ways of writing down the associativity equation. The first is in terms of elements $(ab)c = a(bc)$. Using just the structure morphism μ, one can alternatively rewrite the equation as $\mu \circ_1 \mu = \mu \circ_2 \mu_2$ considered as morphisms of three variables obtained by substitution, where $\circ_i$ means plug into the i–th variable. In this form, one has the following data: (1) an object in a linear monoidal category, we used the monoidal structure to write down $A \otimes A$ and need $A^{\otimes 3}$ for associativity. Moreover, if we are allowed to plug into variables, we will get n–linear maps that is we will use $A^{\otimes n}$. (2) A morphism μ and its iterates and the relations among them given by associativity. Moving to having this data as the value of a functor, the simple data which will be encoded in the Feynman category whose monoidal functors are commutative algebras. This is simply as follows: one basic object $*$ and one basic generating morphism $\pi_2 : * \otimes * \to *$, the proto–multiplication, with the relation of associativity given by a commutative diagram corresponding to the equation $\pi_2 \circ_1 \pi_2 = \pi_2 \circ_2 \pi_2$. Commutativity corresponds to the fact that $\pi_2 \circ (12) = \pi_2$, where (12) is the elementary transposition. The monoidal part of the Feynman category is the category finite sets with surjections with disjoint union as monoidal product, see §2.7.2 and Proposition 2.16 for full details. The morphisms are surjection and the basic morphisms are the surjections $\pi_S : S \to \{*\}$ for a chosen atom $*$. The multiplication is the value of the functor on the surjection $\pi_2 : \{1,2\} \twoheadrightarrow \{1\}$ and the associativity corresponds to the fact that there is only one surjection $\{1,2,3\} \twoheadrightarrow \{1\}$. Note that we now do not have to specify that the functor takes values in a linear category. In general, a functor out of the Feynman category into *any* monoidal target category $\mathcal{C}$ is equivalent to the data of a commutative monoid.

This begs the questions, which we will answer in the text:

(1) What are the natural generalizations of groups, algebras and representations in terms of Feynman categories?
(2) Are there are similar Feynman categories for modules and their generalizations?
(3) What type of operations on algebras translate to the Feynman categories?

The answer to (1) is that there are indeed many Feynman categories naturally generalizing groups and algebras. There are even constructions, like the plus construction, which build more complex Feynman categories from simpler ones. In particular, there are two constructions, which allow one to give more structure to the objects and the morphisms. The first is called decoration and the second indexed enrichment. Decorations, which are a form of Grothendieck (op)–fibration, lead to a factorization system for morphisms in the category of Feynman categories analogous to Galois covers, see §3.2 and [**BK17**]. Other constructions allow one to consider lax–monoidal functors or regular functors instead of strong monoidal functors as representations.

Indexed enrichments are tied to the so–called plus construction, which gives rise to several hierarchies. The most basic one starts with the trivial Feynman category whose representations are objects (§2.7.1). progresses to the category whose representations are associative algebras and the next step is given by non–symmetric operads. Beyond that one finds hyper-operads, which are necessary for the bar and co–bar construction and so on. This may provide a first point of contact and exhibit the naturalness of the notions. To obtain symmetric versions, one can use a forgetful functor which induces a cover by a decoration. In this fashion, there is a boot-strap, which generates a large part of the theory simply from the trivial Feynman category. Another hierarchy starts with a Feynman category $\mathfrak{G}$ based on graphs, see Appendix A. This leads to modular operads, hyper-(modular)-operads, etc., which are intimately related to moduli spaces of curves, and, among other things, lead to twisted modular operads see [**KW17**, 4.1] and [**KWZ12, KL16, BK20a**] as well as §7. In general, the plus construction is a formalization of the fact that the morphisms in a category regarded as having inputs and outputs give rise to flow charts, see also [**KM20**]. For the reader not so familiar with these notions, the self–contained presentation using the approach outlined above gives to a natural construction which makes their existence transparent. For instance, operads naturally appear when considering modules over algebras.

The plus construction is also important in comparing Feynman categories to operadic categories of [**BM15**], see [**BKM**]. One application of these categories is to higher category theory as they produce Batanin's n–operads [**Bat98, Bat17**] which lead to an approach to higher categories. In the context of Feynman categories, the construction goes back to indexing [**KW13**] as a codification and generalization the notion of hyper–operads and twists as introducted in [**GK98**]. The fact that this is related to so–called plus constructions, was explained to us by M. Batanin and C. Berger, which lead to the formulation given in [**KW17**, §3,§4] that is presented below. The origin of plus constructions goes back to [**BD98**], see also [**BB17**] for a plus construction for polynomial monads. Iterations of plus constructions can be found in [**BD98**], under the name of opetopes, and also in [**Bat98, BFSV03, Lei04**].

As to question (2): there are Feynman categories which allow to encode modules (§3.3). It turns out, that in the analogy with groups, algebras and modules are formalized by indexed enrichment using the aforementioned plus construction: the hierarchies are more like ladders on which there are two ways to move: "up" creating a new Feynman category and "down" using the upper rung to define an indexed enrichment. The representations of these indexed enriched versions are then the sought after modules. For representations and modules it is important that these categories can be enriched. Enrichment several different flavors, namely, combinatorial, topological and algebraic. The native constructions are combinatorial in nature as categories are based on sets. The other two are more complicated and are enriched, either in a Cartesian category, which behave very much like $\mathcal{S}et$, or in non–Cartesian category, e.g. a linear ones, such as $\mathcal{V}ect_k$. These are of basic interest in representation theory. In the analogy with groups enrichment over $\mathcal{V}ect_k$ is the transition from group representations to $k[G]$ algebras. We give the details in §4 stressing the intricacies that are presented in the non–Cartesian/linear case.

As far as question (3) is concerned, there are analogues of the bar and cobar constructions (§5), as well as of a dual transform, aka. Feynman transform, which is the cobar on the dual of A. This yields a generalization of Maurer–Cartan equations in the form of Master Equations (§5.3), which are important in deformation theory. There is a topological version of this, the so–called W–construction given in §6. The construction has a cubical nature and the cubical setting gives a natural wall structure. We give the construction for monoids and show that as expected one obtains the cubical decomposition of associahedra which also appear in the stability conditions for A_n type algebras [**KKGJ15**].

Organization of the text. The text is designed to be as self-contained as possible and is aimed at a diverse audience.

We start in §1 with collecting classic results for groups and quiver representations, but reformulated in categorical language. This presentation might be of independent interest as a primer.

The next paragraph, §2, contains the definition of Feynman category introduced as a special type of monoidal category. The representations are then given by strong monoidal functors. The development is parallel to that of §1. The presentation of indexed Feynman categories is new. The section ends with examples which are based on finite sets. Here we provide new details. The representations of these are various kinds of algebras. The group(oid) representations are also included as a basic example. Further examples are provided by graphical Feynman categories. The theory of graphs we use is detailed in Appendix A.

We then turn to various constructions for Feynman categories in §3. These yield Feynman categories whose representations are lax-monoidal or simply functors. At this level the finite set based Feynman categories have FI–modules and (co)–(semi)–simplicial objects as objects. The next operation is that of decoration. It yields the graphical Feynman categories that encode operad–like types, see Table 7. The next construction is the plus construction. Here we give a detailed exposition of the condensed presentation in [**KW17**, §3.6], providing several explicit calculations. The new precision yields gcp–version of the plus–construction, which is a generalization of hyper version contained in [**KW17**, §3.7]. The relationship to indexing is also made more explicit here then previously. A detailed graphical based analysis is given in Appendix B, where we also give a careful discussion of levels.

In §4 we tackle the enriched version. This is technically the most demanding and contains many new details. The bar/co-bar transformation and a dual transformation, aka. Feynman transform along with the master equations are discussed in §5. Traditionally the bar/co-bar adjunction can be used to define resolutions. For this one needs a model structure in general. The relevant details are reviewed in Appendix D. The W–construction is reviewed in §6. Here we also reconstruct the associahedra in their cubical decompositions.

We end the paper with an outlook, §7 that contains further applications as well as speculations about cluster transformations, relations to moduli spaces and 2–Segal objects.

Appendix A also has several details not present in other discussions, such as more details about the composition, the composition of ghost graphs, and grafting into vertices. The presentation of the category of graphs following [**BM08**] is of independent interest as it captures just the correct amount of combinatorics for subtle considerations.

There is an additional Appendix C, which gives the definition of 2–categories double categories and their relationship to Feynman categories and indexing. These can provide a rather technical, but natural, background.

New results. For the reader already familiar with (some of) the notions, there are several new results. The connection to Frobenius reciprocity is new. Algebras receive a full treatment at all levels and the relation to classical results are pointed out along the way. For instance, the examples of §2.7 are partially new and partially given in fuller detail. The treatment of noncommutative sets is entirely novel and provides a new avenue of construction. Tables 1 and 2 are the most exhaustive and detailed up to date.

The role of monoidal units is treated more carefully. First, for the free and nc construction in §3.1 which leads to more precise theorems. Again a more careful treatment of units has lead to the definition and construction of $\mathfrak{F}^{+gcp}$ in §4. The graphical description of $\mathfrak{F}^+, \mathfrak{F}^{+gcp}$ and $\mathfrak{F}^{hyp}$ in Appendix B is new in its level of detail and clarifies the short description in [**KW17**]. This is a key result in the current paper. Key results and definitions in this direction are Proposition 4.11, Definition 4.12, Theorems 4.15 and B.1. The explicit computations of $\mathfrak{F}^+, \mathfrak{F}^{+gcp}$ and $\mathfrak{F}^{hyp}$ are new, cf. Propositions 3.34 and 3.36.

The classification of FI enrichments in §4.4.2 is new and shows how the perspective of Feynman category leads to a natural classification and extension of previous results.

Throughout the paper, The role of indexing is paid more attention to. Especially in the section §2.6.1,§3.3 and Appendix C. The latter appendix also has a full development of the details of the constructions only briefly introduced in [**KW17**] in terms of double categories and 2–categories. The incorporation of holonomy and connections is new. In similar vein, in the section on decoration and covers §3.2, the criterion for being a cover is now given in Proposition 3.7.

Conventions. For a complex $C_\bullet$, we define the shift by $(\Sigma C)_n = (C[1])_n = C_{n+1}$. The effect is that the complex shifts down and suspension $s := \Sigma^{-1}$ shifts the complex up. If $f : C \to D[k]$ them we set $|f| = k$. This means that f is given by a collection of maps $f_n : C_n \to D_{n+k}$ and $|f(c)| = |f||c|$.

1. Representations from a categorical viewpoint

1.1. Representations. A k-representation of a group G is a pair (V, ρ) of a vector space V over k and a morphism of groups $\rho : G \to Aut(V)$, where $Aut(V)$ is the group of k-linear automorphisms of V. This data is neatly organized and generalized as follows.

DEFINITION 1.1. A *groupoid* is a category whose morphisms are all invertible.

EXAMPLE 1.2. Let $\underline{G}$ be the category with one object $*$ and morphisms $Hom_{\underline{G}}(*, *) = G$ where the composition map is given by group multiplication: $f \circ g = fg$. The unit id_* is the unit $e \in G$, the inverses of the morphisms are the inverse group elements g^{-1}, hence this is indeed a groupoid.

DEFINITION 1.3. A *representation of a groupoid* $\mathcal{G}$ is a functor $F : \mathcal{G} \to \mathcal{C}$.

EXAMPLE 1.4. Let k-$\mathcal{V}ect$ be the category of k vector spaces. A functor $F : \underline{G} \to k\text{-}\mathcal{V}ect$ is exactly a k representation of G. Since $\underline{G}$ only has one object $*$, on the object level the functor is completely fixed by $V := F(*)$. For the morphisms, we have an additional morphism $\rho := F : Aut(*) \to Aut(V)$. Thus the functor is determines and is uniquely determined by the pair (V, ρ).

As the example illustrates, one can quickly get generalizations. Groupoid representations are given by collections of objects, automorphisms of them and isomorphisms between them. Another generalization is given by changing the target category $\mathcal{C}$ from $\mathcal{V}ect_k$ to some other category to obtain groupoid representations in different categories.

EXAMPLE 1.5. For any category $\mathcal{C}$, we let $Iso(\mathcal{C})$ be the wide sub–category with $Obj(Iso(\mathcal{C})) = Obj(\mathcal{C})$ and $Mor(Iso(\mathcal{C}))) = Iso(\mathcal{C}) \subset Mor(\mathcal{C})$ the subset of isomorphisms. This is a groupoid sometimes called the underlying groupoid.

In the example above the functor $\rho : \underline{G} \to \mathcal{C}$ actually factors through $Iso(\mathcal{C})$ and more generally so does any functor whose source is a groupoid. Note, a category $\mathcal{V}$ is a groupoid if and only if $Iso(\mathcal{V}) = \mathcal{V}$.

EXAMPLE 1.6. Another typical groupoid is a action groupoid. Let X be a set and G be a group action ρ on X. Let $Mor = G \times X$ and $Obj = X$. Furthermore set $s = \pi_2 : Mor \to Obj : s(g,x) = x$, define $t : Mor \to Obj : t(g,x) = \rho(g)(x)$ and $id : Obj \to Mor : id(x) = (e,x)$ where $e \in G$ is the unit. Lastly, define composition $\circ : Mor_s \times_t Mor \to Obj$ given by $(g, \rho(h)(x)) \circ (h,x) = (gh, x)$. Then this data forms a category and moreover a groupoid.

REMARK 1.7. If one wants to add more geometry or more conditions to a groupoid, one can consider a groupoid internal to a category $\mathcal{C}$. Like a category internal to $\mathcal{C}$ this is a pair of objects Mor and Obj of $\mathcal{C}$ which form the morphisms and objects of a category together with morphisms in $\mathcal{C}$ $s,t : Mor \to Obj$, $id : Obj \to Mor$ and $\circ : Mor_s \times_t Mor \to Obj$ satisfying all the conditions of a category.

1.1.1. *Intertwiners as natural transformations.* Morphisms between representations (V, ρ_V), (W, ρ_W) aka. intertwiners are morphisms $N : V \to W$ such that

$$N \circ \rho_V = \rho_W \circ N \tag{1.1}$$

This equation is also the equation for natural transformations. Recall that functors from $\mathcal{C}$ to $\mathcal{D}$ form a category $Fun(\mathcal{C}, \mathcal{D})$ whose morphisms are natural transformations. Where $Nat(F, F')$ are the natural transformations from F to F' and a natural transformation N is given by a collection of morphisms $N_X : F(X) \to F'(X)$ indexed by the objects of $\mathcal{C}$ that satisfy

$$\forall \phi \in Hom_{\mathcal{C}}(X,Y) : N_X \circ F'(\phi) = N_Y \circ F(\phi) \tag{1.2}$$

EXAMPLE 1.8. In the example where $\mathcal{C} = \underline{G}$, there is only one object $*$ and hence only one morphism $N_* = N$, and the equation becomes (1.1).

1.2. Graphs and quivers. A groupoid gives rise to a graph and vice–versa any graph gives rise to a (free) groupoid, as we will review, see also e.g. [**KKW15**].

We will use the Borisov–Manin definition of graphs and morphisms, [**BM08, KW17**]. Full details are in Appendix A. In this formalism, a graph Γ is a collection $(V, F, \partial, \imath)$, where V is a set of vertices, F is a set of flags aka. half edges, $\partial : F \to V$ assigns a base vertex to each flag and $\imath : F \to F$ is an involution $\imath^2 = id$. Edges are orbits of order two of $\imath$, that is an edge $e = \{f, \imath(f)\}$ comprises two half edges and the orbits of order one, are "unpaired" half edges aka. tails. The set of edges will be denoted by T and that of tails by T. A *directed edge* $\vec{e}$ is a pair $(f, \imath(f))$. Each edge gives rise to two directed edges and by picking the first flag the set of directed edges is in bijection with the flags, that *are not* tails. A path is a sequence of directed edges $\vec{e_i} = (f_i, \imath(f_i))$, such that $\partial(\imath(f_i)) = \partial(f_{i+1})$. The set of all paths on Γ is denoted by $\mathcal{P}(\Gamma)$. A *directed graph*, aka. quiver, is a graph, with a choice of direction for all of its edges.

1.3. Graphs and groupoids. Given a groupoid $\mathcal{G}$ the corresponding graph has $V = Obj(\mathcal{G})$ and flags $F = Mor(\mathcal{G})$, $\partial(\phi) = s(\phi)$ and $\imath(\phi) = \phi^{-1}$. This graph has no tails, and hence the directed edges are in bijections with the flags. Notice that (i) each object has an identity map, thus there is at least one loop at each vertex, (ii) the graph structure does not encode the composition. We do however have a morphisms $\mathcal{P}(\Gamma) = Mor(\mathcal{G})$, by sending the sequence $\vec{e_i}, i = 1, \dots, n$ to $f_n \circ \dots \circ f_1$.

Vice–versa, given a graph Γ without tails, setting $Obj(\mathcal{G}) = V$ and F as the set of directed edges as the basic morphisms, where the source and target maps are given by ∂. One cannot read off relations, but one can simply regard the free category on a given graph Γ to be the category generated by the flags of Γ, that is morphisms $Mor(\mathcal{G}) = \mathcal{P}(\Gamma)$ together with the identity morphisms of all the objects and concatenation being composition of composable paths.

REMARK 1.9. One can use functorial language to say that there is a forgetful functor $\Gamma : \mathcal{C}at_{sm} \to \mathcal{G}raphs^{nt}$ from the category of small category to that of graphs without tails and that this functor has a left adjoint free functor F, $F \dashv G$, see §1.5.1 below.

1.4. Monoids and quiver representations. On can relax the invertibility and see that using the construction of Example 1.2 actually any unital monoid A gives rise to a category $\underline{A}$.

Similarly, relaxing the invertibility, but adding a direction for each edge, a directed graph, gives rise to a category, where the generating morphisms are exactly the directed edges. This is usually known as a quiver.

1.5. Restriction, induction and Kan extensions. Given a morphism of groups $f : H \to G$, in particular a subgroup $H \subset G$, there are two canonical operations for representations. Restriction $res^G_H : Rep(G) \to Rep(H)$ and induction $ind^G_H : Rep(H) \to Rep(G)$.

In the categorical formulation this amounts to the following triangles.

(1.3)
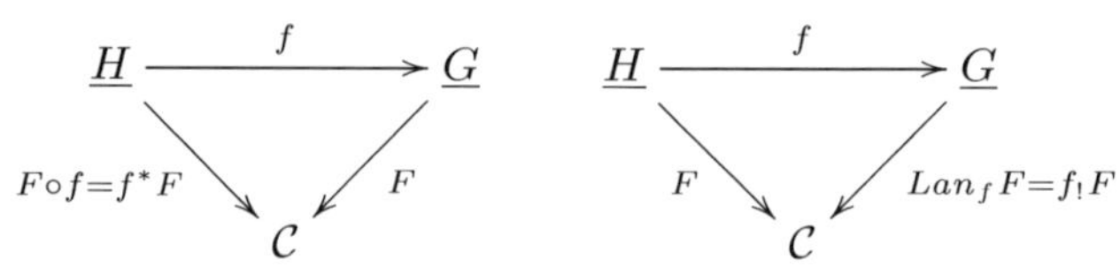

Here pull–back f^* is simply restriction. Induction is more complicated and is given by the so–called left Kan extension.

In general, the situation is that one has a functor $f : \mathcal{D} \to \mathcal{E}$ this gives rise to a morphism in the category of functors and natural transformations $f^* : Fun[\mathcal{E},\mathcal{C}] \to Fun[\mathcal{E},\mathcal{C}]$ by sending $F \in Fun[\mathcal{E},\mathcal{C}]$ to $f^*F = F \circ f$. A left adjoint functor $f_!$, $f_! \dashv f^*$, if it exists if it exists gives a functor in the other direction

$$f_! : Fun[\mathcal{E},\mathcal{C}] \leftrightarrows Fun[\mathcal{E},\mathcal{C}] : f^* \tag{1.4}$$

The left Kan extension, if it exists provides such a left adjoint: $Lan_f = f_!$.

1.5.1. *Adjoint functors.* A functor $F \in Fun(\mathcal{C},\mathcal{D})$ is called the left adjoint to a functor $G \in Fun(\mathcal{D},\mathcal{C})$ if there are natural bijections

$$Hom_{\mathcal{C}}(X, GY) \leftrightarrow Hom_{\mathcal{D}}(FX, Y) \tag{1.5}$$

Typical pairs are $G =$ forget and $F =$ free. E.g. if $\mathcal{C} = \mathcal{S}et$ and $\mathcal{D} = \mathcal{G}roup$. In the case at hand, the functors are $F = f^*$ and $f_!$ which run between the categories indicated in (1.4).

1.5.2. *Left Kan extension.* The Kan extension, gives a left adjoint functor, the putative formula for the point-wise Kan left extension is

$$Lan_f F(Y) = \mathrm{colim}_{(f \downarrow Y)} F \circ s \tag{1.6}$$

if the colimit exists. We will now discuss how to calculate such a beast in the situation above.

First, $(f \downarrow Y)$ is a so–called comma or relative slice category which has as objects pairs $(X, \phi\colon f(X) \to Y)$ where $X \in \mathcal{D}$ and $\phi \in Hom_{\mathcal{E}}(f(X), Y)$. The morphisms from $(X, \phi\colon f(X) \to Y)$ to $(X', \phi'\colon f(X') \to Y)$ are morphisms $\psi : X \to X'$ such that $\phi = \phi' \circ f(\psi)$

(1.7)
$$\begin{array}{ccc} X & & f(X) \xrightarrow{\ \phi\ } Y \\ \psi \downarrow & & f(\psi) \downarrow \quad \nearrow_{\phi'} \\ X' & & F(X') \end{array}$$

Second, the source map s sends $(X, \phi) \mapsto X$ and hence the evaluation at the object (X, ϕ) is $F \circ s(X, \phi) = F(X)$.

Finally in a category $\mathcal{C}$ with direct sums $\bigoplus$ or more generally co–products $\coprod$, we can compute small co–limits as follows. Given an index category $\mathcal{I}$ and a functor $F : \mathcal{I} \to \mathcal{C}$, the co-limit is:

(1.8)
$$\operatorname{colim}_{\mathcal{I}} F = \bigoplus_{X \in Obj(\mathcal{I})} F(X)/\sim$$

where $\sim$ is the equivalence relation induced by $F(X) \ni y \sim F(\phi)(y) \in F(X')$, where $\phi \in Hom_{\mathcal{I}}(X, X')$.

Example 1.10. Consider a set X with a group action $\rho : G \times X \to X$. Set $\mathcal{I} = \underline{G}$ and consider $\mathcal{C} = \mathcal{S}et$. A functor $F : \mathcal{I} \to \mathcal{S}et$ then given by $X = F(*)$ and a morphism $F : G \to Aut(X)$ which is equivalent to the action $\rho : G \times X \to X$. The colimit $\operatorname{colim}_{\mathcal{I}} F = X_G$ is given by the co–invariants of X, that is $X/\sim$ where $x \sim x'$ if there is a $g \in G$ such that $g(x) = x'$. These is exactly $F(*)/\sim$ above. Note that there is a natural map quotient $X \to X_G$

The co-limit is actually more that an object, but it is an object together with morphisms and as such is defined by its universal property. The co-limit $\operatorname{colim}_{\mathcal{I}} F$ is a coherent collection (aka. co-cone) $(C, (\pi_X : F(X) \to C)_{X \in Obj(\mathcal{I})})$ where coherent means that for all $\phi : X \to X' : \pi_{X'} \circ F(\phi) = \pi_X$. The universal property is that if (C', π'_X) is any other co–cone, there is a map $\psi : C \to C'$ which commutes with all the data.

In computing co-limits the following slogan is useful:

> *A co–limit of a given functor can be computed by using an equivalent indexing category.*

Hence one can compute a co–limit using a skeleton, that is an equivalent category that only has one object in each isomorphism class. For a category $\mathcal{F}$, we will denote its skeleton by $sk(\mathcal{F})$.

Example 1.11. Taking the Kan extension as an example, we see that there is a component for each object (X, ϕ) and the equivalence relation is given by the morphisms ψ as in (1.7), that is

(1.9)
$$Lan_f F(Y) = \bigoplus_{(X, \phi: f(X) \to Y)} F(X)/\sim$$

where $F(x) \ni x \sim F(\psi)(x) \in F(X')$. Note that this allows one to omit components (X, ϕ) whenever there is a morphisms ψ with this as a source. This means that only co–final objects, namely those which are not the source of a non–automorphism play a role when computing the limit.

Finally notice that the Kan extension yields a functor. It also has values on morphisms $Lan_f F(\psi : Y \to Y') : Lan_f F(Y) \to Lan_f F(Y')$. This is obtained by mapping the component (X, ϕ) to $(X, \psi \circ \phi)$ using $F(\psi)$.

1.6. Restriction, induction and Frobenius reciprocity. With this setup, we can retrieve classical results. Consider an inclusion of groups $H \hookrightarrow G$. This gives rise to the functor $f : \underline{H} \to \underline{G}$. Given a representation $\rho \in Fun(\underline{G}, \mathcal{V}ect_k)$, we have $res^G_H(\rho) = \rho \circ f$. Furthermore for $\rho \in Fun(\underline{H}, \mathcal{C})$

$$ind^G_H \rho(*) = Lan_f \rho(g) = \bigoplus_{(*,g): g \in G} V/\sim = G \times_H V \tag{1.10}$$

where $V = \rho(*)$, $g \in Hom_{\underline{G}}(*,*) = G$, H acts on the right by multiplication on G and on the left via ρ on V. In terms of the colimit, $g \sim g'$ if there is an h such that $g = g'h$ and the functorial action of h, is given by sending the component of g to that of $g' = gh^{-1}$ using $\rho(h)$ as the morphism on V, which is exactly what the relative product encodes.

The using the equivalence $Fun[\underline{G}, \mathcal{V}ect_k] = k[G]$-*mod*, the adjointness of $ind^G_H = f_! \dashv f^* = res^G_H$, yields one version of *Frobenius reciprocity.*

$$Hom_{k[G]}(ind^G_H \rho, \lambda) \leftrightarrow Hom_{k[H]}(\rho, res^G_H \lambda) \tag{1.11}$$

1.7. Algebra and dual co–algebra structure. Concatenation operation for morphisms gives a partial composition for morphisms of a category $\mathcal{C}$. Let $C = k[Mor(\mathcal{C})]$ then C is an associative algebra with the multiplication

$$\phi \cdot \psi = \begin{cases} \phi \circ \psi & \text{If they are composable} \\ 0 & \text{else} \end{cases} \tag{1.12}$$

there is an approximate unit for the multiplication which is $\sum_X id_X$. This is a unit, if there are only finitely many objects. $\mathcal{C}$ is called of *decomposition finite* if for each ϕ there are only finitely many pairs (ϕ_0, ϕ_1) such that $\phi = \phi_0 \circ \phi_1$. In this case

$$\Delta(\phi) = \sum_{(\phi_0, \phi_1): \phi_0 \circ \phi_1 = \phi} \phi_0 \otimes \phi_1 \tag{1.13}$$

is a co–associative multiplication, see e.g. [**JR79, GCKT20a, GCKT20b**] and has a co–unit

$$\epsilon(\phi) = \begin{cases} 1 & \text{if } \phi = id_X \text{ for some } X \\ 0 & \text{else} \end{cases} \tag{1.14}$$

This is one of the instances, where cutting is simpler than gluing, in the sense that in order to glue, one usually has conditions and hence only partial structures, while when cutting, the cut pieces have no conditions as they could be re–glued.

EXAMPLE 1.12.

(1) For a groupoid $\underline{G}$, C is the group algebra $k[G]$ and the co-multiplication is the dual $\Delta(g) = \sum_{(g_1, g_2): g_1 g_2 = g} g_1 \otimes g_2$. This is the usual co–multiplication one gets for the functions on the group.
(2) For a quiver this is the quiver algebra $k\vec{\Gamma}$ or the path algebra. The algebra is free and hence decomposition finite and thus has a dual co–algebra. The co–product is de–concatenation of paths.

(3) For a groupoid, C is the groupoid algebra and the co–product is the de–composition co–product. If the groupoid is given by a graph Γ there are no relations. The path–algebra is a groupoid algebra and represents the fundamental groupoid. The groupoid may or may not be of finite type. Characteristics which keep it from being of finite type are isomorphisms of infinite order, i.e. loops, infinitely many paths connecting two vertices, or an infinite path connecting two vertices.

2. Feynman categories

Feynman categories are a special type of monoidal category which generalize the groupoids in two ways.

2.1. Monoidal categories. A monoidal category $\mathcal{C}$ is a category with a functor $\otimes : \mathcal{C} \times \mathcal{C} \to \mathcal{C}$. This means the for any two objects X, Y there is an object $X \otimes Y$ and for any two morphisms $\phi \in Hom_{\mathcal{C}}(X, Y), \phi' \in Hom_{\mathcal{C}}(X', Y')$ a morphism $\phi \otimes \phi' \in Hom_{\mathcal{C}}(X \otimes X', Y \otimes Y')$, such that

$$(\phi \otimes \phi') \circ (\psi \otimes \psi') = (\phi \circ \psi) \otimes (\phi' \circ \psi') \tag{2.1}$$

There are several other structures needed for a monoidal category

(1) A unit object $\mathbb{1}$ together with isomorphisms aka. unit constraints $X \otimes \mathbb{1} \stackrel{\lambda}{\to} X \stackrel{\rho}{\leftarrow} \mathbb{1} \otimes X$.
(2) Associativity isomorphisms $A_{XYZ} : (X \otimes Y) \otimes Z \stackrel{\sim}{\to} X \otimes (Y \otimes Z)$
(3) In the case of a *symmetric* monoidal category isomorphisms, aka. commutativity constraints $C_{XY} : X \otimes Y \stackrel{\sim}{\to} Y \otimes X$ with $C_{YX} \circ C_{XY} = id$

that satisfy various compatibility conditions, such as the pentagon axiom, see e.g. [**Kas95**]. A monoidal category is called strict if λ, ρ and A are identities. MacLane's coherence axiom states, that every monoidal category is equivalent (even monoidally) to a strict one [**ML98**].

EXAMPLE 2.1. The most well–known example is $(\mathcal{V}ect_k, \otimes_k)$, here $\mathbb{1} = k$. For $\mathbb{Z}$ graded k vector spaces $\mathcal{V}ect_k^{\mathbb{Z}}$ the usual Koszul sign conventions are commutativity constraints given by $C_{VW}(v \otimes w) = (-1)^{|v||w|} w \otimes v$, where $|v|$ is the $\mathbb{Z}$ degree. This formula also is used in the $\mathbb{Z}/2\mathbb{Z}$ case. The unit is k in degree 0.

Another important example is $(\mathcal{S}et, \amalg)$ where $\amalg$ is disjoint union. One can also consider $(\mathcal{T}op, \times)$.

A strong monoidal functor between two monoidal categories $(\mathcal{C}, \otimes_{\mathcal{C}}) \to (\mathcal{D}, \otimes_{\mathcal{D}})$ is a functor $F : \mathcal{C} \to \mathcal{D}$ and natural isomorphisms $\Phi_{XY} : F(X) \otimes_{\mathcal{C}} F(Y) \stackrel{\sim}{\to} F(X) \otimes_{\mathcal{D}} F(Y)$, $F(\mathbb{1}_{\mathcal{C}}) = \mathbb{1}_{\mathcal{D}}$ and $F(\lambda_{\mathcal{C}}) = \lambda_{\mathcal{D}}$ as well as for ρ and all is compatible with the other constraints.

If the Φ_{XY} are identities the functor is called *strict monoidal*. If they are just natural morphisms, the functor is called *lax monoidal*. A co–lax or *op–lax monoidal* functor has morphisms $\hat{\Phi} : F(X) \otimes_{\mathcal{D}} F(Y) \to F(X) \otimes_{\mathcal{C}} F(Y)$

Strong monoidal functors form a category $Fun_{\otimes}(\mathcal{C}, \mathcal{D})$, the same is the case for the other versions. The natural transformation have to respect the other structure maps.

2.1.1. *Free monoidal categories.* The strict free monoidal category on a category $\mathcal{C}$ is given by objects that are words in objects of $\mathcal{C}$ and morphisms induced by morphisms in $\mathcal{C}$. For the non–strict case, the free monoidal category are bracketed words with associativity morphisms between them and the symmetric version adds commutation isomorphisms. The free monoidal category in both the symmetric and non–symmetric case will be denoted by $\mathcal{V}^{\otimes}$. The strict and non–strict versions are equivalent as (symmetric) monoidal categories, see e.g. [**Kau17**, §2.4] for further discussion. For the examples, we mostly use the strict version, as it has less objects. The non–strict version is more natural when one is dealing with functors into non–strict monoidal categories, see below.

EXAMPLE 2.2. If $\mathcal{V} = \underline{1}$ then the *strict symmetric version* is: $\mathcal{V}^{\otimes} \simeq \mathbb{S}$; viz. the category whose objects are natural numbers $\mathbb{N}_0$ corresponding to the powers $n = *^{\otimes n}$ and $Hom_{\mathcal{V}^{\otimes}}(n,n) = \mathbb{S}_n$ the symmetric group, with all other Hom–sets being empty. Here $\mathbb{1} = 0 = *^{\otimes 0}$, that is the empty word. For the *strict non–symmetric* version: $\mathcal{V}^{\otimes} = \mathbb{N}_0$, that is the discrete category of natural numbers.[1]

The free monoidal category has a universal property. For this notice that there is an inclusion $j : \mathcal{C} \to \mathcal{C}^{\otimes}$ by one letter words. The property can now be phrased as follows, every functor $f : \mathcal{C} \to \mathcal{D}$ into a monoidal category $(\mathcal{D}, \otimes)$ has a lift $f^{\otimes} : \mathcal{C}^{\otimes} \to \mathcal{D}$ as a monoidal functor such that $f = f^{\otimes} \circ j$. This association is functorial and

$$Fun(\mathcal{C}, \mathcal{D}) \simeq Fun_{\otimes}(\mathcal{C}^{\otimes}, \mathcal{D}) \tag{2.2}$$

EXAMPLE 2.3. For instance, we have the $k[G]$-mod $=$ $Fun(\underline{G}, \mathcal{V}ect_k)$ $=$ $Fun_{\otimes}(\underline{G}^{\otimes}, \mathcal{V}ect_k)$. Similarly for $k\vec{\Gamma}$.

2.2. Algebra structure for strict monoidal categories. If $(\mathcal{C}, \otimes)$ is a strict monoidal category there is a unital algebra structure on $C = Mor(\mathcal{C})$ given by $\otimes$. The unit is $id_{\mathbb{1}}$.

REMARK 2.4. Thus on a monoidal category, C has two algebra structures, which are compatible by the intertwining relation (2.1), or if it is decomposition finite. (1) a unital algebra structure given by $\mu = \otimes$ with unit $id_{\mathbb{1}}$ and (2) a co–unital co–algebra structure (Δ, ϵ) given by deconcatenation, see §1.7.

It is *not true in general that these structure from a bi–algebra.* This is the case for non–symmetric Feynman categories, and for the induced structures on isomorphism classes for Feynman categories [**GCKT20b**].

2.3. Feynman categories. Consider a triple $\mathfrak{F} = (\mathcal{V}, \mathcal{F}, \imath)$ of a groupoid $\mathcal{V}$ a (symmetric) monoidal category $\mathcal{F}$ and a functor $\imath : \mathcal{V} \to \mathcal{F}$. By universality of the free (symmetric) monoidal category, there is a functor $\imath^{\otimes} : \mathcal{V}^{\otimes} \to \mathcal{F}$ which factors through $Iso(\mathcal{F})$ since $\mathcal{V}^{\otimes}$ is again a groupoid — words in isomorphisms are isomorphisms. Among the morphisms in $\mathcal{F}$ there are *basic morphisms* $X \to \imath(*)$ which are the objects of the comma category $(\mathcal{F} \downarrow \mathcal{V})$ the morphisms in this category

[1] A category is discrete if the only morphisms being are identity morphisms id_X. This defines a way to identify sets with small discrete categories.

are commutative squares

(2.3)
$$\begin{CD} X @>{\phi}>> \imath(*) \\ @V{\psi}VV @VV{\sigma}V \\ X' @>{\phi'}>> \imath(*') \end{CD}$$

where σ is (necessarily) an isomorphism.

The tensor product induces a morphism $(\mathcal{F} \downarrow \mathcal{V})^{\otimes}$ to the category of arrows $(\mathcal{F} \downarrow \mathcal{F})$. It sends a word in $\phi_i : X_i \to \imath(*_i) : \phi_1 \cdots \phi_n \mapsto \phi_1 \otimes \cdots \otimes \phi_n : X_1 \otimes \cdots \otimes X_n \to \imath(*_1) \otimes \cdots \otimes \imath(*_n) = \imath^{\otimes}(*_1 \cdots *_n)$. The isomorphisms in $(\mathcal{F} \downarrow \mathcal{F})$, are commutative diagrams

(2.4)
$$\begin{CD} X @>{\phi}>> Y \\ @V{\sigma}V{\simeq}V @V{\simeq}V{\sigma'}V \\ X' @>{\phi'}>> Y' \end{CD} \qquad \text{denoted by } (\sigma \Downarrow \sigma')(\phi) : \phi \to \phi'$$

We will abbreviate to $(\sigma \Downarrow \sigma')$, if the source is clear. Alternatively $(\sigma \Downarrow \sigma')$ can be interpreted be a map $\mathrm{Hom}(s(\sigma), s(\sigma') \to \mathrm{Hom}(t(\sigma), t(\sigma')$. These morphisms can also be considered as 2–morphisms in a double category, see Example §C.6 in Appendix C.

DEFINITION 2.5. [**KW17**] A triple $\mathfrak{F}$ as above is a Feynman category if

(i) $\imath^{\otimes} : \mathcal{V}^{\otimes} \to \mathit{Iso}(\mathcal{F})$ yields an equivalence of categories.

(ii) The monoidal product yields an equivalence $(\mathit{Iso}(F \downarrow \mathcal{V}))^{\otimes} \simeq \mathit{Iso}(\mathcal{F} \downarrow \mathcal{F})$.

(iii) Every slice category $(\mathcal{F} \downarrow \imath(*))$ is essentially small.

A Feynman category is called *strict* if the equivalences are identities. Using MacLane's coherence, one can show that every Feynman category is equivalent to a strict one. We call a Feynman category *strictly strict*, if the equivalences become identities when using the strict free monoidal structures. $\mathfrak{F}$ is *skeletal* if $\mathcal{F}$ is.

The first condition says that each object Y decomposes up to isomorphism into a word in $\mathcal{V}$: $Y \simeq \bigoplus_{v \in V} \imath(*_v)$, such a decomposition is unique up to unique isomorphism and all isomorphisms of $\mathcal{F}$ are induced from the (iso)morphisms in $\mathcal{V}$ acting on the letters of the word. This means that each object has a well defined length $|X|$ given by the length of an isomorphic word. The second condition means that every morphisms $\phi : X \to Y$ in $\mathcal{F}$ decomposes isomorphically into a tensor product of basic morphisms according to a decomposition of $Y \simeq \bigoplus_{v \in V} \imath(*_v)$. This decomposition is unique up to unique isomorphism.

(2.5)
$$\begin{CD} X @>{\phi}>> Y \\ @V{\hat{\sigma}}V{\simeq}V @V{\simeq}V{\sigma}V \\ \bigotimes_{v \in V} X_v @>{\bigotimes_{v \in V} \phi_v}>> \bigotimes_{v \in V} \imath(*_v) \end{CD}$$

with $\phi_v : X_v \to \imath(*_v)$.

2.3.1. *Native length and element–type morphisms.* Notice by condition (i) the length of an object $|X|$ = tensor length is well defined. If $X \simeq \bigotimes_{v \in V} *_v$ then $|X| = |V|$. This defines the length of a morphism $\phi : X \to Y$ by $|\phi| = |X| - |Y|$. Isomorphisms necessarily have length 0. There are morphisms of negative length.

These come from the fact that morphisms $\mathbb{1} \to *_v$ are allowed, by the axioms. We will call these morphisms of element–type or simply elements. It follows from the axioms that any morphisms factors as a tensor product into morphisms of positive length and element–type morphisms.

2.3.2. *Non-Sigma version.* Leaving out the "symmetric" in the monoidal categories, one arrives at the notion of a non–Sigma Feynman category. Let $\mathcal{V}$ be a groupoid, $\mathcal{F}$ be a monoidal category, $i : \mathcal{V} \to \mathcal{F}_<$ a functor, $\mathcal{V}^{\otimes}$ the free monoidal category and $\imath^{\otimes} : \mathcal{V}^{\otimes} \to \mathcal{F}_<$ be the induced functor.[2]

DEFINITION 2.6. [**KW17**] A triple $\mathfrak{F}_< = (\mathcal{V}, \mathcal{F}_<, \imath)$ as above is a non–Sigma Feynman category if

(i) $\imath^{\otimes} : \mathcal{V}^{\otimes} \to Iso(\mathcal{F})$ yields an equivalence of categories.
(ii) The monoidal product yields an equivalence $(Iso(F \downarrow \mathcal{V}))^{\otimes} \simeq Iso(\mathcal{F} \downarrow \mathcal{F})$.
(iii) Every slice category $(\mathcal{F} \downarrow \imath(*))$ is essentially small.

Note that now the decompositions (2.5) are unique up to isomorphisms in the letters —permutations are not possible anymore.

2.4. A bi-algebra and Hopf algebras structures for Feynman categories. The following result from [**GCKT20b**] is a surprising feature of Feynman categories.

THEOREM 2.7 ([**GCKT20b**]). *The algebra structure of* §2.2 *and the co–algebra structure of* §1.7 *for a decomposition finite monoidal category* $\mathcal{F}$

(a) *satisfy the bi-algebra equation if* $\mathcal{F} = \mathcal{F}_<$ *belongs to a non–Sigma Feynman category* $\mathfrak{F}_<$.
(b) *induce a bi-algebra structure on the co-invariants* $\mathcal{B}$ *of* C *taken with respect to isomorphisms if* $\mathcal{F}$ *is part of a Feynman category* $\mathfrak{F}$.

In the symmetric case let $\mathcal{C}$ *be the ideal spanned by* $[id_X] - [id_{\mathbb{1}}]$ *in the bialgebra* $\mathcal{B}$ *then*

(c) $\mathcal{C}$ *is a co-ideal.*
(d) *If* $\mathfrak{F}$ *satisfies additional natural conditions listed in* [**GCKT20b**, §1.6] *then* $H = \mathcal{B}/\mathcal{C}$ *is a Hopf algebra.*

For a non–Sigma skeletal strictly strict $\mathfrak{F}_<$ *the corresponding ideal is given by the relations* $\frac{1}{|Aut(X)|} id_X - id_{\mathbb{1}}$. *With a modified co–unit, the quotient* $\mathcal{B}/\mathcal{C}$ *yields a Hopf algebra.*

Examples are the various Hopf algebras of Connes and Kreimer for trees and graphs [**CK00, CK01**], the Hopf algebra of Baues for double loop spaces [**Bau98**] and the Hopf algebra of Goncharov for multiple zeta values.

REMARK 2.8. Note if $\mathfrak{F}$ is a Feynman category with co-algebra $\mathcal{C}$, then $\mathfrak{F}^{op}$ will have the co-algebra structure $\mathcal{C}^{op}$. Thus $\mathfrak{F}^{op}$ although not a Feynman category will also yield a bi-algebra. One can speculate that up to taking the opposite category, the bi-algebra structure is a defining feature.

[2] We will use the notation $\mathcal{F}_<, \mathfrak{F}_<$ to indicate that these are non–symmetric, aka. ordered, versions.

2.5. Representations of Feynman categories, aka. $\mathcal{O}ps$. Fix a (symmetric) monoidal target category $(\mathcal{C}, \otimes)$. We define:

$$\mathcal{F}\text{-}\mathcal{O}ps_{\mathcal{C}} := Fun_{\otimes}(\mathcal{F}, \mathcal{C}) \text{ and } \mathcal{V}\text{-}\mathcal{M}ods_{\mathcal{C}} := Fun(\mathcal{V}, \mathcal{C}) \tag{2.6}$$

where $Fun_{\otimes}$ means strong monoidal functors. We will denote such functors by $\mathcal{O}$ and suppress $\mathcal{C}$ when it is fixed. The functors $\mathcal{F}\text{-}\mathcal{O}ps$ are "representations". They are usually operators or operations, which is why we call them $\mathcal{O}ps$. One such functor gives an $\mathcal{F}$-operation, that is an $\mathcal{F}$-*op* or an *op* for short.

2.5.1. *Intertwiners and monoidal category structure.* Using natural transformations as morphisms both $\mathcal{F}\text{-}\mathcal{O}ps$ and $\mathcal{V}\text{-}\mathcal{M}ods$ are categories. These natural transformations correspond to intertwiners. This yields the natural definition of equivalence of $\mathcal{O}ps$ and $\mathcal{M}ods$ as isomorphic objects in these categories.

$\mathcal{O}ps$ and $\mathcal{M}ods$ are symmetric monoidal categories for the level–wise tensor product. That is for $\mathcal{O}, \mathcal{P} \in \mathcal{F}\text{-}\mathcal{O}ps$ or $\mathcal{M}ods$

$$(\mathcal{O} \otimes \mathcal{P})(X) := \mathcal{O}(X) \otimes \mathcal{P}(X) \tag{2.7}$$

The monoidal unit is given by the trivial functor $\mathcal{T}$, which is defined by $\mathcal{T}(X) = \mathbb{1}_{\mathcal{C}}$ and $\mathcal{T}(\phi) = id_{\mathbb{1}_{\mathcal{C}}}$. The unit, associativity and commutativity constraints are those induced from $\mathcal{C}$.

2.5.2. *A second monoidal structure.* Due to the fact that in the setting of Theorem 2.7, $\mathcal{B}$ is a bi–algebra there is an additional monoidal structure on the co–completion of $\mathcal{O}ps$, which has as of yet not been explored. For the co-completion and universal operations see [**KW17**, §6].

2.5.3. *Free $\mathcal{O}ps$ and monadicity.* There is a forgetful functor $G : \mathcal{F}\text{-}\mathcal{O}ps_{\mathcal{C}} \to \mathcal{V}\text{-}\mathcal{M}ods_{\mathcal{C}}$ were $G(\mathcal{O}) = \imath^*\mathcal{O} = \mathcal{O} \circ \imath$. This functor is strong symmetric monoidal functor.

THEOREM 2.9 ([**KW17**, Theorem 1.5.3.]). *The functor G has a left adjoint (free) functor $F \dashv G$, which is lax symmetric monoidal.*

There is another way to understand the operations as an algebra over a triple or a monad. Given a pair of adjoint functors $F: \mathcal{C} \leftrightarrows \mathcal{D}: G$, there is the endofunctor $T = G \circ F \in Fun(\mathcal{D}, \mathcal{D})$, which is a unital monoid as follows:

1. There is a natural transformation $\mu: T \circ T \to T$ given by the structure morphism of adjoint functors $\epsilon : F \circ G \to id_{\mathcal{D}}$ (here $id_{\mathcal{D}}$ is the identity functor): $T \circ T = (G \circ F) \circ (G \circ F) = G \circ (F \circ G) \circ F \xrightarrow{\epsilon} G \circ id_{\mathcal{C}} \circ F = T$.
2. The other structure map of the adjunction $\eta : id_{\mathcal{D}} \to F \circ G = T$ yields a unit for T: $T = T \circ id_{\mathcal{D}} \xrightarrow{\epsilon} T \circ T \xrightarrow{\mu} T$ is the identity transformation. Likewise for the left unit equation.

An algebra over a triple (T, μ, ϵ) is an object $\mathcal{M}$ in $\mathcal{D}$ together with a transformation $\rho : T\mathcal{M} \to \mathcal{M}$ that is associative $\mu \circ \rho = \rho \circ \rho : T^2\mathcal{M} \to \mathcal{M}$. The T–algebras in $\mathcal{D}$ form a category denoted by $\mathcal{D}^T$. In the case at hand, $\mathcal{M} \in \mathcal{V}\text{-}\mathcal{M}ods$ is a $\mathcal{V}$ module and ρ gives the operation of T on $\mathcal{M}$.

THEOREM 2.10 ([**KW17**, Corollary 1.5.5]). *The adjunction $F \dashv G$ is monadic, that is $(\mathcal{V}\text{-}\mathcal{M}ods)^T = \mathcal{F}\text{-}\mathcal{O}ps$.*

The image of F in $\mathcal{F}\text{-}\mathcal{O}ps$ are the free $\mathcal{F}\text{-}\mathcal{O}ps$ and these are equivalent to the so–called Kleisly category $(\mathcal{V}\text{-}\mathcal{M}ods)_T$.

2.6. The category of Feynman categories. Feynman categories again form a category. A morphism of Feynman categories $\mathfrak{F} = (\mathcal{V}, \mathcal{F}, \imath)$ to $\mathfrak{F}' = (\mathcal{V}', \mathcal{F}', \imath')$ is a pair $\mathfrak{f} = (v, f)$, $v \in Fun(\mathcal{V}, \mathcal{V}')$, $f \in Fun_{\otimes}(\mathcal{F}, \mathcal{F}')$ such that $f \circ \imath = \imath' \circ v$. Theorem 3.10 give an important structural theorem by establishing a factorization system.

2.6.1. *Indexed Feynman categories.*

DEFINITION 2.11. A Feynman category $\mathfrak{F} = (\mathcal{V}, \mathcal{F}, \imath)$ indexed over a Feynman category $\mathfrak{B} = (\mathcal{V}_{\mathcal{B}}, \mathcal{B}, \imath_{\mathcal{B}})$ is a morphism of Feynman categories $\mathfrak{b} = (v_b, f_b) : \mathfrak{F} \to \mathfrak{B}$ whose underlying functor $f_b : \mathcal{F} \to \mathcal{B}$ is surjective on objects.

An indexing is called strong, if it is bijective on objects and surjective on morphisms. A strong indexing is strict, if induces an equivalence of $\mathcal{V} \simeq \mathcal{V}_{\mathcal{B}}$.[3]

REMARK 2.12. Let $\mathfrak{b} = (v_b, f_b) : \mathfrak{F} = (\mathcal{V}, \mathcal{F}, \imath) \to \mathfrak{B} = (\mathcal{V}_{\mathcal{B}}, \mathcal{B}, \imath_{\mathcal{B}})$ be an indexing, then

(i) Morphisms decompose fiberwise:

$$\mathrm{Hom}_{\mathcal{F}}(X, Y) = \amalg_{\phi \in \mathrm{Hom}_{\mathcal{B}}(f_b(X), f_b(Y))} f_b^{-1}(\phi) \tag{2.8}$$

(ii) Composition and the monoidal product are partially defined fiberwise

$$\begin{aligned} f_b^{-1}(\phi) \times f_b^{-1}(\psi) &\overset{\circ}{\dashrightarrow} f_b^{-1}(\phi \circ \psi) \\ f_b^{-1}(\phi) \times f_b^{-1}(\psi) &\overset{\otimes}{\dashrightarrow} f_b^{-1}(\phi \otimes \psi) \end{aligned} \tag{2.9}$$

These two partial products are associative, satisfy the interchange relation (2.1) and are compatible with the groupoid action of $Iso((\mathcal{B} \downarrow \mathcal{B})$ lifted to $Iso(\mathcal{F} \downarrow \mathcal{F})$.

If the indexing is strong, then these products are fully defined.

(iii) Any invertible $\hat{\sigma} \in Mor(\mathcal{V})$ has $f_b(\hat{\sigma}) \in Mor(V_{\mathcal{B}})$. And for $\sigma \in Mor(\mathcal{V})$: $f_b^{-1}(\sigma) = f_b^{-1}(\sigma)^{\times} \amalg \overline{f_b}^{-1}(\sigma)$, where the first set is made up of all the invertible elements in the fiber. If the indexing is strong, then the fiber has exactly one element: $f_b^{-1}(\sigma) = \sigma$.

(iv) There are unit elements

$$id_X \in f_b^{-1}(id_{f_b(X)}) \tag{2.10}$$

(v) The monoidal unit, since native length is preserved and a monoidal unit is unique up to isomorphism, $\mathbb{1}_{\mathcal{F}} \in f_b^{-1}(\mathbb{1}_{\mathcal{B}}) = \mathscr{G}$, where $\mathscr{G}$ is a discrete groupoid. If the indexing is strong then $f_b^{-1}(\mathbb{1}_{\mathcal{B}}) = \mathbb{1}_{\mathcal{F}}$.

Examples of indexing are given by decoration, see §3.2 and enrichment see §3.3.

REMARK 2.13. Using the fact that a monoidal category is a two–category with one object, see Appendix C, one can rephrase Remark 2.12 as saying that for a strong indexing f_b^{-1} is a lax monoidal lax 2–functor to $\underline{\mathcal{S}et}$. This relationship is the basis of indexed enrichment, see §4, where $\mathcal{S}et$ is allowed to be replaced by some other symmetric monoidal category.

---

[3]Similar conditions are necessary to obtain morphisms of the associative Hopf algebras [**GCKT20b**, §1.7].

2.6.2. *Pull–back/push–forward adjunction (restriction, induction and Frobenius reciprocity).* There is a natural pull–back or restriction for $\mathcal{O}ps$: $\mathfrak{f}^* : \mathcal{F}'\text{-}\mathcal{O}ps_{\mathcal{C}} \to \mathcal{F}\text{-}\mathcal{O}ps_{\mathcal{C}}$ given by $\mathfrak{f}^*\mathcal{O} = \mathcal{O} \circ f$, which is again a strong symmetric monoidal functor.

THEOREM 2.14 ([**KW17**, Theorem 1.6.2]). *The functor $\mathfrak{f}^*$ has a left adjoint $\mathfrak{f}_! \dashv \mathfrak{f}^*$ which is symmetric monoidal.*

The formula is again given by a left Kan extension. $\mathfrak{f}^*\mathcal{O} = Lan_f\mathcal{O}$. What is not obvious and is proven in *loc. cit.* is that this Kan extension yields a *monoidal* functor.

2.7. Examples. We will go through several examples. These examples are part of the fundamental ladder mentioned in the introduction whose base is the trivial Feynman category. The next level is given by finite sets and their variations. The different Feynman categories we discuss are collected in Table 1, their non–Sigma analogues are in Table2. The corresponding $\mathcal{O}ps$ are collected in Table 3.

2.7.1. *Trivial Feynman category.* More generally, the trivial Feynman category on a groupoid $\mathcal{V}$ is $\mathfrak{V} = (\mathcal{V}, \mathcal{V}^{\otimes}, \jmath)$. It has the following properties:

(1) $\mathcal{V}^{\otimes}\text{-}\mathcal{O}ps_{\mathcal{C}} \simeq \mathcal{V}\text{-}\mathcal{M}ods_{\mathcal{C}}$, by the universal property of the free monoidal category.
(2) For $\mathcal{V} = \underline{1}$, we will denote $\mathfrak{V}$ by $\mathfrak{F}^{triv}$. We have $\mathcal{V}^{\otimes}\text{-}\mathcal{O}ps_{\mathcal{C}} \simeq \mathcal{V}\text{-}\mathcal{M}ods_{\mathcal{C}} = Obj(\mathcal{C})$. This is *the* trivial Feynman category.
(3) If $\mathcal{V} = \underline{G}$ and $\mathcal{C}$ is k–linear then $\mathcal{V}^{\otimes}\text{-}\mathcal{O}ps_{\mathcal{C}} \simeq \mathcal{V}\text{-}\mathcal{M}ods_{\mathcal{C}} = k[G] - mods$ in $\mathcal{C}$.
(4) If we consider the inclusion $\imath : \underline{H} \to \underline{G}$. Then $i^* = res^G_H$ and $i_! = ind^G_H$. The adjointness of the functors is Frobenius reciprocity in the form (1.11).
(5) More generally, given any Feynman category $\mathfrak{F} = (\mathcal{V}, \mathcal{F}, \imath)$ we can consider $\mathfrak{V}$ and the morphism given by $\mathfrak{i} = (id, \imath^{\otimes})$. The using the isomorphism $\jmath^* : \mathcal{V}^{\otimes}\text{-}\mathcal{O}ps_{\mathcal{C}} \xrightarrow{\sim} \mathcal{V}\text{-}\mathcal{M}ods_{\mathcal{C}}$, $\mathfrak{i}_! \circ (\jmath^*)^{-1} = F$ and $\jmath^* \circ \mathfrak{i}^* = G$ are the adjoint pair of the free and forgetful functor in Theorem 2.9. Thus showing that this is a special case of Theorem 2.14.

In general there may be more basic morphisms apart from those coming from $\mathcal{V}$. In particular there may be basic morphisms $\mathbb{1} \to \imath(*)$ and $X \to \imath(*)$

2.7.2. *Finite Sets.* Consider the symmetric monoidal category $(\mathcal{F}in\mathcal{S}et, \amalg)$ whose unit is $\mathbb{1} = \emptyset$, consider the inclusion functor $\imath : \underline{1} \to \mathcal{F}in\mathcal{S}et$ that sends $*$ to the atom $\{*\}$. Then $\mathfrak{FinSet} = (\underline{1}, \mathcal{F}in\mathcal{S}et, \imath)$ is a Feynman category. The axioms are satisfied:

(i) $\underline{1}^{\otimes} \simeq \mathbb{S} = Iso(sk(\mathcal{F}in\mathcal{S}et)) \simeq Iso(\mathcal{F}in\mathcal{S}et))$ where $sk(\mathcal{F}in\mathcal{S}et)$ is the skeleton of $\mathcal{F}in\mathcal{S}et$ whose objects are the sets $\underline{n} = \{1, \dots, n\}$, $n \in \mathbb{N}_0$ with $\underline{0} = \emptyset$.
(ii) Given any morphisms $S \to T$ between finite sets, we can decompose it using fibers as.

$$
\begin{array}{ccc}
S & \xrightarrow{\quad f \quad} & T \\
{\scriptstyle =}\downarrow & & \downarrow{\scriptstyle =} \\
\amalg_{t \in T} f^{-1}(t) & \xrightarrow{\quad \amalg f_t \quad} & \amalg_{t \in T}\{*\}
\end{array}
\tag{2.11}
$$

where f_t is the unique map $f^{-1}(t) \to \{*\}$. Note that this map exists even if $f^{-1}(t) = \emptyset$. This shows the condition (ii), since any isomorphisms of this decomposition must preserve the fibers.

(iii) The slice category $(\mathcal{F}in\mathcal{S}et \downarrow *)$ is equivalent to its skeleton $\mathbb{S}$.

REMARK 2.15. We can also regard a skeletal version of $\mathcal{F}in\mathcal{S}et$, this category has as objects the sets $\underline{n} = \{1, \dots, n\}$ with all morphisms between them. The disjoint union is $\underline{n} \amalg \underline{m} = \underline{n+m}$ with the unit $\underline{0} = \emptyset$. The isomorphisms are $\mathbb{S}_n$ for $\underline{n}$, that is $Iso(sk(\mathcal{F}in\mathcal{S}et)) = \mathbb{S}$. This yields the strictly strict skeletal Feynman category $(\underline{\mathbf{1}}, sk(\mathcal{F}in\mathcal{S}et), \imath)$.

$\mathfrak{F}in\mathfrak{S}et$ has the Feynman subcategories $\mathfrak{F}\mathfrak{S} = (\underline{\mathbf{1}}, FS, \imath)$ and $\mathfrak{F}\mathfrak{I} = (\underline{\mathbf{1}}, FI, \imath)$, where the maps are restricted to be surjections resp. injections; see Table 1. This means that none of the fibers are empty in the surjective case and or all of the fibers are empty or singletons in the injective case.

PROPOSITION 2.16. *The following Feynman categories have $\mathcal{V}$-$\mathcal{M}ods_{\mathcal{C}} = Obj(\mathcal{C})$ and the following $\mathcal{O}ps$:*

(1) *$\mathfrak{F}\mathfrak{S}$: FS-$\mathcal{O}ps_{\mathcal{C}}$ is equivalent to the category of non–unital commutative monoids in $\mathcal{C}$.*
(2) *$\mathfrak{F}\mathfrak{I}$: FI-$\mathcal{O}ps_{\mathcal{C}}$ are equivalent to pointed objects in $\mathcal{C}$.*
(3) *$\mathfrak{F}in\mathfrak{S}et$: $\mathcal{F}in\mathcal{S}et$-$\mathcal{O}ps$ are unital commutative monoids.*

PROOF. The statement about $\mathcal{V}$-$\mathcal{M}ods$ is clear, as $Fun(\mathcal{T}, \mathcal{C}) = Obj(\mathcal{C})$. For the first statement, let $\mathcal{O} \in FS$-$\mathcal{O}ps_{\mathcal{C}}$ and set $C := \mathcal{O}(\imath(*))$. By compatibility with the tensor product, up to equivalence, we may assume that $\mathcal{O}$ is strict and replace FS, FI or $\mathcal{F}in\mathcal{S}et$ with its skeleton. In all cases, the objects are the sets $\underline{n} = \{1, \dots, n\}$ with $\underline{0} = \emptyset = \mathbb{1}$. Thus up to equivalence, $\mathcal{O}$ is fixed on objects as $\mathcal{O}(\underline{n}) = C^{\otimes n}$ with the symmetric group acting by permuting the tensor factors using the commutativity constraints in $\mathcal{C}$.

The basic maps in FS are the unique surjections $\pi_n : \underline{n} \twoheadrightarrow \underline{1}$. Set $\mu = \mathcal{O}(\pi_2) : C^{\otimes 2} \to C$. Then (C, μ) is a commutative non–unital monoid in $\mathcal{C}$. The multiplication μ is associative as $\pi_2 \circ (\pi_2 \amalg id) = \pi_3 = \pi_2 \circ (id \amalg \pi_2)$ and hence $\mathcal{O}(\pi_2 \circ (\pi_2 \amalg id)) = \mu \circ (\mu \otimes id) = \mu \circ (id \otimes \mu) = \mathcal{O}(\pi_2 \circ (id \amalg \pi_2))$. For the commutativity let τ_{12} be the transposition that interchanges 1 and 2, then $\pi_2 = \pi_2 \circ \tau_{12}$ and hence $\mu_2 = \mathcal{O}(\pi_2) = \mathcal{O}(\pi_2 \circ \tau_{12}) = \mathcal{O}(\pi_2) \circ \mathcal{O}(\tau_{12}) = \mu_2 \circ C_{CC}$ where C_{CC} is the commutativity constraint.

The basic morphisms for FI are $i : \emptyset = \mathbb{1} \to \underline{1}$ and $id_1 : \underline{1} \to \underline{1}$. Any injection can be written as a tensor product of these two maps. The map $\eta := \mathcal{O}(i) : \mathcal{O}(\mathbb{1}) = \mathbb{1}_{\mathcal{C}} \to \mathcal{O}(\underline{1}) = C$ makes C into a pointed object. The values of $\underline{1}$ and i determines the functor $\mathcal{O}$ uniquely up to isomorphism.

Finally, the morphisms in $\mathcal{F}in\mathcal{S}et$ are generated by id_1, π_2 and i using both the monoidal structure and concatenation. There is one more relation, that is $\pi_2 \circ (id_1 \amalg i) = id_1$, where we have tacitly used a strict unit constraint $\underline{1} = \underline{1} \amalg \emptyset$. Applying $\mathcal{O}$, we see that $\mathcal{O}(\pi_2 \circ (id_1 \amalg i)) = \mu \circ (id \otimes \eta) = id = \mathcal{O}(id_1)$ again suppressing unit constraints. The fact that η is a left identity follows from commutativity. □

REMARK 2.17. Judging by the name we chose for these categories, one could expect that to see find FS and FI algebras and indeed $Fun(FI, \mathcal{C})$ are FI–algebras and $Fun(FS, \mathcal{C})$ are FS algebras. By definition, however, $\mathcal{O}ps$ are *monoidal* functors and not ordinary functors. But, there is a free monoidal construction, see §3.1 which to every Feynman category $\mathfrak{F}$ associated a Feynman category $\mathfrak{F}^{\boxtimes}$ with $\mathcal{F}^{\boxtimes}$-$\mathcal{O}ps_{\mathcal{C}} = Fun_{\otimes}(\mathcal{F}^{\otimes}) = Fun(\mathcal{F}, \mathcal{C})$, and this way, we obtain FI–algebras as $\mathcal{O}ps$.

FC:	underlying $\mathcal{F}$	definition
$\mathfrak{FinSet}$	$\mathcal{F}in\mathcal{S}et$	Finite sets and set maps
$\mathfrak{FS}$	FS	Finite sets and surjections
$\mathfrak{FI}$	$\mathcal{I}nj$	Finite sets and injections
$\mathfrak{NC}Set$	$\mathcal{NC}Set$	Finite sets and set map with orders on the fibers aka. noncommutative sets
Δ_+S	Δ_+S	Augmented crossed simplicial group
$\mathfrak{FS}^<$	$FS^<$	Finite sets and surjections with orders on the fibers

TABLE 1. Set based Feynman categories Feynman categories. $\mathcal{V} = \underline{1}$ is trivial.

non-Σ FC	underlying $\mathcal{F}$	definition
$\mathfrak{F}_<\mathfrak{S}$	$\mathcal{F}in\mathcal{S}et_<$	Ordered finite sets and order preserving maps.
$\mathfrak{F}_<\mathfrak{S}$,	OS	Ordered finite sets and order preserving surjections
$\mathfrak{F}_<\mathfrak{I}$	OI	Ordered finite sets and order preserving injections
Δ_+	Δ_+	Augmented Simplicial category, Skeleton of $\mathcal{F}in\mathcal{S}et_<$
$\mathfrak{FI}^{op}_{*,*}$	$OI^{op}_{*,*}$	Subcategory of Δ_+^{op} of double base–point preserving injections

TABLE 2. Set based non-Σ Feynman categories. $\mathcal{V} = \underline{\mathbf{1}}$ is trivial.

2.7.3. *Ordered finite sets.* In the non–Σ case, a basic example is $\mathfrak{FinSet}_< = (\underline{\mathbf{1}}, \mathcal{F}in\mathcal{S}et_<, \imath)$, where $\mathcal{F}in\mathcal{S}et_<$ is the category of ordered finite sets with order preserving maps with $\amalg$ as monoidal structure; the order of $S \amalg T$ is lexicographic, S before T. The functor $\imath$ is given by sending $*$ to the atom $\{*\}$. Viewing an order on S as a bijection to $\{1, \dots, |S|\}$, we see that $\mathbf{N}_0$ is the skeleton of $Iso(\mathcal{F}in\mathcal{S}et_<)$. The diagram (2.11) translates to this situation, and we obtain a non–Σ Feynman category. The skeleton of this Feynman category is the strictly strict Feynman category $(\underline{\mathbf{1}}, \Delta_+, \imath)$, where Δ_+ is the augmented simplicial category and $\imath(*) = [0])$. Restricting to order preserving surjections and injections, we obtain the Feynman subcategories $\mathfrak{FS}_< = (\underline{\mathbf{1}}, OS, \imath)$ and $\mathfrak{FI}_< = (\underline{\mathbf{1}}, OI, \imath)$. We can also restrict the skeleton of $\mathcal{F}in\mathcal{S}et_<$ given by Δ_+ and the subcategory of order preserving surjections and injections. See Tables 2. In Δ_+ the image of $*^{\otimes n}$ under $\imath^{\otimes}$ will be the set $\underline{n}$ with its natural order.

NB: to make contact with the standard notation of n–simplices, $[n] = \underline{n+1}$, so that $[0] = \underline{1}$ and $[-1] = \underline{0} = \emptyset$ with the monoidal structure $[n] \amalg [m] = [n] * [m] = [n+m+1]$, where $*$ is the join operation.

PROPOSITION 2.18. *The following Feynman categories have $\mathcal{V}$-$\mathcal{M}ods_{\mathcal{C}} = Obj(\mathcal{C})$ and he following $\mathcal{O}ps$:*

(1) *For $\mathfrak{F}_<\mathfrak{S}$: the OS-$\mathcal{O}ps_{\mathcal{C}}$ is equivalent to the category of non–unital associative monoids in $\mathcal{C}$.*
(2) *For $\mathfrak{F}_<\mathfrak{I}$: the OI-$\mathcal{O}ps_{\mathcal{C}}$ are equivalent to pointed objects in $\mathcal{C}$.*
(3) *For $\mathfrak{FinSet}_<$: the $\mathcal{F}in\mathcal{S}et_<$-$\mathcal{O}ps$ are pointed unital associative monoids.*

PROOF. The proof is as above, save the action of the symmetric groups, which is not present. Hence there is no commutativity condition. For the unit, since there is no commutativity, we have two relations between π_2 and i : $\pi_2 \circ (id_1 \amalg i) = \pi_2 \circ (i \amalg id_1) = id_1$ giving the left and right unit equations. □

REMARK 2.19. Again, at this point the $\mathcal{F}$-$\mathcal{O}ps$ are monoidal functors not simply functors, but see §3.1 below.

2.7.4. *Hybrids.* To obtain the symmetric Feynman category whose $\mathcal{O}ps$ are associative algebras or unital associative algebras one has to consider ordered sets with set maps and orders on the fibers. $Aut(n)$ acts trivially on the morphism π_n, which was the reason for the commutativity. To remedy the situation, we notice that on an ordered $(S, <)$, $Aut(S)$ acts transitively on the orders of S. Thus adding an order to the fibers of π_n, the different orders will prevent from obtaining the same map by pre–composing with elements of $Aut(S)$.

Let $\mathcal{NC}Set$ (noncommutative sets), cf. [**Lod98, PR02, So03**] be the category whose objects are finite sets. And whose morphisms from S to T are pairs $(f, <_f)$ where $f : S \to T$ and $<_f$ is a collection of orders $<_{f^{-1}(t)}, t \in T$ on the fibers $f^{-1}(t)$ of f. Composition is given by lexicographic composition of orders. For $S \xrightarrow{g} T \xrightarrow{f} U$, $(f \circ g)^{-1}(u) = f^{-1}(g^{-1}(u)) = \amalg_{s \in g^{-1}(u)} f^{-1}(s)$ so that every and the order is given by $t' <_{f \circ g} t$ if t and t' are in the same fiber $f^{-1}(s)$ and $t' <_f t$, or if t' is in the fiber of s' and t is in the fiber of s and $s' <_g s$. Since isomorphisms in $\mathcal{F}inSet$ have one element fibers, they remain isomorphisms in $\mathcal{NC}Set$.

The skeleton of this category is known by the name of augmented crossed simplicial group defined by the symmetric groups $\Delta_+ S$ (aka. $\Delta\Sigma)_+$), [**FL91**]. In the simplicial notation $Aut([n]) = \mathbb{S}_{n+1}$.

We let $\mathfrak{NCSet}$ be the Feynman category $(\underline{\mathbf{1}}, \mathcal{NC}Set, \imath)$ and $\mathfrak{FS}_<$ the Feynman subcategory whose morphisms have underlying maps that are surjections.

It is easy to check that these are Feynman categories. They are also examples of enriched Feynman categories as discussed in §4. It is also obtained from a plus construction.

PROPOSITION 2.20. *The category $\mathcal{NC}Set$-$\mathcal{O}ps_{\mathcal{C}}$ and respectively $\mathfrak{FS}_<$-$\mathcal{O}ps_{\mathcal{C}}$ are equivalent to unital associative monoids (aka. algebras) in $\mathcal{C}$ and to the category of possibly non–unital associative monoids (aka. algebras) in $\mathcal{C}$ respectively.*

There is an embedding $i : \mathfrak{FS}^< \to \mathfrak{NCSet}$, i^* forgets the unit and $i_!$ is the free adjunction of a unit to an algebra.

PROOF. As above, on objects, the monoidal functors $\mathcal{O}$ are fixed by the value $\mathcal{O}(\underline{1}) =: A$ up to equivalence, since then up to equivalence $\mathcal{O}(S) = A^{\otimes S}$. Starting with surjections, we see that these are generated up to isomorphism by the $\pi_n : \underline{n} \twoheadrightarrow \underline{1}$ and a choice of order on the fiber, that is a choice of order on $\underline{n}$. Let $\mu := \mathcal{O}(\mu_2, 1 < 2)$ then $\mu : A \otimes A \to A$ yields the multiplication. Associativity follows directly. If we are in $\mathcal{F}inSet_<$, then we add the inclusion $i : \emptyset \to \underline{1}$ as a generating morphism. The unique fiber is empty and has the empty order. As before $\eta := \mathcal{O}(i) : \mathbb{1}_{\mathcal{C}} \to A$ provides the unit. This yields the functor from $\mathcal{O}ps$ to algebras exhibiting the equivalence.

For the other direction, notice that if $\tau_{1,2} : \underline{2} \to \underline{2}$ exchanges 1 and 2 $(\pi_2, 2 < 1) = (\pi_2, 1 < 2) \circ \tau_{1,2})$. More generally transpositions generate $\mathbb{S}_n$, which acts transitively on the orders of the fiber of π_n. Hence, the identity map $id_{\underline{1}}$, $(\pi_2, 1 < 2)$ generate all surjections up to isomorphism, which are permutations of the orders of the fibers. The latter are generated by transpositions. These maps together with i generate all maps, thus fixing their values yields a functor in the reverse direction. Here one uses, that transpositions are mapped to commutativity constraints; hence e.g. $\mathcal{O}(\pi_2, 2 < 1) = \mathcal{O}((\pi_2, 1 < 2) \circ \tau_{1,2}) = \mu \circ C_{A,A} = \mu^{op}$.

$\mathfrak{F}$	$\mathcal{F}$-$\mathcal{O}ps$ equivalent to
$\mathcal{F}in\mathcal{S}et$	unital commutative monoids/algebras
$\mathfrak{F}\mathfrak{S}$	commutative monoids/algebras
$\mathfrak{NC}Set$	unital associative monoids/algebras
$\mathfrak{F}\mathfrak{S}_<$	associative monoids/algebras
$\mathfrak{F}\mathfrak{I}$	pointed objects.

TABLE 3. Feynman categories based on finite sets and their $\mathcal{O}ps$

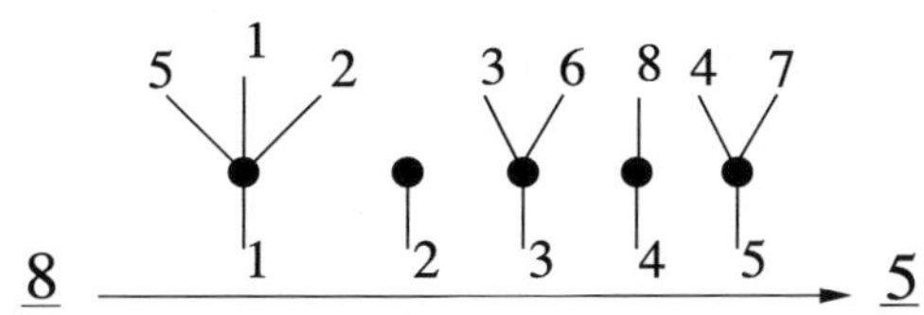

FIGURE 1. A morphism f from $\underline{8} = \{1, \ldots, , 8\}$ to $\underline{5} = \{1, \ldots, 5\}$ with the orders and fibers: $2 < 1 < 5$ on $f^{-1}(1)$, $\emptyset$ on $f^{-1}(2)$, $3 < 6$ on $f^{-1}(3)$, 8 on $f^{-1}(4)$, and $7 < 4$ on $f^{-1}(3)$.

The last statement is straightforward. □

The following is straightforward:

PROPOSITION 2.21. *$\mathfrak{b}(v_b, f_b) : \mathfrak{NC}Set \to \mathcal{F}in\mathcal{S}et$ with $f_b : \mathcal{NC}Set \to \mathcal{F}in\mathcal{S}et$ given by the identity on objects and defined on morphisms as the forgetful functor $f_b : (f, <_f) = f$ is a strong cover, but not strict.*

The pull–back f_b^* is the inclusion of unital commutative algebras into unital algebras. The push–forward is the Abelianization.

For the non–unital versions mutatis mutandis the same results hold for the restriction of $\mathfrak{b} : \mathfrak{F}\mathfrak{S}^< \to \mathfrak{F}\mathfrak{S}$. □

We show in Lemma 4.17 that this is an indexed enrichment. Going one level higher, the enrichment is by the associative operad, which can be obtained via a push–forward along a forgetful map from a plus construction, see Lemma 3.37. The relation between the two is in Remark 4.18.

2.7.5. *Graphical interpretation.* A convenient graphical notation to write down a map with ordered fibers is given by planar planted corollas.

First, the fibers of a morphism $f : S \to \{t\}$ give a planted corolla $*_{S \amalg \{t\}, t}$, that is a corolla with flags $S \amalg \{t\}$ and root t. Vice–versa, any morphism $S \twoheadrightarrow T$ can then be encoded by a forest of planted corollas $\amalg_{t \in T} *_{f^{-1}(t) \amalg \{t\}, t}$. Note that empty fibers correspond to 0-ary corollas. The map is a surjection, if there are no zero-ary corollas, and an injection, if all the corollas are either 1-ary or 0-ary.

An order on the fibers $<_f$ extends to an order $<$ on $f^{-1}(t) \amalg \{t\}$, by considering t to be the first element. That is the fibers can be viewed as planar planted corollas $*_{S \amalg \{t\}, t, <}$. And, any morphism in $\mathcal{F}in\mathcal{S}et_<$ can be written as an forest of planar planted corollas indexed by $t \in T$. An example is given in Figure 1.

2.7.6. *Graphical Feynman categories.* There is a Feynman category $\mathfrak{G} = (\mathcal{C}rl, \mathcal{A}gg, \imath)$ whose technical definition with all details is in Appendix A.2. It uses the technical framework of [**BM08, KW17**] which is also given in Appendix A to be self–contained.

$\mathfrak{F}$	$\mathfrak{F}^{\boxtimes}$-$\mathcal{O}ps_{\mathcal{C}}$	$\mathfrak{F}^{\boxtimes}$-$\mathcal{O}ps_{\mathcal{C}^{op}}$
Δ_+	augmented co–simplicial objects	augmented simplicial objects
$\mathcal{F}_{<\mathfrak{I}}$	co–semi–simple objects	semi–simple objects
$\mathfrak{F}\mathfrak{I}$	FI–algebras	FI-co-algebras
$\mathfrak{F}\mathfrak{S}$	FS–algebras	FS–co–algebras

TABLE 4. Examples of $\mathfrak{F}^{\boxtimes}$ whose $\mathcal{O}ps$ in $\mathcal{C}$ and $\mathcal{C}^{op}$ are familiar notions

To give the basic structure without all the details, we note that $\mathcal{C}rl$ is equivalent to a the category $Iso(\mathcal{F}inSet)$ of finite sets and isomorphisms. If S is a finite set, then it determines an object $*_S$ of $\mathcal{C}rl$ which is a corolla with S with external flags and any bijection $S \leftrightarrow S'$ yields an isomorphism $*_S \leftrightarrow *_{S'}$. Consequently $Iso(\mathcal{A}gg) \simeq Iso(\mathcal{F}inSet)^{\otimes}$. Its elements are collections or aggregates of corollas. The morphisms between aggregates are rather complicated. They are morphisms of the aggregates thought of as graphs without edges as defined in [**BM08**, **KW17**]. Given a morphism between aggregates, ϕ, there is an underlying graph, the ghost graph of ϕ, which is denoted by $\mathbb{\Gamma}(\phi)$. The ghost graph does not fix the morphism uniquely. It does fix the isomorphism class of a basic morphism —that is morphism whose target is a single corolla. The extra data of for a basic morphism is given by an identification of the tails of the ghost graph with the tails of the target corolla.

A *graphical Feynman category* is a Feynman category indexed over $\mathfrak{G}$.

The $\mathcal{O}ps$ for graphical Feynman categories include all the known operad types as well as rather new ones, see Table 7 in Appendix A.5. Moreover all these examples can be obtained via the operations below, especially decoration, and taking sub–Feynman categories, aka. restriction.

3. Constructions and examples

3.1. Functors and lax monoidal functors as $\mathcal{O}ps$. By definition $\mathcal{F}$-$\mathcal{O}ps_{\mathcal{C}}$ are strong monoidal functors $Fun(\mathcal{F},\mathcal{C})$. This begs the question if there are constructions of Feynman categories to obtain lax monoidal functors $Fun_{lax\otimes}(\mathcal{F},\mathcal{C})$ or simply functors $Fun(\mathcal{F},\mathcal{C})$ as $\mathcal{O}ps$, which is indeed the case:

THEOREM 3.1 ([**KW17**, §3]). *Given a Feynman category $\mathfrak{F}$, there are Feynman categories $\mathfrak{F}^{\boxtimes} = (\mathcal{V}^{\boxtimes}, \mathcal{F}^{\boxtimes}, \imath^{\boxtimes})$ and $\mathfrak{F}^{nc} = (\mathcal{V}^{nc}, \mathcal{F}^{nc}, \imath^{nc})$[4] such that*

$$\mathcal{F}^{\boxtimes}\text{-}\mathcal{O}ps = Fun(\mathcal{F},\mathcal{C}) \tag{3.1}$$

$$\mathcal{F}^{nc}\text{-}\mathcal{O}ps = Fun_{lax\otimes}(\mathcal{F},\mathcal{C}) \tag{3.2}$$

The original statements are in [**KW17**, §3]. We give the constructions below, filing in some details concerning units.

EXAMPLE 3.2. As announced, $\Delta_+^{\boxtimes}$-$\mathcal{O}ps$ are augmented simplicial objects, $FI^{\boxtimes}$-$\mathcal{O}ps$ are FI-modules etc., see Table 3.2.

3.1.1. *Free construction $\mathfrak{F}^{\boxtimes}$.* For this $\mathcal{F}^{\boxtimes}$ is the free (symmetric) category on $\mathcal{F}$. We use $\boxtimes$ for the new free monoidal structure, which we also call outer. $\mathcal{V}^{\boxtimes} = \mathcal{V}^{\otimes} \simeq \mathcal{F}$. The basic morphisms "are" the morphisms of $\mathcal{F}$: $(\mathcal{F}^{\boxtimes} \downarrow \mathcal{V}^{\boxtimes}) \simeq (\mathcal{F}^{\boxtimes} \downarrow \mathcal{F}) = (\mathcal{F} \downarrow \mathcal{F})$ under the equivalence $\imath^{\otimes} : \mathcal{V}^{\otimes} \to \mathcal{F}$ and by the definition of the free (symmetric) monoidal category.

[4]nc stands for non–connected

In the free monoidal category $\mathcal{F}^{\boxtimes}$ there is a free unit $\mathbb{1}_{\boxtimes}$ which can be different from $\mathbb{1}_{\otimes}$, thus for $\hat{\mathcal{O}} \in \mathcal{F}^{\boxtimes}\text{-}\mathcal{O}ps_{\mathcal{C}}$, we have that $\hat{\mathcal{O}}(\mathbb{1}_{\boxtimes}) \simeq \mathbb{1}_{\mathcal{C}}$, but no condition on $\mathbb{1}_{\otimes}$.

EXAMPLE 3.3. Examples are given by the Feynman category indexed over finite sets.

3.1.2. *NC-construction.* Here again $\mathcal{V}^{nc} = \mathcal{V}^{\otimes}$ and $Obj(\mathcal{F}^{nc}) = Obj(F^{\boxtimes})$, but the basic morphisms $(\mathcal{F}^{nc} \downarrow \mathcal{V})$ are defined as $Hom_{\mathcal{F}^{nc}}(\boxtimes_i X_i, Y) = Hom_{\mathcal{F}}(\bigotimes_i X_i, Y)$. This effectively adds the data of functor $\mu : \mathcal{F} \boxtimes \mathcal{F} \to \mathcal{F}$, that is a natural family of morphisms $\mu_{X,Y} : (X \boxtimes Y) = X \otimes Y$, and a morphisms $\epsilon : \mathbb{1}_{\boxtimes} \to \mathbb{1}_{\otimes}$, compatible with the unit constraints, to the morphisms of $\mathcal{F}^{\boxtimes}$. The data of ϵ was not addressed separately in [**KW17**, §3.1].

The construction of $\mathcal{O}ps$ is as follows: If $\hat{\mathcal{O}} \in \mathcal{F}^{nc}\text{-}\mathcal{O}ps_{\mathcal{C}}$, one defines $\mathcal{O} \in Fun_{lax\otimes}(\mathcal{F}\mathcal{C})$ as $\mathcal{O}(X) := \hat{\mathcal{O}}(X)$ and the same on morphisms $\hat{\mathcal{O}}(\phi) = \mathcal{O}(\phi)$ for $\phi : X \to Y$.

The two structural morphisms are defined as follows:

$$\mathcal{O}(X) \otimes \mathcal{O}(Y) = \hat{\mathcal{O}}(X) \otimes \hat{\mathcal{O}}(Y) \simeq \hat{\mathcal{O}}(X \boxtimes Y) \overset{\hat{\mathcal{O}}(\mu_{X,Y})}{\to} \hat{\mathcal{O}}(X \otimes Y) = \mathcal{O}(X \otimes Y) \tag{3.3}$$

and

$$\hat{\mathcal{O}}(\mathbb{1}_{\boxtimes}) \simeq \mathbb{1}_{\mathcal{C}} \overset{\hat{\mathcal{O}}(\epsilon)}{\to} \hat{\mathcal{O}}(\mathbb{1}_{\otimes}) \tag{3.4}$$

yields the structural unit morphism for the underlying lax-monoidal functor.

Vice–versa, using the structure for the free monoidal category, we can extend a $\mathcal{O} \in \mathcal{F}^{nc}\text{-}\mathcal{O}ps$ functor to all the morphisms of $\mathcal{F}^{\boxtimes}$ by using the functor underlying $\mathcal{O}$ and then extending it to the free (symmetric) monoidal category by the universal property as $\mathcal{O}^{\boxtimes}$. Then one only needs to define $\mathcal{O}^{nc}$, we only need the values on $\mu_{X,Y}$ and ϵ, which are fixed by the equations (3.3) and (3.4).

REMARK 3.4.

(1) It is most natural to take ϵ to be an isomorphism and moreover to identify $\mathbb{1}_{\boxtimes}$ and $\mathbb{1}_{\otimes}$. An example is taking the empty forest to be given by an empty tree, or more generally, the empty sentence (the empty collection of words) is identified with the empty word.
(2) The nc–construction plays a crucial role in the connection to Hopf algebras [**GCKT20a, GCKT20b**].
(3) An example of nc–construction first appeared in [**KWZ12**].

3.2. Decorations, covers and factorizations of morphisms in the category of Feynman categories. Given an $\mathcal{O} \in \mathcal{F}\text{-}\mathcal{O}ps_{\mathcal{S}et}$ there is a new Feynman category $\mathfrak{F}_{\text{dec }\mathcal{O}}$ defined in [**KL16**]. This decoration is a variation of the Grothendieck (op)–fibration construction. It also establishes a theory of covers that is compatible with Galois covers in the sense of Grothendieck [**BK17**, §3].

The objects of $\mathcal{F}_{\text{dec}\mathcal{O}}$ are pairs (X, a_x) with $X \in Obj(\mathcal{F})$ and $a_x \in \mathcal{O}(X)$ and morphisms are given by

$$Hom_{\mathcal{F}_{\text{dec}\mathcal{O}}}((X, a_x), (Y, a_y)) = \{\phi \in Hom_{\mathcal{F}}(X, Y) | \mathcal{O}(\phi) a_x = a_y\}$$

Likewise, the objects of $\mathcal{V}_{\text{dec}\mathcal{O}}$ are pairs $(*, a_*)$ with $* \in Obj(\mathcal{V})$ and $a_* \in \mathcal{O}(\imath(*))$ with the morphisms analogous to the ones above. The inclusion $\imath_{\text{dec}\mathcal{O}}$ is given by $(*, a_*) \mapsto (\imath(*), a_*)$.

REMARK 3.5. If $\mathcal{O} \in \mathcal{F}\text{-}\mathcal{O}ps_{\mathcal{C}^{op}}$ the condition reads $\mathcal{O}(\phi)(a_y) = a_x$.

THEOREM 3.6 ([**KL16**, Theorem 4.1]). *Given a functor $\mathcal{O} \in \mathfrak{F}\text{-}\mathcal{O}ps_{\mathcal{S}et}$ then $\mathfrak{F}_{\mathrm{dec}\mathcal{O}} = (\mathcal{V}_{\mathrm{dec}\mathcal{O}}, \mathcal{F}_{\mathrm{dec}\mathcal{O}}, \imath_{\mathrm{dec}\mathcal{O}})$ as above is a Feynman category and there is a morphism of Feynman categories $\mathfrak{p}_{\mathcal{O}} = (p_{\mathcal{V}}, p) : \mathfrak{F}_{\mathrm{dec}\mathcal{O}} \to \mathfrak{F}$, which forgets the decoration, i.e. $p((X, a_x)) = X$ and $p_{\mathcal{V}}$ is its restriction to $\mathcal{V}$.*

This construction is functorial in $\mathcal{O}$ and with respect to morphisms of Feynman categories;,i.e. the following diagrams commute. For a morphism (natural transformation) $\eta : \mathcal{O} \to \mathcal{P}$ in $\mathfrak{F}\text{-}\mathcal{O}ps$ there is a commutative diagram

$$\begin{array}{ccc} \mathfrak{F}_{\mathrm{dec}\mathcal{O}} & \longrightarrow & \mathfrak{F}_{\mathrm{dec}\mathcal{P}} \\ & \searrow^{\mathfrak{p}_{\mathcal{O}}} \quad {}^{\mathfrak{p}_{\mathcal{P}}}\swarrow & \\ & \mathfrak{F} & \end{array} \tag{3.5}$$

where the upper arrow is given by $(X, a_X) \to (X, \eta_X(a_X))$, where $\eta_X : \mathcal{O}(X) \to \mathcal{P}(X)$ is the natural transformation.

Given a morphism of Feynman categories $\mathfrak{f} = (v, f) : \mathfrak{F} \to \mathfrak{F}'$, there is a commutative diagram

$$\begin{array}{ccc} \mathfrak{F}_{\mathrm{dec}\mathcal{O}} & \xrightarrow{\mathfrak{f}_{\mathcal{O}}} & \mathfrak{F}'_{\mathrm{dec}\ f_!(\mathcal{O})} \\ \downarrow{\scriptstyle \mathfrak{p}_{\mathcal{O}}} & & \downarrow{\scriptstyle \mathfrak{p}_{f_!(\mathcal{O})}} \\ \mathfrak{F} & \xrightarrow{\mathfrak{f}} & \mathfrak{F}' \end{array} \tag{3.6}$$

The maps $\mathfrak{p}$ in the theorem above called covers, viz. $\mathfrak{f} : \mathfrak{G} \to \mathfrak{F}$ is a cover if $\mathfrak{G} = \mathfrak{F}_{\mathrm{dec}\mathcal{O}}$ for some $\mathcal{O} \in \mathcal{F}\text{-}\mathcal{O}ps$ and $\mathfrak{f} = \mathfrak{p} : \mathfrak{F}_{\mathrm{dec}\mathcal{O}} \to \mathfrak{F}$.

PROPOSITION 3.7. *A morphism $\mathfrak{f} = (v, f)$ of Feynman categories is a cover if*

(1) *Any morphism whose source is in the image of v respectively f, has a lift.*
(2) *Any lift of a morphism in the image of v respectively f is uniquely determined by its source.*

PROOF. Given $\mathfrak{f} : \mathfrak{F}_{\mathrm{dec}\ \mathcal{O}} \to \mathfrak{F}$, we verify the two conditions. The objects in the image of f are the $X \in Obj(\mathcal{F})$ with $\mathcal{O}(X) \neq \emptyset$. If X is in the image of f, i.e. $\mathcal{O}(X)$ is not empty then given any $\phi : X \to Y$, $\mathcal{O}(Y) \supset \mathcal{O}(\phi)(X) \neq \emptyset$, so that Y is in the image of f as well. Moreover, for any $a_x \in \mathcal{O}(X)$ fix $a_y := \mathcal{O}(\phi)(a_x) \in O(Y)$ then $\phi : (X, a_x) \to (Y, a_y)$ is a lift of ϕ and any lift of ϕ is uniquely fixed by the choice of by a_x.

Vice–versa, given any morphism $\mathfrak{f} : \mathfrak{F}' \to \mathfrak{F}$ satisfying (1) and (2), gives rise to a functor $\mathcal{O} \in \mathcal{F}\text{-}\mathcal{O}ps$, such that $\mathcal{F}_{\mathrm{dec}\ \mathcal{O}} = \mathcal{F}'$. On objects $\mathcal{O}(X) = f^{-1}(X)$, that is the set of fibers, aka. elements. This will be $\emptyset$ if X is not in the image. Given a morphism $\phi \in Hom_{\mathcal{F}}(X, Y)$ in the image fix a $\hat{X} \in \mathcal{O}(X)$ using (1) and (2) there is a unique lift $\hat{\phi}(\hat{X}) \in Hom_{\mathcal{F}'}(\hat{X}, \hat{Y})$ with $\hat{Y} = t(\hat{\phi}(\hat{X}))$. Define $\mathcal{O}(\phi) : \mathcal{O}(X) \to \mathcal{O}(Y)$ by $\mathcal{O}|_{\hat{X}} := \hat{\phi}(\hat{X})$. If $\phi \in Hom_{\mathcal{F}}(X, Y)$ and X not in the image of f, that is $\mathcal{O}(X) = \emptyset$, then $\mathcal{O}(\phi)$ is the unique map with source $\emptyset : \mathcal{O}(X) = \emptyset \to \mathcal{O}(Y)$. Finally, we check that the functor is (symmetric) monoidal. We have that $\mathcal{O}(X \otimes Y) = f^{-1}(X \otimes Y) = \{Z | f(Z) = X \otimes Y\}$ using that $\mathfrak{f}$ is a morphism of Feynman categories,we can decompose Z and we have $f^{-1}(X \otimes Y) = \{(Z', Z'') | f(Z') = X, F(Z'') = Y\} = \mathcal{O}(X) \times \mathcal{O}(Y)\}$.

Decorating with this functor, we get $\mathcal{F}_{\mathrm{dec}\mathcal{O}}$ with objects $(X, \hat{X})$ and morphism $\hat{\phi} : (X, \hat{X}) \to (Y, \hat{Y})$ lifting ϕ. The isomorphism is given by sending $(X, \hat{X})$ to $\hat{X}$, the inverse is fixed by $X = f(\hat{X})$.

The determination of the groupoid part is analogous and the inclusion is then clearly the restriction. □

EXAMPLE 3.8. Table 8 contains shows examples decorations for graphical Feynman categories $\mathfrak{G}$.

There is a second type of standard morphism, which is called connected.

DEFINITION 3.9. A morphism of Feynman categories $\mathfrak{f} : \mathfrak{F}' \to \mathfrak{F}$ is connected if $f_!(\mathcal{T}_{\mathcal{F}'}) = \mathcal{T}_{\mathcal{F}}$ where $\mathcal{T}_{\mathcal{F}'} : \mathcal{F}' \to \mathcal{S}et$ and $\mathcal{T}_{\mathcal{F}} \to \mathcal{S}et$ are the trivial $\mathcal{O}ps$ to $\mathcal{S}et$ with the Cartesian monoidal product $\times$.

The two sets of morphisms form an orthogonal factorization system in the sense of [**BK17**], where these types of morphisms are linked to comprehension schemes [**Law70**] and a general theory of Galois type covers. The following theorem follows from [**BK17**, Proposition 2.3]:

THEOREM 3.10. *Any morphism of Feynman categories* $\mathfrak{f} : \mathfrak{G} \to \mathfrak{F}$ *has a unique factorization as* $\mathfrak{f} = \mathfrak{p} \circ \mathfrak{i}$ *where* $\mathfrak{i}$ *is connected and* $\mathfrak{p}$ *is a cover.*

REMARK 3.11.

(1) A cover $\mathfrak{p} = (v, p) : \mathfrak{F}' \to \mathfrak{F}$ is isomorphic to the decoration by $p_!(\mathcal{T}_{\mathcal{F}'}) \in \mathcal{F}\text{-}\mathcal{O}ps$, [**KL16**].
(2) The decoration construction is also intimately tied to cyclic operads, modular operads and moduli spaces; see [**BK17, BK20a**] and §7.1 and Appendix A.5.2 below.
(3) The existence of a factorization follows already from [**KL16**], the uniqueness requires a finer analysis.

3.2.1. *Covers, connected morphisms and indexing.*

LEMMA 3.12.

(1) *The forgetful functor* $\mathfrak{p} : \mathfrak{F}_{\mathrm{dec}\mathcal{O}} \to \mathfrak{F}$ *is an indexing if* $\mathcal{O}(X) \neq \emptyset$ *for all* X. *Thus restricting to the full image of* p, *we obtain a indexing, which is not strong in general.*
(2) *A strong indexing is a cover, if and only if it is the an isomorphism and hence strict.*

PROOF. The image of a cover are precisely those X for which $\mathcal{O}(X) \neq \emptyset$. If a cover is a strong indexing, then there is only one object in the inverse image and thus every morphism has a unique left. Hence, the cover is an isomorphism. □

REMARK 3.13.

(1) Any indexing factors as a connected morphism and a cover.
(2) By the above, we see that a strong indexing is connected.
(3) Strong and strict indexings give rise to enrichments discussed below, ¶4.

There are interesting connected morphisms which are not strong. These typically arise from inclusions given by restrictions, see §A.5 in the Appendix for examples.

3.2.2. $\mathcal{F}_{\mathrm{dec}\mathcal{O}}$-*Ops*. In [**BK17**], we showed

PROPOSITION 3.14. *The* $\mathcal{F}_{\mathrm{dec}\mathcal{O}}$-*Ops are* $\mathcal{F}$-*Ops over* $\mathcal{O}$, *that is* $\mathcal{P} \in \mathcal{F}$-*Ops with a natural transformation* $\mathcal{P} \to \mathcal{O}$. □

3.3. Modules and plus construction. Remarkably, Feynman categories often can be used to encode modules as well as algebras. As constructed above Feynman categories can be used to "encode" algebras, see Table 3. One can ask the question if there is a way to encode modules over a given algebra or more generally modules over $\mathcal{F}$-*Ops*. We will give the answer in two parts. Here we fist consider the case of Feynman categories over $\mathcal{S}et$, in §4 we will then deal with more general types of modules. In particular, one would like to consider modules in linear categories. This is possible along the same lines presented here, but technically more involved. This is why we postpone it to the next section. In this section, we will fix the target category $\mathcal{C}$ to be $(\mathcal{S}et, \times)$. The arguments generalize in a straightforward fashion to a Cartesian target category $\mathcal{C}$.

EXAMPLE 3.15 (Paradigmatic example). Consider an associative monoid A, then there is an $\mathcal{A} \in FS^{<}$-$\mathcal{O}ps_{\mathcal{S}et}$ such that $A = \mathcal{A}(*)$. The set–modules, aka. set–algebras, over the associative monoid A are sets M with structure maps $\rho : A \times M \to A, (a, m) \mapsto am$ that satisfy $a_1(a_2m) = (a_1a_2)m$. The morphisms of modules are intertwiners. If A is unital, with unit $1 \in A$ then there is another condition for modules: $1m = m$. In this case A is actually the value of $\mathcal{A} \in NC\mathcal{S}et$-$\mathcal{O}ps_{\mathcal{S}et}$ and

Assuming that A is unital, we can consider $\underline{A}$ cf. §1.4. The category of A–modules in $\mathcal{S}et$ is $Fun(\underline{A}, \mathcal{S}et) \simeq Fun_{\otimes}(\underline{A}^{\otimes}, \mathcal{S}et)$ and intertwiners are natural transformations. To separate out the isomorphisms, splits as a disjoint union $A = A^{\times} \amalg A^{red}$ where $A^{\times} = G(A)$ are the invertible elements, then $\mathcal{V}_{\mathcal{A}} = \underline{A^{\times}}$ and $\mathcal{F} = \underline{A}^{\otimes}$ together with the natural inclusion form a Feynman category $\mathfrak{F}^{triv}_{\mathcal{A}} = (\underline{A^{\times}}, \underline{A}^{\otimes}, \imath)$.

Let $\mathfrak{V}_{\mathcal{A}}$ be the trivial Feynman category on $\mathcal{V}_{\mathcal{A}}$. There is a natural functor $\mathfrak{F}^{triv}_{\mathcal{A}} \to \mathfrak{V}_{\mathcal{A}}$ which is a strong indexing. It is identity on $A^{\times}$ and sends A^{red} to id_*. The indexing is strict, if A is reduced that is $A^{\times} = 1$.

The basic results in the theory, see [**KW17**, §§3.6,3.7] are that there is a plus construction $\mathfrak{F}^{+}$ for a given Feynman category and that there is a quotient of it $\mathfrak{F}^{hyp}$ which is called the hyper construction. The latter is important for twisting as in [**GK98**], see §3.3.3, §3.3.4 and is needed in §5.. Being more careful with the units, we give a new construction $\mathfrak{F}^{+gcp}$ which allows one to define modules via indexed enrichments, see [**KW17**, §4.1] and §4, especially §4.4.1 below.

In the example above $(\mathfrak{F}^{triv})^{+} = \mathfrak{F}\mathfrak{S}^{<}, (\mathfrak{F}^{triv})^{+gcp} = \mathfrak{NCSet}, (\mathfrak{F}^{triv})^{hyp} = \mathfrak{F}^{triv}$ and $\mathcal{A} \in \mathcal{NCS}et$-$\mathcal{O}ps_{\mathcal{S}et}$ gives rise to $\mathfrak{F}^{triv}_{\mathcal{A}}$ indexed over $\mathfrak{V}_{\mathcal{A}}$. If A is reduced, then $\mathfrak{V}_{\mathcal{A}} = \mathfrak{F}^{triv}$.

We now describe these construction adding more details to the condensed presentation in [**KW17**, §3]. A graphical version of these constructions is given in Appendix B. This graphical treatment provides a solid combinatorial language to write out the proofs in detail, improving the level of precision over that of [**KW17**]. These, however, use the full strength of graph formalism of [**BM08,KW17**], which is reviewed in Appendix A. With this in mind, we relegated the more detailed proofs to the appendix to not hinder the flow.

3.3.1. *A look ahead.* For clarity, we will deal first with the case of $\mathcal{C} = \mathcal{S}et$ and relay the subtleties of enrichment to §4. In full generality, for any split $\mathcal{O} \in$

$\mathcal{F}^{+gcp}$-$\mathcal{O}ps_{\mathcal{E}}$ there is an indexed enriched Feynman category $\mathfrak{F}_{\mathcal{O}}$ enriched over $\mathcal{E}$ whose $\mathcal{V}_{\mathcal{O}}$ is a freely enriched groupoid. With this construction, we can defined the sought after Feynman category for modules.

DEFINITION 3.16. An algebra over $\mathcal{O} \in \mathcal{F}^{+gcp}$-$\mathcal{O}ps$, aka. $\mathcal{O}$–module, is an element of $\mathcal{F}_{\mathcal{O}}$-$\mathcal{O}ps$.

Thus we can define modules over an $\mathcal{O} \in \mathcal{F}$-$\mathcal{O}ps_{\mathcal{C}}$ if $\mathfrak{F} = \tilde{\mathfrak{F}}^+$, for some $\tilde{\mathfrak{F}}$ and $\mathcal{O}$ lifts to $\mathfrak{F}^{+gcp}$-$\mathcal{O}ps_{\mathcal{C}}$.

3.3.2. *Plus construction.* Fix a Feynman category $\mathfrak{F} = (\mathcal{V}, \mathcal{F}, \imath)$ and define a new Feynman category as follows. Set $\mathcal{V}^+ = Iso(\mathcal{F} \downarrow \mathcal{V})$, that is basic morphisms and their isomorphisms. The objects of $\mathcal{F}^+$ are the morphisms of $\mathcal{F}$: $Obj(\mathcal{F}^+) = Mor(\mathcal{F})$ and $Iso(\mathcal{F}^+) = Iso(\mathcal{F} \downarrow \mathcal{F})$. The isomorphism are the $(\sigma \Downarrow \sigma') : \phi \to \phi' = \sigma' \circ \phi \circ \sigma^{-1}$ of (2.4). There is a natural inclusion of $\imath^+ : Iso(\mathcal{F} \downarrow \mathcal{V}) \to Iso(\mathcal{F} \downarrow \mathcal{F})$.

The other morphisms obtained by decomposing the source morphism into basic morphisms and then composing these basic morphisms using concatenation and tensor products to obtain the target morphism. More precisely, consider $\phi := \phi_0 \otimes \phi_1 \otimes \cdots \otimes \phi_n$ such that $\psi = \phi_0 \circ (\phi_1 \otimes \cdots \otimes \phi_n) \in (\mathcal{F} \downarrow \mathcal{V})$ is well defined. There will be one basic generating morphism $\phi \to \psi$ for such a pair denoted by $\gamma(\phi_0; \phi_1, \ldots, \phi_n)$.

A general morphism in $\mathcal{F}^+$ is a concatenation of tensor products, generating morphisms and isomorphisms modulo the relations of a monoidal category, that is associativity, units, interchange and equivariance under the action of isomorphisms given above.

PROPOSITION 3.17. *$\mathfrak{F}^+ = (\mathcal{V}^+, \mathcal{F}^+, \imath^+)$ is a Feynman category.*

PROOF. Condition (i) for $\mathfrak{F}^+$ is the condition (ii) for $\mathfrak{F}$. For condition (ii) fix any morphism $\Phi : \phi \to \psi$. We first show the existence of the decomposition (2.5). By condition (ii) for $\mathfrak{F}$, there exist isomorphisms $(\sigma \Downarrow \sigma') : \phi \to \phi' = \bigotimes_{v \in V} \phi_v$ and $(\tau \Downarrow \tau') : \psi \simeq \psi' = \bigotimes_{w \in W} \psi_w$. Thus we can assume that both ϕ and ψ are decomposable. The statement then follows from the Theorem B.1 in Appendix B. A short version is that any iteration of morphisms is given by a flow chart with input the source of ψ and the output, the target of ψ. The tensor product acts as disjoint union on the flow charts. Decomposing the target decomposes the flow charts by following the sources upwards. Such a chart is connected since otherwise the target would not lie in $\mathcal{V}$ and thus the decomposition is into connected components, with the compositions along these connected components yielding the ψ_v. The last axiom holds due to the axiom (iii) for $\mathfrak{F}$. □

LEMMA 3.18. *Strictifying $\mathcal{F}$, we have $\mathcal{V}^+$ is equivalent to $Iso(\imath^{\otimes} \downarrow \imath)$, $Iso(\mathcal{F}^+)$ is equivalent to $Iso(\imath^{\otimes} \downarrow \imath^{\otimes}) \simeq Mor(\mathcal{V})^{\otimes}$. Assuming strict associativity constraints, this is generated by commutativity constraints and morphisms of $\mathcal{V}$. $Mor(\mathcal{F}^+)$ is generated by these isomorphisms and the morphisms $\phi = \phi_0 \otimes \cdots \otimes \phi_n \to \psi = \phi_0 \circ (\phi_1 \otimes \cdots \otimes \phi_n) \in (\imath^{\otimes} \downarrow \imath)$.*

PROOF. Note that by axiom (ii) $Iso(\mathcal{F} \downarrow \mathcal{F})$ is equivalent to $(Iso(\mathcal{F} \downarrow \mathcal{V}))^{\otimes}$, and by the definition above $\mathcal{F}^+$ can be made strict by considering only objects in $(\mathcal{F} \downarrow \mathcal{V})^{\otimes} \simeq (\imath^{\otimes} \downarrow \imath)^{\otimes}$ together with their isomorphism, which are fixed by axiom (ii), and the generating morphisms $\gamma(\phi_0; \phi_1, \ldots, \phi_n) : \phi = \phi_0 \otimes \cdots \otimes \phi_n \mapsto \psi = \phi_0 \circ (\phi_1 \otimes \cdots \otimes \phi_n) \in (\imath^{\otimes} \downarrow \imath)$. □

The notation γ was chosen to be reminiscent of operadic compositions. Note that in the general case, there is a condition of composability.

COROLLARY 3.19. *$\mathcal{F}^+$-$\mathcal{O}ps_{\mathcal{C}}$ is equivalent to the category of strict monoidal functors $\mathcal{D}$ on the strictification, which are uniquely determined by the data:*

(1) *(Groupoid rep) A functor $\mathcal{D}$ from $\mathcal{V}^+ = Iso(\mathcal{V}^{\otimes} \downarrow \mathcal{V})$ to $\mathcal{C}$ which is given by:*
   (a) *(Object data): $\mathcal{D} : Obj(\mathcal{V}^{\otimes} \downarrow \mathcal{V}) \to Obj(\mathcal{C})$. That is an object $\mathcal{D}(\phi)$ of $\mathcal{C}$ for each basic morphism ϕ.*
   (b) *(Iso data): Actions of the isomorphisms $(\sigma \Downarrow \sigma')$. That is a left action of isomorphisms of $Mor(\mathcal{V})^{\otimes}$ and a right action of $Mor(\mathcal{V})$ on the $\mathcal{D}(\phi)$.*
(2) *(Composition data) Morphisms $\mathcal{D}(\gamma) : \mathcal{D}(\phi_0) \otimes \bigotimes_{i=1}^n \mathcal{D}(\phi_i) \to \mathcal{D}(\psi)$ where $\psi = \phi_0 \circ (\phi_1 \otimes \cdots \otimes \phi_n) \in (\mathcal{F} \downarrow \mathcal{V})$.*

Note that the groupoid data states that if $\phi' \simeq \phi$ via $(\sigma \Downarrow \sigma')$ in $(\mathcal{F} \downarrow \mathcal{F})$, then $\mathcal{D}((\sigma \Downarrow \sigma')) : \mathcal{D}(\phi) \xrightarrow{\sim} \mathcal{D}(\phi')$. In particular, if $\sigma, \sigma' \in Mor(\mathcal{V})$ are in same isomorphism class, that is are maps between elements of $\mathcal{V}$ which are in the same connected component of $\mathcal{V}$, then $\mathcal{D}(\sigma) \simeq \mathcal{D}(\sigma')$.

The corollary allows us to compute the first examples, which are the start of a ladder of complexity.

PROPOSITION 3.20.

(1) *For $\mathfrak{F}^{triv}$, the $\mathcal{O}ps$ are equivalent to associative monoids, aka. algebras.*
(2) *For $\mathcal{F}in\mathcal{S}et^+$ the $\mathcal{O}ps$ are equivalent to operads and for $\mathfrak{F}\mathfrak{S}^+$ to operads without constants.*
(3) *For $\mathcal{F}in\mathcal{S}et^+_<$ the $\mathcal{O}ps$ are equivalent to non–symmetric operads and for $\mathfrak{F}\mathfrak{S}^+_<$ non–symmetric to operads without constants.*

For the reader unfamiliar with operads the latter two statements can serve as a definition, see (3.7) below.

PROOF. To calculate the groupoid data: $\mathcal{V} = \underline{\mathbf{1}}$, $\mathcal{V}^+ = Iso(\mathcal{V}^{\otimes} \downarrow \mathcal{V}) = Mor(\mathcal{V}) = id_*$. So the groupoid part of $\mathcal{D}$ is fixed by an object $A := \mathcal{D}(id_*)$. The composition data is morphism $\mu : \mathcal{D}(\gamma(id_*; id_*) : \mathcal{D}(id_* \amalg id_*) = A \otimes A \to \mathcal{D}(id_* \circ id_*) = \mathcal{D}(id_*) = A$. The associativity of μ follows from the associativity $(id_* \circ \mathrm{id}_*) \circ id_* = id_* \circ (id_* \circ id_*)$.

For $\mathcal{F}in\mathcal{S}et^+$, $\mathcal{V} = \underline{\mathbf{1}}, \mathcal{V}^{\otimes} = \mathbb{S}$ and $\mathcal{V}^+ = Iso(\mathbb{S} \downarrow \{*\})$ has as objects the maps $\underline{n} \to \{*\}$ where n may be zero, where $\underline{0} = \emptyset$. Since $\{*\}$ is a one-element set, there is a unique map $\pi_n : \underline{n} \to \{*\}$ and hence the objects of $\mathcal{V}^+$ are given by $\mathbb{N}_0$. As the action of isomorphisms on the target $\underline{n}$ is trivial, the isomorphisms are exactly the isomorphism of the source $\underline{n}$ and hence $Aut(\pi_n) = \mathbb{S}_n$ and there are no isomorphisms between π_n and π_m for $n \neq m$. Thus $\mathcal{V}^+ \simeq \mathbb{S}$, and the groupoid data is a functor $\mathcal{O} : \mathbb{S} \to \mathcal{C}$ that is an $\mathbb{S}$–module which is a collection of objects $\mathcal{O}(n) = \mathcal{O}(\pi_n)$ together with an action of $\mathbb{S}_n$ on $\mathcal{O}(n)$, were we reverted from the notation $\mathcal{D}$ to the generic $\mathcal{O}$ for elements of $\mathcal{O}ps$, which also conforms with the usual operad notation. The composition data is given by the maps, which define an operad. Denoting $\mathcal{O}(\gamma(\pi_m;\pi_{n_1},\dots,\pi_{n_m}))=\gamma_{m;n_1,\dots,n_m}$

$$\begin{array}{ccc} \mathcal{O}(m)\otimes\mathcal{O}(n_1)\otimes\cdots\otimes\mathcal{O}(n_m) & \xrightarrow{\gamma_{m;n_1,\dots,n_m})} & \mathcal{O}(n) \\ \| & & \| \\ \mathcal{O}(\pi_m \amalg \pi_{n_1} \amalg\cdots\amalg \pi_{n_m}) & & \mathcal{O}(\pi_n) \end{array} \tag{3.7}$$

where $n=\sum n_i$. The associativity gives a condition on γ as does the $\mathbb{S}_n$ actions. These are spelled out in any text on operads, see e.g. [**MSS02, Kau04**] or [**GCKT20a**, §2.2.5] for a formula using indexing.

If we only have surjections, the map π_0 is missing and hence there is no $\mathcal{O}(0)$, in other words, there are no constant terms.

If we have an order on the fibers of the maps, then the objects of $\mathcal{V}^+$ are $(\pi_n,<_{\underline{n}})$ with isomorphisms acting transitively on the orders, that is we have as a groupoid the objects $\{\mathbb{S}_n\}, n\in\mathbb{N}_0$ with $\mathbb{S}_n$ acting on $\mathbb{S}_n$ as automorphisms via the regular representation. Its skeleton is simply $\mathbb{N}_0$ as a discrete category. A representative for each isomorphism class is given by $(\pi_n,<)$ where $<$ is the standard order on $\underline{n}$. Thus the groupoid part is simply given by a sequence of objects $\mathcal{O}(n)$. Composition is as above where γ is fixed above and it has to satisfy the condition of associativity. □

REMARK 3.21. One can ask about the pseudo–operad structure. Using the subset with $\phi_i=\pi_n$, and all other $\phi_j=\pi_1$ we obtain maps

$$\gamma_{n;1,\dots,,1,m,,\dots,1}:\mathcal{O}(n)\otimes\mathcal{O}(1)\otimes\cdots\otimes\mathcal{O}(1)\otimes\mathcal{O}(m)\otimes\mathcal{O}(1)\otimes\cdots\otimes\mathcal{O}(1)\to\mathcal{O}(n+m-1) \tag{3.8}$$

To obtain the usual pseudo–operad maps $\circ_i$, we at this time lack an operadic unit.

If we have a unit $u:\mathbb{1}\to\mathcal{O}(1)$ which is a unit for the operation γ:

$$\gamma_{1;n}\circ(u\otimes id_{\mathcal{O}(n)}) = id_{\mathcal{O}(n)} \tag{3.9}$$

$$\gamma_{m;1,\dots,1}\circ(id_{\mathcal{O}(m)}\otimes u^{\otimes m}) = id_{\mathcal{O}(m)} \tag{3.10}$$

where we tacitly used the unit constraints. Thus if there is a unit and then we can define

$$\begin{aligned} \circ_i:\mathcal{O}(n)\otimes\mathcal{O}(m) &\simeq \mathcal{O}(n)\otimes\mathbb{1}\otimes\cdots\otimes\mathbb{1}\otimes\mathcal{O}(m)\otimes\mathbb{1}\otimes\cdots\otimes\mathbb{1} \\ &\xrightarrow{id_{\mathcal{O}(n)}\otimes u\otimes\cdots\otimes u\otimes id_{\mathcal{O}(m)} u\otimes\cdots\otimes u} \mathcal{O}(n)\otimes\mathcal{O}(1)\otimes\cdots\otimes\mathcal{O}(1)\otimes\mathcal{O}(m)\otimes\mathcal{O}(1)\otimes\cdots\otimes\mathcal{O}(1) \\ &\xrightarrow{\gamma_{n;1,\dots,1,m,1,\dots,1}} \mathcal{O}(n+m-1) \end{aligned} \tag{3.11}$$

and using (3.9) and (3.10), we can recover γ from (3.11) by using iterated $\circ_i$ operations and associativity.

However, if $\mathcal{O}$ is reduced, which means that $\mathcal{O}(1)=\mathbb{1}$ the composition data factors trough unit constraints then $\mathcal{O}$ is automatically unital. In the general case, this is formalized by the definition below.

DEFINITION 3.22. A groupoid compatible pointing for a functor $\mathcal{D}\in\mathcal{F}^+\text{-}\mathcal{O}ps$ is a collection of elements $u_\sigma:\mathbb{1}_\mathcal{C}\to\mathcal{D}(\sigma)$, $\sigma\in Mor(\mathcal{V})$, which satisfy $\mathcal{D}(\sigma)\circ u_{\sigma'}=u_{\sigma\circ\sigma'}$ and are compatible with groupoid action and the composition data $\mathcal{D}(\gamma)$. I.e. the following diagrams commute:

$$
(3.12)\qquad \begin{array}{ccc}
\mathcal{D}(\phi_0)\otimes\mathcal{D}(\sigma_1)\otimes\cdots\otimes\mathcal{D}(\sigma_n) & \xrightarrow{\mathcal{D}(\gamma)} & \mathcal{D}(\phi_0\circ(\sigma_1\otimes\cdots\otimes\sigma_n)) \\
\uparrow{\scriptstyle id\otimes u_{\sigma_1}\otimes\cdots\otimes u_{\sigma_n}} & & \simeq\uparrow{\scriptstyle \mathcal{D}((\sigma_1^{-1}\otimes\cdots\otimes\sigma_n^{-1}\Downarrow id))} \\
\mathcal{D}(\phi_0)\otimes\mathbb{1}\otimes\cdots\otimes\mathbb{1} & \xleftarrow[\simeq]{\text{unit constraints}} & \mathcal{D}(\phi_0)
\end{array}
$$

where the right morphisms is given by the groupoid data and

$$
(3.13)\qquad \begin{array}{ccc}
\mathcal{D}(\sigma_0)\otimes\mathcal{D}(\phi_1) & \xrightarrow{\mathcal{D}(\gamma)} & \mathcal{D}(\sigma_0\circ\phi_1) \\
\uparrow{\scriptstyle u^{\sigma_0}\otimes id} & & \simeq\uparrow{\scriptstyle \mathcal{D}((id\Downarrow\sigma_0))} \\
\mathbb{1}\otimes\mathcal{D}(\phi_1) & \xleftarrow[\simeq]{\text{unit constraint}} & \mathcal{D}(\phi_1)
\end{array}
$$

A functor $\mathcal{D}$ is called reduced if $\mathcal{D}(\sigma)\simeq\mathbb{1}$ for all $\sigma\in Mor(\mathcal{V})$.

A functor $\mathcal{D}$ and a choice of groupoid compatible pointing, is called *groupoid compatibly pointed (gcp)* functor.

A functor $\mathcal{D}\in\mathcal{F}^+$-$\mathcal{O}ps$ is a *hyper–functor* if it is reduced and gcp using the identification $\mathcal{D}(\sigma)\simeq\mathbb{1}$.

REMARK 3.23.

(1) Due to the groupoid data, to check that $\mathcal{D}$ is reduced it suffices to check $\mathcal{D}(id_*)\simeq\mathbb{1}$ for a set of representatives $*$ for the isomorphism classes of $\mathcal{V}$.
(2) Due to the compatibility with the action of the groupoid, the u_σ are also already fixed by a choice of the u_{id_*} where $*$ runs through representative of the isomorphism classes of objects of $\mathcal{V}$. Concretely, if $\sigma:*\to*'$ an isomorphism, then $\mathcal{D}((id\Downarrow\sigma))\circ u_{id_*}=u_\sigma$.
(3) From (3.13) and (3.12) it follows that $\mathcal{D}(\gamma(\sigma;\sigma'))\circ(u_\sigma\otimes u_{\sigma'})=u_{\sigma\circ\sigma}$.

EXAMPLE 3.24. For $\mathfrak{F}^{triv}$, we retrieve the motivating Example 3.15. The $(\mathfrak{F}^{triv})^+=\mathcal{O}ps$ are associative monoids in $\mathcal{C}$. There is again only one isomorphism in the skeleton of $\mathcal{V}$ namely id_*. Gcp means that the monoids are unital and reduced means that $M\simeq\mathbb{1}$. Thus a hyper–functor is trivial.

For $\mathcal{F}in\mathcal{S}et$, we retrieve Remark 3.21. The $\mathcal{F}in\mathcal{S}et^+$-$\mathcal{O}ps$ are operads, the gcp respectively reduced $\mathcal{F}in\mathcal{S}et^+$-$\mathcal{O}ps$ are unital operads, respectively the reduced operads. The hyper functors are unital reduced operads.

REMARK 3.25. For a gcp functor $\mathcal{D}$, we can define the analogue of the $\circ_i$ of (3.11).

DEFINITION 3.26. For a gcp functor $\mathcal{D}$, let $\phi:*_1\otimes\cdots\otimes*_n\to Y$ and $\psi:*'_1\otimes\cdots\otimes*'_n\to*_i$, then define:

$$
(3.14)\quad \circ_i:\mathcal{D}(\phi)\otimes\mathcal{D}(\psi)\simeq\mathcal{D}(\phi)\otimes\mathbb{1}\otimes\cdots\otimes\mathbb{1}\otimes\mathcal{D}(\psi)\otimes\mathbb{1}\otimes\cdots\otimes\mathbb{1}
$$
$$
\xrightarrow{id_{\mathcal{D}(\phi)}\otimes u_{id_{*_1}}\otimes\cdots\otimes u_{*_{i-1}}\otimes id_{\mathcal{D}(\psi)}u_{*_{i+1}}\otimes\cdots\otimes u_{*_n}} \mathcal{D}(\phi)\otimes\mathcal{D}(id_{*_1})\otimes\cdots\otimes\mathcal{D}(id_{*_{i-1}})\otimes\mathcal{D}(\psi)
$$
$$
\otimes\mathcal{D}(id_{*_{i+1}})\otimes\cdots\otimes\mathcal{D}(id_{*_n})
$$
$$
\xrightarrow{\gamma}\mathcal{D}(\phi\circ id_{*_1}\otimes\cdots\otimes id_{*_{i-1}}\otimes\phi)
$$

REMARK 3.27. The difference between this and the equation (3.7) is that there are possibly many objects in $\mathcal{F}$ and hence more identities. Using the general compatibility conditions (3.19) and (3.20), we can recover the γ from the $\circ_i$. Thus the $\circ_i$ give another generating set which is "local" in the sense that it involves only two morphisms and equivalently only one edge in the graphic description of Appendix B.

3.3.3. *Signs as hyper–functors.* For a finite-dimensional k vector space V of dimension n, let $\det(V)$ be the graded vector space $\det(V) = \Sigma^{-n}\Lambda^n V$; this is the one-dimensional top exterior power of V, concentrated in degree $-n$. If S is a finite set, let $\det(S) = \det(k^S)$. There is a natural action of $Aut(S)$ on $\det(S)$. Choosing an order on S, thereby identifying the set $Aut(S)$ with $\mathbb{S}_n$ as an $\mathbb{S}_n$-module $\Sigma^{-|S|}\text{sign}(\mathbb{S}_{|S|})$, where sign is the sign representation.

EXAMPLE 3.28 (Signs: $\mathfrak{K}$). For $\mathfrak{G}$ (see Appendix A, especially §A.2), we define $\mathfrak{K} \in \mathcal{A}gg^+\text{-}\mathcal{O}ps$ as follows.

$$\mathfrak{K}(\phi) = \det(E_{ghost}(\phi)) \tag{3.15}$$

The composition is given by $\mathfrak{K}(\phi_0) \otimes \mathfrak{K}(\phi_1) \otimes \cdots \otimes \mathfrak{K}(\phi_n) = \det(E_{ghost}(\phi)) \otimes \det(E_{ghost}(\phi_1)) \otimes \cdots \otimes \det(E_{ghost}(\phi_n)) \to \det(E_{ghost(\phi_0) \amalg E_{ghost}(\phi_1)} \amalg \cdots \amalg E_{ghost}(\phi_n)) = \det(E_{ghost}(\phi_0 \circ (\phi_1 \otimes \cdots \otimes \phi_n))$ with the identification according to (A.1), see also Lemma A.7. Since an isomorphism σ does not have any edges, $\mathfrak{K}(\sigma) = \mathbb{1} = k$ and the composition is simply given by the unit constraints. Hence $\mathfrak{K}$ is a hyper–functor. This generalizes [**GK98**].

As the twist is a hyper–functor, for the compositions data it is enough to give the data of the $\circ_i$ which are simply $\mathfrak{K}(\phi) \circ_i \mathfrak{K}(\psi) = \det(E_{ghost}(\phi) \amalg E_{ghost}(\psi))$ and check that these are appropriately associative, which is straightforward.

The fact that this gives a hyper-functor basically boils down to the fact that the ghost graph uniquely determines the isomorphism class of a basic morphism in $\mathcal{A}gg$.

EXAMPLE 3.29 (Homology twist). We similarly define Det via $Det(\phi) = \det(H_1(\mathbb{\Gamma}(\phi)))$.

3.3.4. *$\mathcal{V}$–twists and suspension.* There are special types of hyper–functors called $\mathcal{V}$–twists, see [**KW17**, §4.2.1] generalizing cobordism twist of [**GK98**]. Consider $Pic(\mathcal{C})$, that is the full subcategory of tensor invertible elements in $\mathcal{C}$ and an $\mathfrak{L} \in \mathcal{V}\text{-}\mathcal{O}ps_{Pic(\mathcal{C})}$.

Given a hyper–functor $\mathcal{D}$ one defines the $\mathfrak{L}$ twist as

$$\mathcal{D}_{\mathfrak{L}}(\phi) = \mathfrak{L}(t\phi)^{-1} \otimes \mathcal{D}(\phi) \otimes \mathfrak{L}(s(\phi)) \tag{3.16}$$

where again s, t are the source and target maps and the composition uses the morphism $\mathfrak{L}(X)^{-1} \otimes \mathfrak{L}(X) \simeq \mathbb{1}$.

EXAMPLE 3.30 (Suspensions). One of the most important twists is given by using suspension. These are defined for graph based Feynman categories. In particular for the operadic and modular operad categories.

There are two interesting versions. The first is the naive suspension, Σ which takes values $\Sigma(*_S) = \Sigma\mathbb{1}$.

The second, in the genus marked case, is $\mathfrak{s}$ given by $\mathfrak{s}(*_{S,g}) = \Sigma^{-2(g-1)-|S|}\text{sign}_{\mathbb{S}_{|S|}}$. Without the genus marking $g = 0$.

There is a fundamental relation

$$\mathfrak{K} \simeq Det_{\mathfrak{s}\Sigma} \tag{3.17}$$

This states that if $\mathbb{\Gamma}(\phi)$ is contractible, the $\mathfrak{K}$ is simply the suspension $\mathfrak{s}\Sigma$. This is precisely the odd structure for the bar complex, see §5.1. But, *if the underlying graph has topology, the mere suspensions do not suffice.*

This is one of the basic mantras explained in detail in [**KWZ12**].

3.3.5. *Gcp version and hyper version of* $\mathfrak{F}^+$. For the different types of functors in Definition 3.22 there are Feynman categories through which these functors factor.

For this one adjoins morphisms and mods out by relations. The paradigmatic example is $\mathcal{F}in\mathcal{S}et$ which is generated by FS and the morphism $i : \emptyset \to \{*\}$, which satisfies the relation $\pi_{S_+} \circ (id_S \amalg i) = \pi_S$, where $\pi_S : S \to \{*\}$ is the unique surjection, $S_+ = S \amalg \{*\}$ and $id_S \amalg i : S = S \amalg \emptyset \to S$.

DEFINITION 3.31. We define $\mathfrak{F}^{+gcp} = (\mathcal{V}^{+gcp}, \mathcal{F}^{+gcp}, \imath^{+gcp})$ as follows: $\mathcal{V}^{+gcp} = \mathcal{V}^+$, $Iso(\mathcal{F}^{+gcp}) = Iso(\mathcal{F}^+)$ and $\imath^{+gcp} = \imath^+$. But for $\mathcal{F}^{+gcp}$, we first freely adjoin a morphism $i_\sigma : \mathbb{1}_{\mathcal{F}^+} = id_{\mathbb{1}_{\mathcal{F}}} \to \sigma$ for all $\sigma \in \mathcal{V}^+$, and then mod out by the relations

$$(\sigma \Downarrow \sigma')(i_\tau) = i_{\sigma' \circ \tau \circ \sigma} \tag{3.18}$$

implementing the compatibility with the groupoid structure of $\mathcal{V}^+$. The compatibility with the generating morphisms is forced by modding out by the relations postulating that the following diagrams commute

$$\begin{array}{ccc} \phi_0 \otimes \sigma_1 \otimes \cdots \otimes \sigma_n & \xrightarrow{\gamma} & \phi_0 \circ (\sigma_1 \otimes \cdots \otimes \sigma_n) \\ \uparrow{\scriptstyle id \otimes i_{\sigma_1} \otimes \cdots \otimes i_{\sigma_n}} & & \simeq \uparrow {\scriptstyle (\sigma_1^{-1} \otimes \cdots \otimes \sigma_n^{-1}, id)} \\ \phi_0 \otimes \mathbb{1}_{\mathcal{F}^+} \otimes \cdots \otimes \mathbb{1}_{\mathcal{F}^+} & \xrightarrow[\text{unit constraints}]{\simeq} & \phi_0 \end{array} \tag{3.19}$$

where the right morphisms is given by the groupoid data and

$$\begin{array}{ccc} \sigma_0 \otimes \phi_1 & \xrightarrow{\gamma} & \sigma_0 \circ \phi_1 \\ \uparrow{\scriptstyle i_{\sigma_0} \otimes id} & & \simeq \uparrow {\scriptstyle (id \Downarrow \sigma_0)} \\ \mathbb{1}_{\mathcal{F}^+} \otimes \phi_1 & \xleftarrow[\simeq]{\text{unit constraint}} & \phi_1 \end{array} \tag{3.20}$$

$\mathfrak{F}^{hyp} = (\mathcal{V}^{hyp}, \mathcal{F}^{hyp}, \imath^{hyp})$ is defined as follows: $\mathcal{F}^{hyp}$ is the quotient of $\mathcal{F}^{+gcp}$ by the relation that i_σ is invertible. $\mathcal{V}^{hyp}$ is the full sub–groupoid of $\mathcal{V}^+$ whose objects are non–isomorphisms and $\imath^{hyp}$ is the restriction of i^+.

PROPOSITION 3.32. *Both* $\mathfrak{F}^{+gcp}$ *and* $\mathfrak{F}^{hyp}$ *are Feynman categories. Moreover any gcp functor factors through* $\mathfrak{F}^{+gcp}$ *and any hyper–functor factors through* $\mathfrak{F}^{hyp}$.

PROOF. The last statement is straightforward. The first statement for for $\mathfrak{F}^{+gcp}$ this is again straightforward, as we are only adding element–type morphisms and the relations preserve decomposability. For $\mathfrak{F}^{hyp}$ this follows from a two–step argument: First, inverting the i_σ, we see that in $\mathcal{F}^{hyp}$ the full subcategory spanned by the elements $\sigma \in Mor(\mathcal{V})$ and $\mathbb{1}$ is equivalent to a discrete category. There is a unique morphism between any two objects which is an isomorphism: $\mathrm{Hom}_{\mathcal{F}^{hyp}}(\sigma, \sigma') = \{i_{\sigma'} \circ i_\sigma\}$, $\mathrm{Hom}(\mathbb{1}, \sigma) = \{i_\sigma\}$, $\mathrm{Hom}(\sigma, \mathbb{1}) = \{i_\sigma^{-1}\}$ and $\mathrm{Hom}(\mathbb{1}, \mathbb{1}) = id_{\mathbb{1}}$. Contracting this isomorphism class to $\mathbb{1}$, we obtain an equivalent category that is clearly a Feynman category with vertices $\mathcal{V}^{hyp}$. □

REMARK 3.33. From (3.20) and (3.19) it follows that $\gamma(i_\sigma \otimes i_{\sigma'}) = i_{\sigma\circ\sigma'}$. Also, the analog of Remark 3.23 applies using i_σ in lieu of the u_σ.

3.4. Computations for the plus construction: Realizing the first ladder.

In this section, we will apply the explicit graphical presentation for the plus construction of Appendix B and starting from $\mathfrak{F}^{triv}$ construct the categories corresponding to various types of algebras in a first step and in a second step the Feynman categories for various forms of operads.

3.4.1. *From objects to monoids, via the plus construction.* Using Theorem B.1, we can compute $(\mathfrak{F}^{triv})^+$ and $(\mathfrak{F}^{triv})^{+gcp}$ to obtain the first rung of the fundamental ladder.

In $\mathcal{C}rl_{(\mathfrak{F}^{triv})^+}$, see §B.1.2, the vertex color is necessarily id_* as it is the only possible morphism and all the flag colors are $*$. The decorations $(\boldsymbol{\sigma}, \boldsymbol{\imath})$ for morphisms are necessarily given by identities. There is only one objects in $\mathcal{C}rl_{(\mathfrak{F}^{triv})^+}$ whose isomorphism group is trivial. Hence the objects of $(\mathcal{C}rl_{(\mathfrak{F}^{triv})^+})^{\otimes}$ are the $*_{id_n}$, with $id_n =: id_{*^{\otimes n}} = (id_*)^{\otimes n}$ whose automorphisms are $\mathbb{S}_n$,. Identifying n with $\mathrm{id}_*^{\otimes n}$ automorphisms $Iso(\mathcal{A}gg_{(\mathfrak{F}^{triv})^+}) \simeq (\mathcal{C}rl_{(\mathfrak{F}^{triv})^+})^{\otimes} = \mathbb{S}$

A general morphism in $\mathcal{A}gg_{(\mathfrak{F}^{triv})^+}$, see §B.1.3, is given by decorated forests whose trees are linear, i.e. they all have bi–valent vertices, since id_* is the only possible decoration. The map g_V is a surjection $\underline{n} \twoheadrightarrow \underline{m}$. Let $|g_V^{-1}(i)| = n_i$. Writing a 2–regular, aka. linear, tree with n vertices as $\phi\, n$, we see that the morphisms in $\mathcal{A}gg_{(\mathfrak{F}^{triv})^+}$ have underlying trees $\phi\, n_1 \cdots \phi\, n_m$. The vertices in each $\phi\, n_i$ are ordered from the root to the top, thus giving an order on the fibers $g_V^{-1}(i)$, as each vertex has a unique outgoing flag. The extra data identifies the root, that is the only outgoing flag, of $\phi\, n_i$ with the target ϕ corresponding to i in $\underline{m}$.

A basic morphism, is thus given by one linear tree with an ordering of its vertices $(\phi\, n, <)$. Here $<$ can alternatively be thought of as an identification of the vertices of the tree with the factors of id in $(id_*)^{\otimes n}$, that is a bijection $\underline{n} \leftrightarrow \underline{n}$ aka. an order on $\underline{n}$. The isomorphisms are given by permuting the (source) vertices. By pre–composing, they act as permutations on the vertices of the linear tree and hence transitively on the linear orders on the fibers.

Considering $(\mathfrak{F}^{triv})^{+gcp}$, there is one added morphism $i_{id_*} : \emptyset \to \phi$. It is convenient to introduce the notation $\phi\, u$ the ghost-graph of the morphism i_{id_*}. There are the relations $(\pi_2, 1<2) \circ (id, 1) \amalg i = (\pi_2, 1<2) \circ i \amalg (id, 1) = (\pi_2, 1<2)$. Note that $(\pi_2, 2<1) = (\pi_2, 1<2) \circ \tau_{12}$ and thus we also have the unit equation regardless of the position and order of the unit insertion.

An example is given in Figure 2.

PROPOSITION 3.34. *As Feynman categories* $(\mathfrak{F}^{triv})^+ \simeq \mathfrak{F}\mathfrak{S}^{>}$ *and* $(\mathfrak{F}^{triv})^{+gcp} \simeq$ $\mathfrak{N}\mathfrak{C}Set$.

PROOF. The morphism $\mathfrak{f}$ of Feynman categories consists of the functor v given by $\phi = \phi\, id_* \to 1$ and the functor f given by $\phi^{\otimes n} \to \underline{n}$ and on basic morphisms $f(\phi\, n, <) = (\pi_n, <), n \geq 1$ where the order of $\phi\, n$ is given by the height of the vertex i and the order on $\underline{n}$ is the order induced by this. In the case of $(\mathfrak{F}^{triv})^{+gcp}$ we also have $f(\phi\ 0 = i_{id_*}) = i : \emptyset \to \underline{1}$, as a generator with the appropriate relations. □

REMARK 3.35. This gives another natural interpretation of $\mathcal{N}\mathcal{C}Set$ and ΔS complementary to [**Lod98, FL91, PR02, So03**].

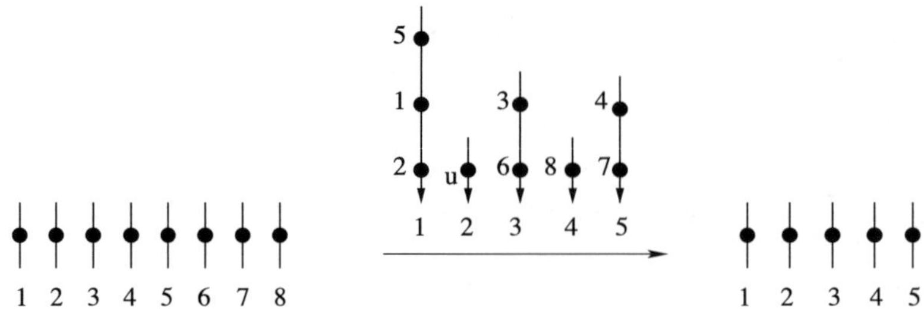

FIGURE 2. A morphism given by a marked linear forest. This yields the same morphism in $(\mathfrak{F}^{triv})^{+gcp} = \mathfrak{NCSet}$ given in Figure 1. The underlying forest is ⧫ 3 •| 0 •| 2 •| 1 •| 2, which fixes the surjection $\underline{8} \twoheadrightarrow \underline{5}$.

3.4.2. *Operads and non–Sigma operads via the plus construction: The second rung.* One can now apply the plus construction again that is compute $((\mathfrak{F}^{triv})^+)^+ = \mathfrak{FS}^+_> \simeq \mathfrak{O}^{\neg\Sigma}$. We can also obtain operads through the plus construction. Let $\mathfrak{O}$ be the Feynman category for operads, $\mathfrak{O}^{\neg\Sigma}$ that for non–Sigma, operads and let $\mathfrak{O}_{unital}$ and $\mathfrak{O}^{\neg\Sigma}_{unital}$ the ones for unital operads and non–Sigma operads. The unital operads are sometimes called May operads and the corresponding Feynman category was called $\mathfrak{F}_{May}$ in [**KW17**]. The latter have an adjoined unit morphism $u :\to \mathcal{O}(1)$ which satisfies the unit equations, see [**KW17**, §2.2], see also Table 7. We will also consider $\mathfrak{O}_0$ and $\mathfrak{O}_0^{\neg\Sigma}$ whose $\mathcal{O}ps$ whose $\mathcal{O}(1) \simeq \mathbb{1}$.

PROPOSITION 3.36. *We have the following identifications:*

$$\mathfrak{FinSet}^+ \simeq \mathfrak{O} \qquad \mathfrak{FinSet}^{+gcp} \simeq \mathfrak{O}_{unital} \qquad \mathfrak{FinSet}^{hyp} = \mathfrak{O}_0$$
$$\mathfrak{NCSet}^+ \simeq \mathfrak{O}^{\neg\Sigma} \qquad \mathfrak{NCSet}^{+gcp} \simeq \mathfrak{O}^{\neg\Sigma}_{unital} \qquad \mathfrak{NCSet}^{hyp} = \mathfrak{O}_0^{\neg-\Sigma}$$

Restricting to surjections, one obtains the Feynman categories for the same representations, but without constants, i.e. $\mathcal{O}(0)$. *In particular,* $\mathfrak{FS}^{hyp} = \mathfrak{O}^{red}$ *that is the Feynman category whose* $\mathcal{O}ps$ *are reduced operads, no* $\mathcal{O}(0)$ *and* $\mathcal{O}(1) \simeq \mathbb{1}$.

PROOF. Consider $\mathfrak{FinSet}^+$. There is only one flag color $\underline{1}$. Up to isomorphism, the basic morphisms are given by $\pi_n : \underline{n} \to \underline{1}$, where $\pi_0 = i : \emptyset \to \underline{1}$. These give a smaller, but equivalent subcategory, known as the biased version. These give rise to the corollas $*_{\pi_n}$, which are the isomorphism classes of objects in $\mathcal{C}rl_{\mathfrak{FinSet}^+}$. The automorphisms in $\mathcal{C}rl_{\mathfrak{FinSet}^+}$ of $*_{\pi_n}$ are $\mathbb{S}_n$ as $\pi_n \circ \sigma = \pi_n$ and σ acts trivially on the flag labeling up to equivalence, since (B.1) holds for any permutation. Thus, replacing $\mathcal{C}rl_{\mathfrak{FinSet}^+}$ by a skeleton, we have $\mathcal{C}rl^+_{\mathfrak{FinSet}} \simeq \mathbb{S}$.

The objects in $\mathcal{A}gg_{\mathcal{F}in\mathcal{S}et^+}$ are then forests of $*_{\pi_n}$ or more generally $*_{\pi_S}$, where $\pi_S : S \twoheadrightarrow \underline{1}$. The only color isomorphism is $id_{\underline{1}}$, thus the data $(\boldsymbol{\sigma}, \boldsymbol{\imath})$ is trivial. Thus, up to isomorphism, the basic morphisms in $\mathcal{A}gg_{\mathfrak{FinSet}}$ are exactly the graph morphisms $\phi : *_{\underline{n}_1} \amalg \cdots \amalg *_{\underline{n}_m} \to *_{\underline{n}}$ whose underlying ghost graph is a tree. This implies that $n = \sum_i n_i$ The arity of vertex decorated by π_n is n and hence knowing the arity fixes this decoration. The flag colors, the edge and tail decorations are all redundant as well. This means that we have obtained the skeletal version of $\mathfrak{O}$.

More generally allowing for the morphisms $\pi_S : S \twoheadrightarrow \{r\}$, the vertices $*_{\pi_S}$ will be or arity $|S|$ and have incoming flag set in bijection with S and an outgoing flag corresponds to $\{r\}$. Hence we have $*_{\pi_S}$ is completely determined by the rooted corolla $*_{S\amalg\{r\},t}$. And we obtain $\mathfrak{FinSet}^+ = \mathfrak{O}$.

In $\mathfrak{FinSet}^{+gcp}$ there is an extra morphism $u := i_{id_{\underline{1}}} : \emptyset \to id_{\underline{1}}$ which provides a unit-element morphism, [**KW17**, 2.2]. It is straightforward to check that for $\mathcal{O} \in \mathfrak{FinSet}^{+gcp}\text{-}\mathcal{O}ps$, $\mathcal{O}(u)$ is a unit.

In the case of $\mathfrak{NCSet}$, consider the objects of $\mathcal{C}rl_{\mathfrak{NCSet}^+}$. In the biased version, that is using only the sets $\underline{n}$, the objects are $*_{\pi_n,<_n}$, where $<_n$ is an order on n. The permutations act transitively on the orders of the decorated corollas: $(p \Downarrow id)(*_{\pi_n,<_n}) = *_{\pi_n,p(\circ <_n)}$, were $p(\circ <_n)$ is the permutation of §B.1.2. Thus there are no automorphisms and a skeletal version of $\mathcal{C}rl_{\mathfrak{NCSet}^+}$ is the discrete category $\mathbb{N}_0$ whose objects are the natural numbers with only identity morphisms. This means that up to isomorphism, there is a unique ordered vertex of arity n for each n which can be represented by the planar, planed corolla $*_{\underline{n}\amalg\{r\},r}$. Proceeding to $\mathcal{A}gg_{\mathfrak{NCSet}^+}$, we see that the planar, planted forest underlying the morphism uniquely determines the morphism up to isomorphism. Thus, we have objects given by planar, planted corollas and basic morphisms given by graph morphisms whose ghost graph is a tree. This tree becomes planar if one pulls back the orders from the orders on the source using the identification of the vertices of the graph and the source. Finally, compatibility states that the order of the leaves of the target corolla is that of the leaves of the tree. This is the description of $\mathcal{O}^{\neg\Sigma} = \mathcal{O}_{\text{dec } Assoc}$, see [**KL16**] and Appendix A.5.2 below. □

LEMMA 3.37. *Forgetting the order yield a forgetful functor $\mathfrak{p} : \mathfrak{O}^{\neg\Sigma} \to \mathfrak{O}$. This morphism is a cover with decoration $p_!(\mathcal{T}) = Assoc$.*

Furthermore $Assoc((*_{S\amalg\{r\}}, r)) = \{$all orders $<_S$ on $S\}$, with the usual composition: $\gamma((*_{S\amalg\{r_0\}\cdot,r_0,<_S}; (*_{T_1\amalg\{r_1\}}, r_n, <_n), \dots, (*_{T_1\amalg\{r_1\}}, r_1, <_1)) = *_{T\amalg\{r_0\},r_0,<_T}$ where $T = T_1 \amalg \dots \amalg T_n$ with the order $<_T = <_1 \amalg \dots \amalg <_n$.

This cover restricts to the non–unital and reduced $\mathcal{O}(1)$ and reduced cases.

PROOF. This is contained in [**KL16**]. The key steps are: to define $\mathfrak{p} = (p_{\mathcal{V}}, p)$, functor $p_{\mathcal{V}}$ is given simply by $(*_{S\amalg\{r\}}, r, <_S) \mapsto (*_{S\amalg\{r\}}, r)$ on objects. On morphisms of $\mathfrak{O}^{\neg\Sigma}$, p is given by forgetting the planar structure on the forests. This is a cover, as it is surjective on objects and morphisms and every morphism has a unique lift once the target is fixed. This latter is the case as the planar structure on the forest is completely fixed by the planar structure of the source corollas. On objects computing $p_!(*_{S\amalg\{r\}}, r)$ yields the fiber that is the set $\{(*_{S\amalg\{r\}}, r, <_S)\}$ or simply the set of orders on S. Composition of the linear orders is that according to the forests. I.e. lifting the planar structure of the source corollas yields a planar structure on each tree underlying a morphism and this in turn yields the given order of the leaves, which are the set T. □

4. Modules and enriched Feynman categories

4.1. Enriched categories. To consider modules of algebras, or more generally if $\mathcal{O} \in \mathcal{F}\text{-}\mathcal{O}ps_{\mathcal{C}}$ their modules will need Feynman categories enriched in $\mathcal{C}$. The general reference is [**Kel82**], to which we refer for full details. We will give the salient features here.

Recall that a category enriched $\mathcal{F}$ in a symmetric monoidal category $(\mathcal{E}, \otimes)$ has a class of objects $Obj(\mathcal{F})$ and morphisms $\mathrm{Hom}(X,Y) \in Obj(\mathcal{E})$ together with a associative unital composition maps which are morphisms in $\mathcal{E}$: $\circ : \mathrm{Hom}(Y,Z) \otimes_{\mathcal{E}} \mathrm{Hom}(X,Y) \to \mathrm{Hom}(X,Z)$ and units $id_X : \mathbb{1}_{\mathcal{E}} \to \mathrm{Hom}(X,X)$. Likewise for a monoidal structure, all structure morphisms are morphisms and objects $\mathcal{E}$, in particular:

$$\otimes : \underline{\mathrm{Hom}}(X,Y) \otimes_{\mathcal{E}} \underline{\mathrm{Hom}}(Z,W) \to \underline{\mathrm{Hom}}(X \otimes Z, Y \otimes W) \tag{4.1}$$

is a morphism in $\mathcal{E}$.

EXAMPLE 4.1. Consider the category $\underline{G_k}$ enriched over $\mathcal{V}ect_k$ which has one object $*$ and morphism set $k[G]$ with composition on basis elements given by $\circ(g,h) = g \circ h$.

EXAMPLE 4.2 (Internal Hom). A monoidal category can be enriched over itself. The standard example is $\mathcal{V}ect_k$, since $\underline{\mathrm{Hom}}(V,W)$ again has the structure of a vector space. More generally, $\mathcal{E}$ can be enriched over itself if it is closed. This means that it an internal hom, $\underline{\mathrm{Hom}}(X,Y) \in E$, and $\underline{\mathrm{Hom}}(X \otimes Y, Z) \simeq \underline{\mathrm{Hom}}(X, \underline{\mathrm{Hom}}(Y,Z))$ functorially, see [**Kel82**, §1.6] for details.

One can then consider functors between two categories $\mathcal{C}$ and $\mathcal{D}$ enriched over the same $\mathcal{E}$ in straightforward formalism. In particular, $\underline{\mathrm{Hom}}_{\mathcal{C}}(X,Y) \to \underline{\mathrm{Hom}}_{\mathcal{D}}(\mathcal{O}(X),\mathcal{O}(Y)) : \phi \mapsto \mathcal{O}(\phi)$ is a morphism in $\mathcal{E}$.

EXAMPLE 4.3. Functors $\mathcal{O}$ from $\underline{G_k}$ to $\mathcal{V}ect_k$ thought of as enriched over $\mathcal{V}ect_k$ are $k[G]$-modules. Let $M = \mathcal{O}(*)$ the fact that the map $\mathrm{Hom}(*,*) = k[G] \to \underline{\mathrm{Hom}}(M,M)$ is a morphism in $\mathcal{V}ect_k$ means that the action is k-linear, i.e. the action is given by $\mu : k[G] \otimes_k M \to M$.

4.1.1. *Freely enriched categories.* Given an enriched category one can define an underlying category by defining the underlying morphism via $Hom_{\mathcal{S}et}(X,Y) := \mathrm{Hom}_{\mathcal{E}}(\mathbb{1}_{\mathcal{E}}, \underline{\mathrm{Hom}}(X,Y))$. This is actually a 2–functor, see [**Kel82**], which has an adjoint, called free enrichment. We will use the notation $F_{\mathcal{E}}$ for the free enriched version of $\mathcal{E}$. E.g. if $\mathcal{E} = \mathcal{V}ect_k$ then $Hom_{\mathcal{F}_{\mathcal{E}}}(X,Y)$ is the free vector space on $Hom_{\mathcal{F}}(X,Y)$. If $\mathcal{E} = \mathcal{T}op$ the $Hom_{\mathcal{F}_{\mathcal{E}}}(X,Y)$ is $Hom_{\mathcal{F}}(X,Y)$ with the discrete topology.

4.1.2. *Cartesian vs. linear enriched.* There are basically two types, Cartesian enriched and linearly enriched. Cartesian enriched means that $\otimes_{\mathcal{E}}$ is also a Cartesian product like in $\mathcal{T}op$[5]. Linear means that one is at lead $\mathcal{A}b$ enriched, and $\otimes_{\mathcal{E}}$ is "bi–linear". Typical examples for $\mathcal{E}$ are $\mathcal{V}ect_k$, dg-$\mathcal{V}ect_k$, etc..

There are basically no big modifications to Feynman categories in the Cartesian enriched case. In the linear case, there are necessary modifications as the notion of a groupoid becomes unavailable. Note that $GL(V) \subset End(V)$ is a subspace, but not a linear subspace.

4.2. Modifications in case of enrichment. In this sub-paragraph, we will collect the modifications that are necessitated in the enriched case, especially in that of linear enrichment.

4.2.1. *Cartesian enriched Feynman categories.* Generally in the enriched case axiom (ii) is be replaced by the rather technical axiom (ii').

(ii') The pull-back of preserves $\imath^{\otimes\wedge} : [\mathcal{F}^{op}, Set] \to [\mathcal{V}^{\otimes op}, Set]$ *restricted to representable pre–sheaves* is monoidal.

[5]Fixing a convenient topological category.

The monoidal structure on pre–sheaves is given by Day convolution $\circledast$, thus (ii') means that

$$(4.2)\quad \imath^{\otimes\wedge}Hom_{\mathcal{F}}(\,\cdot\,, X\otimes Y) = Hom_{\mathcal{F}}(\imath^{\otimes}\,\cdot\,, X\otimes Y) = \imath^{\otimes\wedge}Hom_{\mathcal{F}}(\,\cdot\,, X)\circledast \imath^{\otimes\wedge}Hom_{\mathcal{F}}(\,\cdot\,, Y)$$

Using the definition of the Day convolution the right hand side of (4.2) becomes the co–end condition:

$$(4.3)\quad \imath^{\otimes\wedge}Hom_{\mathcal{F}}(\,\cdot\,, X)\circledast \imath^{\otimes\wedge}Hom_{\mathcal{F}}(\,\cdot\,, Y) = Hom_{\mathcal{F}}(\imath^{\otimes}\,\cdot\,, X)\circledast Hom_{\mathcal{F}}(\imath^{\otimes}\,\cdot\,, Y)$$
$$= \int^{Z,Z'} Hom_{\mathcal{F}}(\imath^{\otimes}Z, X)\times Hom_{\mathcal{F}}(\imath^{\otimes}Z', Y)\times Hom_{\mathcal{V}^{\otimes}}(\,\cdot\,, Z\otimes Z')$$

The co–end formula expresses the "bi–linearity" of composition [**Aus74**, **ML98**].

Just like condition (ii), the smallness condition (iii) should be modified in the enriched case as (co)limits become so–called indexed (co)–limits, see [**Kel82**].

(iii') For all $*\in\mathcal{V}$, the indexing functors $\tilde{\imath}^{\otimes}(*) := Hom_{\mathcal{F}}(\imath^{\otimes}*, -)$ are essentially small.

The indexing functor takes care of the "linearity" of morphisms.

Definition 4.4. A Feynman category $\mathfrak{F}$ enriched over a Cartesian $\mathcal{E}$ is a triple $(\mathcal{V},\mathcal{F},\imath)$ of a category $\mathcal{F}$ enriched over $\mathcal{E}$ and an enriched category $\mathcal{V}$ which satisfy the enriched version of the axioms of Definition 2.5. That is (i), (ii') and (iii') as given above

4.2.2. *Linear enrichment/weak Feynman categories/index enriched Feynman categories.* As there is no good notion of groupoid, in the linear case, the axiom (i) has to be modified to (i').

Definition 4.5. A weak Feynman category is a triple $(\mathcal{W},\mathcal{F},\imath)$, where both $\mathcal{W}$ and $\mathcal{F}$ are categories enriched over $\mathcal{E}$, $\imath:\mathcal{W}\to\mathcal{F}$ is a functor enriched over $\mathcal{E}$, $\mathcal{F}$ is symmetric monoidal enriched over $\mathcal{E}$, and $\mathcal{W}$ symmetric monoidal tensored over $\mathcal{E}$ satisfying: (i') $\imath^{\otimes}$ is essentially surjective, and (ii') and (iii') as above.

This notion is closely related to Getzler's patterns, see [**Get09**, **KW17**, **BKW18**].

Definition 4.6. An indexed enriched Feynman category is a weak Feynman category $\mathfrak{F}=(\mathcal{V},\mathcal{F},\imath)$ indexed over a $\mathcal{S}et$ Feynman category $\mathfrak{B}=(\mathcal{V}_{\mathcal{B}},\mathcal{B},\imath_{\mathcal{B}})$, such that $\mathcal{V}=(\mathcal{V}_{\mathcal{B}})_{\mathcal{E}}$.

Assumption 4.7. From now an, we assume that $\mathcal{C}=\mathcal{E}$ is enriched over itself and has a co–product. For representations, one is interested in Abelian categories $\mathcal{C}$, which is why we denote the co–product by $\oplus$.

4.3. Enrichment functors. Enrichment functors are the generalization of Remark 2.13. We refer to Appendix C for the two–categorical notions.

Definition 4.8. A weak enrichment functor for a Feynman category $\mathfrak{F}$ is a lax 2–functor $\mathcal{F}\to\underline{\mathcal{E}}$ with is strictly monoidal, see Remark 2.13. An enrichment functor is a weak enrichment functor that also satisfies $\mathcal{D}(\sigma)=\mathbb{1}$.

4.3.1. *Indexed enriched Feynman categories over $\mathfrak{F}$ and $\mathfrak{F}^{hyp}$.* The following is proved in [**KW17**, Proposition 4.1.2, Theorem 1.4.1] connecting the plus construction and enrichment.

THEOREM 4.9.

(1) *There is a 1–1 correspondence between indexed enriched Feynman categories over $\mathfrak{F}$ and enrichment functors.*
(2) *There is a 1–1 correspondence between enrichment functors and $\mathfrak{F}^{hyp}$-$\mathcal{O}ps_{\mathcal{E}}$.*

Denote the indexed enriched Feynman category of $\mathfrak{F}$ corresponding to $\mathcal{D} \in \mathfrak{F}^{+}$-$\mathcal{O}ps_{\mathcal{E}}$ by $\mathfrak{F}_{\mathcal{D}}$, then monadicity holds for the weak Feynman category $\mathfrak{F}_{\mathcal{D}}$.

The correspondence is represented by the formula (4.4) which is a generalization of (2.8).

$$\underline{\mathrm{Hom}}_{\mathcal{F}_{\mathcal{D}}}(X,Y) = \bigoplus_{\phi \in \mathrm{Hom}_{\mathcal{F}}(X,Y)} \mathcal{D}(\phi) \tag{4.4}$$

with composition

$$\mathcal{D}(\phi) \otimes \mathcal{D}(\psi) \to \mathcal{D}(\phi \circ \psi) \tag{4.5}$$

REMARK 4.10. The first two parts the statement for $\mathcal{E} = \mathcal{S}et$ is contained in Remarks 2.12 and 2.13. The general proof of these statements is similar. also follows from the definitions, in particular Definition C.12 and Corollary 3.19. Note that the condition that composition with isomorphism is strict, i.e. it is given by unit constraints from [**KW17**] follows from Definition C.12 condition (3) together with the groupoid action via Remark 3.33. This is why we could remove this extra condition in the definition of an enrichment functor.

4.3.2. *Generalizing to $\mathfrak{F}^{+gcp}$.* The results and constructions are analogous to Example 3.15.

Relaxing the condition of an enrichment functor to a weak enrichment functor, we obtain the generalization of Theorem 4.9 (1).

PROPOSITION 4.11. *Weak enrichment functors are in 1–1 correspondence with $\mathfrak{F}^{+gcp}$-$\mathcal{O}ps_{\mathcal{E}}$.*

PROOF. This is straightforward as in Remark 4.10 using Definitions 3.22, C.12 and Proposition C.15. □

There is a generalization of the results of Theorem 4.9 (2) to $\mathfrak{F}^{+gcp}$-$\mathcal{O}ps$. In the enriched case, we have to be careful about splitting.

DEFINITION 4.12. We call $\mathcal{D} \in \mathcal{F}^{+}$-$\mathcal{O}ps_{\mathcal{C}}$ split, if each for all $\sigma \in Mor(\mathcal{V})$.

(1) $\mathcal{D}(\sigma) = \mathcal{D}(\sigma)^{\times} \oplus \mathcal{D}^{red}(\sigma)$.
(2) Any invertible $\phi \in \mathcal{D}(\sigma)$ with $\phi^{-1} \in \mathcal{D}(\sigma^{-1})$ is in $D(\sigma)^{\times}$.
(3) $\bigoplus_{\sigma \in Mor(\mathcal{V})} \mathcal{D}(\sigma) = \mathcal{G}(\mathcal{D})_{\mathcal{E}}$ that is the free enrichment of a set–groupoid $\mathcal{G}(\mathcal{D})$.

A $\mathcal{D} \in \mathcal{F}^{gcp+}$-$\mathcal{O}ps_{\mathcal{C}}$ is called split if it satisfies the conditions above and furthermore The morphism $\mathcal{D}(i_{\sigma}) : \mathbb{1} \to \mathcal{D}(\sigma)$ is split. I.e. $\mathcal{D}(\sigma)^{\times} = \mathbb{1} \oplus \overline{\mathcal{D}}(\sigma)^{\times}$, where the first summand is $im(\mathcal{D}(i_{\sigma}))$.

A weak enrichment functor is split, if the corresponding functor $\mathcal{D} \in \mathfrak{F}^{+gcp}$-$\mathcal{O}ps$ is split.

REMARK 4.13.

(i) Condition (3) means that $Obj(\mathscr{G}(\mathcal{D})) = Obj(\mathcal{V})$, but $Mor(\mathscr{G}(\mathcal{D})) = \amalg_{\sigma \in Mor(\mathcal{V})} \amalg_{\sigma_i \in I_\sigma} \sigma_i$ and $\mathcal{D}(\sigma)^\times = \bigoplus_{\sigma_i \in I_\sigma} \mathbb{1}$ and there is are composition morphisms $I_\sigma \times I_{\sigma'} \to I_{\sigma\sigma'}$ for composable σ, σ', such that fixing an element in I_σ or an element in $I_{\sigma'}$ the composition morphism is a bijection.

(ii) In the $+gcp$ case, all the $\mathcal{D}(\sigma)^\times = \mathbb{1} \oplus \bar{\mathcal{D}}(\sigma)^\times$ split with the first component being $\mathcal{D}(i_\sigma)$. In the language above I_σ is a pointed set $(I_\sigma, 0)$ and there is an inclusion $\mathcal{V} \to \mathscr{G}(\mathcal{D})$, with the image of σ being $0 \in I_\sigma$. In particular, there is an involution, $\bar{}: I_\sigma \to I_{\sigma^{-1}}$ for which for $\sigma : X \to Y$ the composition $\mathbb{1}_{\sigma_i^{-1}} \otimes \mathbb{1}_{\sigma_i} \to \mathbb{1}_{id_X}$ where id_X is the base point of I_{id_X} .

(iii) Any $\mathcal{D} \in \mathcal{F}^{hyp}\text{-}\mathcal{O}ps_{\mathcal{C}}$ is split.

ASSUMPTION 4.14. From now on, we will assume that all functors from $\mathcal{F}^+$ and $\mathcal{F}^{+gcp}$ are split.

In the case $\mathcal{E} = \mathcal{S}et$ the splitting is simply given by $\mathcal{D}(\sigma) = \mathcal{D}(\sigma)^\times \amalg \mathcal{D}(\sigma)^{red}$

Given a split $\mathcal{D} \in \mathcal{F}^{+gcp}$, set $\mathcal{V}_{\mathcal{B}} := \mathscr{G}(\mathcal{D})$ and let $\mathcal{F}_B$ be the trivially extended monoidal category along the projection $j : \mathcal{V}_{\mathcal{B}} \to \mathcal{V}$ given by $I_\sigma \mapsto \sigma$. This means the $Obj(\mathcal{F}_{\mathcal{B}}) = Obj(\mathcal{F})$. To give the morphisms, note that $Mor(\mathcal{F})$ is a $\mathcal{V}^\otimes$-$\mathcal{V}^\otimes$ bi–module with the action $(\sigma \Downarrow \sigma')$. This action is extended to bi–module action of $\mathcal{V}_{\mathcal{B}}$ by $(\sigma_i \Downarrow \sigma'_j)(\phi) = (\sigma \Downarrow \sigma')(\phi)$.

There is the natural inclusion $i_{\mathcal{B}} : \mathcal{V}_{\mathcal{B}} \to \mathcal{F}_{\mathcal{B}}$. Set $\mathfrak{B}(\mathcal{D}) = (\mathcal{V}_{\mathcal{B}}, \mathcal{F}_{\mathcal{B}}, \imath_{\mathcal{B}})$.

THEOREM 4.15. *For a split weak enrichment functor* $\mathcal{D} : \mathfrak{F}^+$ *there is weak Feynman category* $\mathfrak{F}_{\mathcal{D}}$ *indexed over the Feynman category* $\mathfrak{B}(\mathcal{D})$.

PROOF. The fact that $\mathfrak{F}_{\mathcal{D}}$ is indexed over $\mathfrak{F}_{\mathcal{D}}$ is clear. The fact that $\mathfrak{B}$ is a Feynman category follows in a straightforward fashion similar to [**KW17**, Theorem 4.1.4]. □

COROLLARY 4.16. *A split weak enrichment functor* $\mathcal{D}$ *for* $\mathfrak{F}$ *lifts to an enrichment functor* $\tilde{\mathcal{D}}$ *over* $\mathfrak{B}(\mathcal{D})$.

The values of $\tilde{\mathcal{D}}$ on morphisms are $\tilde{\mathcal{D}}(\sigma_i, \phi) = \mathbb{1}_{\sigma_i} \otimes \mathcal{D}(\phi) \simeq \mathcal{D}(\phi)$ where σ_i is the isomorphism corresponding to $i \in I_\sigma$, $\mathbb{1}_{\sigma_i}$ is the corresponding component of $\mathscr{G}(\mathcal{D})$ and the morphism is given by pre–composing the morphism $\mathcal{D}(\sigma) \otimes \mathcal{D}(\phi) \to \mathcal{D}(\phi)$ with the inclusion of $\mathbb{1}_{\sigma_i} \to \mathcal{D}(\sigma)$.

4.3.3. *$\mathfrak{NC}Set$ as an indexed enriched Feynman category over $\mathfrak{FinSet}$.* We know that $\mathfrak{FinSet}^{+gcp} = \mathfrak{O}$, so enrichment functors will be operads. Let $Assoc \in \mathfrak{O}\text{-}\mathcal{O}ps_{\mathcal{S}et}$ be the associative operad as in Lemma 3.37, then $Assoc \in \mathfrak{O}^{hyp}$, since $Assoc(*_{\{s,t\},t})$ has only one element. The following is now straightforward.

LEMMA 4.17. $\mathfrak{NC}Set = \mathfrak{FinSet}_{Assoc}$ *is indexed enriched over* $\mathfrak{FinSet}$. □

REMARK 4.18. We now have two description of $\mathfrak{O}^{\neg\Sigma}$. Using the Lemma 4.17 above, Proposition 3.36 and Lemma 3.37

$$(\mathfrak{FinSet}_{Assoc})^{+gcp} = (\mathcal{F}in\mathcal{S}et^{+gcp})_{\text{dec } Assoc} \tag{4.6}$$

This is part of a general statement, see §4.5.1 and [**KM20**].

4.3.4. *Enriching quivers.* Quivers give a generalization of Example 4.3. As an example consider of a simple quiver $Q : \bullet_1 \to \bullet_2$. That is the category has two isomorphisms $\mathrm{id}_{\bullet_1}$ and $id_{\bullet_2}$ together with a morphism $\phi : \bullet_1 \to \bullet_2$. This give rise to a Feynman category $\mathfrak{F}_Q$, where $\mathcal{V}$ is the discrete category with objects $\bullet_1, \bullet_2$ and $\mathcal{F} = Q^{\otimes}$.

PROPOSITION 4.19. *The weak enrichments of $\mathfrak{F}_Q$ are in 1–1 correspondence with $(A, B, {}_AM_B)$ of two unital split, in the sense of split enrichment functors, algebras and a bi–module.*

PROOF. We can define a class of enrichment functors by giving two groups and a bi–module over them $\mathcal{D}$ will have values $\mathcal{D}(id_{\bullet_1}) = A$, $\mathcal{D}(id_{\bullet_2}) = B$ and $\mathcal{D}(\phi) = {}_AM_B$. The algebra structure comes from the compositions $id_{\bullet_i} \circ id_{\bullet_i} = id_{\bullet_i}$ The bi–module structure comes from the compositions $\phi \circ id_{\bullet_1} = id_{\bullet_2} \circ \phi = \phi$. □

If the functors are weak or not depends on the algebras A and B.

EXAMPLE 4.20. A particular example over $\mathcal{S}et$ is the choice $A = GL(V), B = GL(W), Hom(V, W)$, where $V, W \in \mathcal{V}ect_k$. Notice that this is not enriched over k as discussed above. It is also a weak indexing. With the base category $\mathfrak{B}$ having two objects with automorphisms $GL(V)$ and $GL(W)$ respectively and one morphism between the two objects, i.e. the trivial A-B module.

REMARK 4.21. Going over to finite graphs, this type of example is tied to the quantum graph symmetries [**KKW15, KKWK16**], where the enrichment now takes values in C^* algebras and has applications to material science, [**KWK18**].

4.3.5. *Twists.* One reason one uses the categories $\mathfrak{F}_{\mathcal{D}}$ is to obtain the necessary sign twist for the bar and co–bar constructions. In particular, the twists of §§3.3.3 and 3.3.4 are important for the transforms in §5.

There is also a nice interpretation of twisting the triples for indexed Feynman categories, which we will not describe in detail here, but refer to [**KW17**, Proposition 4.1.7]. We wish to note, that the $\mathcal{V}$–twists modify the triples in an isomorphic way, hence one obtains isomorphisms between $\mathcal{F}_{\mathcal{D}}$-$\mathcal{O}ps$ and $\mathcal{F}_{\mathcal{D}_{\mathfrak{L}}}$-$\mathcal{O}ps$, which is what one is used to in the algebra case, see §5.1 and in general [**KWZ12**] for relevant examples.

4.4. Modules. With this preparation, we can finally define modules for $\mathfrak{F}^{+gcp}$-$\mathcal{O}ps$. This generalizes the definition of [**KW17**], where the modules were only defined for $\mathfrak{F}^{hyp}$-$\mathcal{O}ps$.

DEFINITION 4.22. Given a $\mathcal{D} \in \mathfrak{F}^{+gcp}$-$\mathcal{O}ps_E$, $\mathcal{D}$-modules in a monoidal category $\mathcal{C}$ enriched over $\mathcal{D}$ are $\mathfrak{F}_{\mathcal{D}}$-$\mathcal{O}ps_{\mathcal{C}}$.

4.4.1. *Modules over an associative algebra.* We can now do the construction of Remark 3.15 full justice, that is we can consider modules over an algebra and not just a monoid. The example also exhibits all the features above.

Let A be an associative algebra over k, that is the value of $\mathcal{A} \in \mathfrak{F}\mathfrak{S}^{>}$-$\mathcal{O}ps_{\mathcal{V}ect_k} = (\mathfrak{F}^{triv})^+$. We consider the category $\mathcal{F}_A$ whose objects are the natural numbers and whose morphisms are given by $\underline{\mathrm{Hom}}_{\mathcal{F}_A}(n, n) = A^{\otimes n}$ and $\underline{\mathrm{Hom}}_{\mathcal{F}_A}(n, m) = 0$ for $n \neq m$ with permutation action and unit $\mathbb{1} = k = A^{\otimes 0}$. This is the free symmetric category on the category enriched over k, $\mathcal{V}_A$, which has one object $*$ and morphisms $Hom_{\mathcal{V}_A} = A$. Thus $Fun_{\otimes}(\mathcal{F}_A, \mathcal{E}) = Fun(\mathcal{V}_A, \mathcal{E}) = A\text{-}mod_{\mathcal{E}}$ that is the category of

A–modules in $\mathcal{E}$. Here we assume that $\mathcal{E}$ is also enriched over $\mathcal{V}ect_k$ and the monoidal functors are over $\mathcal{V}ect_k$.

Now, if $\mathcal{A}$ is split as a functor in $(\mathfrak{F}^{triv})^+$-$\mathcal{O}ps$, then $A = A^\times \oplus A$ and any invertible element lies in $A^\times = G_k = k[G]$ for some group G.

If A is unital that is $\mathcal{A} \in NCS\text{-}\mathcal{O}ps_{\mathcal{V}ect_k} = (\mathfrak{F}^{triv})^{+gcp}$, being pointed by the unit and $A^\times = \mathbb{1} \oplus \bar{A}^\times$, which is the inclusion of $e \in k[G]$. In this case, $\mathfrak{B}_{\mathcal{A}} = \mathfrak{V}_{\mathcal{A}}$. Again, hyper–means that A is reduces, that is $\bar{(A)}^\times = 0$ and $A^\times = \mathbb{1}$. In this case indeed $\mathfrak{B} = \mathfrak{F}^{triv}$ and $\mathcal{F}_{\mathcal{A}}$ is indexed enriched over $\mathfrak{F}^{triv}$.

REMARK 4.23. This suggests the study of group like elements of a Hopf algebra as a replacement for the condition of having an underlying set–groupoid. This will be explored in the future.

4.4.2. *Enrichments of FI, FI_G and FI_d.* Consider a functor $\mathcal{D} \in FI^{+gcp}$-$\mathcal{O}ps_{\mathcal{E}}$. There are two generating morphisms in FI: id_* and $i : \emptyset \to 1$. Let $\mathcal{D}(id_*) = A = A^\times \oplus \bar{A}$ and $\mathcal{D}(i) = M$. The morphism i_{id_*} provides $\mathbb{1} \to A^\times$.

PROPOSITION 4.24. *The weak indexed enrichments of FI are in 1–1 correspondence with pairs (A, M) where A is a unital algebra (monoids) and M is an A–module.*

PROOF. The composition $id_* \circ id_* = id_*$ provides the multiplication map $\mu : \mathcal{D}(id_*) \otimes \mathcal{D}(id_*) = A \otimes A \to A = \mathcal{D}(id_*)$, the unit is provided by $\mathbb{1} \to \mathcal{D}(id_*) = A$ from the $+gcp$ data, while the composition $id_* \circ i = i$ provides the module map $\rho : \mathcal{D}(id_*) \otimes \mathcal{D}(i) = A \otimes M \to M = \mathcal{D}(i)$. □

REMARK 4.25.

(1) As $\mathfrak{F}\mathfrak{I} \subset \mathcal{F}in\mathcal{S}et$ we have that $\mathfrak{F}\mathfrak{I}^{+gcp} \subset \mathfrak{F}in\mathfrak{S}et^{+gcp} = \mathfrak{O}_{unital}$ and hence $\mathfrak{F}\mathfrak{I}^{+gcp}$-$\mathcal{O}ps$ are unital operads with only $\mathcal{O}(1)$ and $\mathcal{O}(0)$. It is well known that this pair is a pair of a unital algebra and a module over it.
(2) Considering $\mathfrak{F}\mathfrak{I}^+$-$\mathcal{O}ps$, one arrives at a non–unital algebra and a module over it.

Two special cases over $\mathcal{S}et$ have been considered in [**SS19, SS17**]. Here we generalize these in two ways. There is a general solution yielding the two construction as special cases and this can be performed in any enriched setting, that is also k–linearly.

Let us briefly review the two constructions. For the first the category G–maps is studied. Objects are finite sets with a morphism between R and S given by a pair of maps: and injection $f : R \to S$ and a map $\rho : R \to G$. Composition is given by $(g, \sigma) \circ (f, \rho) = (g \circ f, \tau)$ with $\tau(x) = \sigma(f(x))\rho(x)$.

The category FI_d is given as follows: again the objects are finite sets and morphisms given by pair (f, m), where $f : R \to S$ is an injection, but now $m : S \setminus f(R) \to \{1, \ldots, d\}$. For the composition $(f, m) \circ (g, n) = (f \circ g, p)$ with $R \xrightarrow{g} S \xrightarrow{f} T$, where p is defined on $T \setminus f \circ g(R) = f(T \setminus g(R)) \amalg T \setminus g(S)$ as $m \circ f^{-1} \amalg n$.

The extra data is compatible with orders and defines the categories OI_G and OI_d.

PROPOSITION 4.26. *Restricting A to be group G and M to be trivial, we recover the category FI_G . Restricting A to be trivial and $M = \{1, \ldots, d\}$ then we recover the category FI_d.*

Similar results hold in the ordered cases OI_G and OI_d.

PROOF. In general, decomposing S as $\amalg_{s\in S}*$, according to the decomposition (2.5) a morphism in $Hom_{FI_{\mathcal{D}}}(R,S)$ is given by a tensor product of pairs $(id_*,g),a\in A)$ and $(i,m\in M)$, with a factor of id_* for each element in the image of R and a factor of i for each element not in the image. If A is G and M is trivial, then an injection $R\to S$ is given by a tensor product of pairs $(id_*,g),g\in G)$ and $(i,1)$. The data of the (id_*,g) is equivalent to the data of a map $R\to g$ which assigns g to the factor of id_* corresponding to r. Composition gives $(id_*,g)\circ(id_*,h)=(id_*,gh)$, where $(id<h)$ is the factor that corresponds to $f(x)$ if (id_*,x) corresponds to x. and $(id_*,g)\circ(i,1)=(i,1)$, since the action is trivial $\rho(g)1=1$. Thus the two categories coincide.

If A is trivial and $M=\{1,\dots,d\}$ then a morphism $R\to S$ is given by tensor factors of $(id_*,1)$ and (i,j) with $j\in\{1,\dots,d\}$. This data is encoded in the injection $f:R\to S$ and a morphism on $S\setminus f(R)$. Composition now reads $(id_*,1)\circ(i,j)=(i,j)$ which is the first part of the formula for p. The second part is simply the insertion of factors of i for the elements missed by g.

The proofs for the ordered cases is analogous. $\square$

Note that the case of $A=G$ is one in which $A^{red}=\emptyset$. An in this case the indexing, $A^\times=G$ is not reduced. The (weak) Feynman category is actually indexed over itself.

4.5. Decoration. In the case of decoration in the linear case, we need the modification that is spelled out in [**KL16**, §2.2].

DEFINITION 4.27. For a fixed choice of $\jmath: Iso(\mathcal{F})\to V^{\otimes}$ realizing the equivalence of condition (i):

The objects of $\mathcal{F}_{\mathrm{dec}\mathcal{O}}$ are tuples $(X,a_{v_1},\dots,a_{v_{|X|}})$, where for $\jmath(X)=\bigotimes_{v\in I}*_v=*_{v_1}\otimes\dots\otimes *_{v_{|X|}}$, $a_{v_i}\in\mathcal{O}(*_{v_i})$.

The morphism of $\mathcal{F}_{\mathrm{dec}\mathcal{O}}$ are given by the subset

$$Hom_{\mathcal{F}_{\mathrm{dec}\mathcal{O}}}((X,a_{w_1},\dots,a_{w_{|X|}}),(Y,b_{v_1},\dots,b_{v_{|Y|}}))\subset Hom_{\mathcal{F}}(X,Y)$$

of those morphisms $\phi:X\to Y$, such that if $\bigotimes_v\phi_v$ is the decomposition of ϕ according to the diagram (2.5), with $\phi_v:X_v=\bigotimes_{w\in I_v}\imath(*_w)\to\imath(*_v)$, then $\mathcal{O}(\phi_v)(\bigotimes_{w\in I_v}(a_w))=b_v$.

The monoidal structure is given by

$$(X,a_{w_1},\dots,a_{w_{|X|}})\otimes(Y,b_{v_1},\dots,b_{v_{|Y|}}))=(X\otimes Y,a_{w_1},\dots,a_{w_{|X|}},b_{v_1},\dots,b_{v_{|Y|}})$$

and the commutativity constraints are given by those of $\mathcal{F}$ on the first component and the respective permutations on the others.

The result can also be made into enriched category, if $\mathcal{F}$ is tensored over $\mathcal{E}$.

DEFINITION 4.28. Assuming the conditions above, the category $\mathcal{F}_{\mathrm{dec}\mathcal{O}}$ as an enriched category over $\mathcal{E}$ has objects $X\otimes\mathcal{O}(X)$ and formally the same set of morphisms $\mathcal{F}$–$Hom_{\mathcal{F}_{\mathrm{dec}\mathcal{O}}}(X\otimes\mathcal{O}(X),Y\otimes\mathcal{O}(Y))=Hom_{\mathcal{F}}(X,Y)$. A morphisms ϕ via tensoring becomes the morphism $\phi\otimes\mathcal{O}(\phi)$ in $\mathcal{C}$. Its symmetric monoidal structure is given by $(X\otimes\mathcal{O}(X))\otimes_{\mathcal{F}_{\mathrm{dec}\mathcal{O}}}(Y\otimes\mathcal{O}(Y))=(X\otimes_{\mathcal{F}}Y)\otimes\mathcal{O}(X\otimes Y)$, with composition of morphisms and symmetries given by the isomorphism $(X\otimes Y)\otimes\mathcal{O}(X\otimes Y)\simeq(X\otimes\mathcal{O}(X))\otimes(Y\otimes\mathcal{O}(Y))$ in $\mathcal{C}$ provided by the strong symmetric monoidal structure of $\mathcal{O}$. $\mathcal{V}_{\mathrm{dec}\mathcal{O}}$ is likewise defined by objects $(V\otimes\mathcal{O}(\imath(V)))$ with $V\in\mathcal{V}$ and the morphisms of $\mathcal{V}$. The inclusion is given by $\imath(V\otimes\mathcal{O}(\imath(V)))=(\imath(V)\otimes\mathcal{O}(\imath(V)))$.

Similar modification allow for the decoration of enriched Feynman categories with functors to $\mathcal{C}$ which is also enriched over $\mathcal{E}$.

4.5.1. *The relation between decoration and plus construction.* With the notion of enrichment for decoration, we can state a Proposition generalizing Remark 4.18 that will be proven in [**KM20**]

PROPOSITION 4.29. *In general* $(\mathfrak{F}^{+gcp})_{\text{dec } \mathcal{O}} = (\mathfrak{F}_{\mathcal{O}})^{+gcp}$.

5. Bar, co–bar, Feynman transforms, & master equations

In analogy with (co)–algebras, there are three transforms we will consider for $\mathcal{F}$-$\mathcal{O}ps$: the bar–, the cobar transform and the Feynman transform aka. dual transform. The bar–cobar transforms yield a pair of adjoint functors. These transforms serve a dual purpose. One is to give resolutions, the other is to give deformations through master–equations, which generalize the Maurer–Cartan equation.

5.1. Motivating example: Algebras.

5.1.1. *Bar transform.* If $(A, \epsilon, d_A, |\cdot|)$ is an augmented associative algebra dg, then the bar transform is the dg–co–algebra given by the free co–algebra $BA = T\Sigma^{-1}\bar{A}$, where $\bar{A} = ker(\epsilon)$ together with differential from algebra structure. The usual notation for an element in BA is $a_0|a_1|\dots|a_n$ it will have degree $\sum_{i=0}^n(|a_i| + 1) = \sum_i |a_i| + n + 1$. In this notation the co–product is

$$\begin{aligned} \Delta(a_0|\dots|a_n) &= 1 \otimes a_0|\dots|a_n + a_0|\dots|a_n \otimes 1 + \bar{\Delta}(a_0|\dots|a_n) \\ \bar{\Delta}(a_0|\dots|a_n) &= \sum_{i=1}^n a_0|\dots|a_{i-1} \otimes a_i|\dots|a_n \end{aligned} \tag{5.1}$$

Using the differentials

$$\begin{aligned} \partial_i(a_0|a_1|\dots|a_n) &= a_0|\dots|a_{i-1}a_i|\dots|a_n \\ \partial_B(a_0|a_1|\dots|a_n) &= \sum_{i=1}^n (-1)^{\sum_{j=1}^{i-1}(|a_j|+1)}\partial_i(a_0|a_1|\dots|a_n) \end{aligned} \tag{5.2}$$

The total differential on BA is $d_B + d_A$.

5.1.2. *Odd version.* Notice that looking at ΣBA, the degree of an element $a_0|\dots|a_n$ is n which is the number of bars. As a mnemonic one can put $deg(|) = 1$. In the shifted version the co–product reduced co–product $\bar{\Delta}$ has degree -1 as there is one bar less. This type of odd structure lies at the heart of the story for deformations. It actually gives the right degree to the Hochschild complex, see [**KWZ12**] for a detailed exposition.

5.1.3. *Cobar and bar/cobar.* Likewise let $(C, \eta, |\cdot|)$ be an associative co–augmented connected dg–co–algebra, $\bar{C} = coker(\eta)$. The co–bar transform is the dg–algebra $\Omega C := Free_{alg}(\Sigma^{-1}\bar{C})$ together with a differential coming from co–algebra structure dual to the structures above. The bar–cobar transform ΩBA is a resolution of A.

5.1.4. *Dual and Feynman transforms.* For the dual or Feynman transform consider a finite–dimensional algebra A or a graded algebra with finite dimensional pieces and let $\check{A}$ be its (graded) dual co–algebra. Then the dual or Feynman transform of A is $FA := \Omega\check{A}$ together with a differential from multiplication. Now, the double Feynman transform FFA a resolution.

There is particular interest in the differential for deformations. In particular one considers Maurer-Cartan elements which satisfy the equation $dm + \frac{1}{2}[m, m] = 0$.

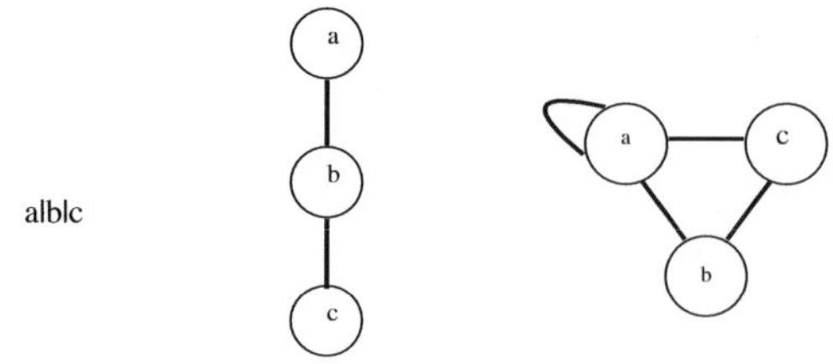

FIGURE 3. The sign mnemonics for the bar construction, traditional version with the symbols | of degree 1, the equivalent linear tree with edges of degree 1, and a more general graph with edges of degree 1. Notice that in the linear rooted case there is a natural order of edges, this ceases to be the case for more general graphs

5.2. Transforms for Feynman categories. As before ,one can ask the question of how much of the structure of these transforms can be pulled back to the Feynman category side. The answer is: "Pretty much all of it". We shall not discuss all the details which can be found in [**KW17**], but will give an overview following[**Kau17**].

We start with a general overview and a discussion of the necessary structures. The result of the transforms BA or FA is actually an odd version of a (co)–free co–algebra or an odd algebra with a (co)differential.

Algebras are a general case of elements of $\mathcal{F}$-$\mathcal{O}ps$ and hence the transforms will be defined for inputs $\mathcal{O} \in \mathcal{F}\text{-}\mathcal{O}ps_{\mathcal{C}}$. The co–algebra output means that one is going to the opposite category $\mathcal{C}^{op}$ for the target category in the output of the transforms. The construction will be free constructions, which, however, also have the extra structure of an additional (co)differential. The "oddness" is necessary for the signs that are needed in order for the differentials to square to zero. In general, this means that one needs an odd version $\mathfrak{F}^{odd}$ of the Feynman category $\mathfrak{F}$. In order to define this odd version, one needs to make assumptions on the Feynman category and fix a presentation. The transform will transform an *op* into a new *op* for the odd version of the Feynman category $\mathfrak{F}^{odd}$ either in $\mathcal{C}^{op}$ or $\mathcal{C}$. In the graphical case, this is achieved by the twist by $\mathfrak{K}$, see Example 3.28. The equation (3.16) then neatly explains the relation between odd version and the shifts. The twist by $\mathfrak{K}$ means that each edge gets degree 1, which is exactly the convention that $deg(|) = 1$ in the bar construction; see Figure 3.

Thus the resulting Feynman category is actually a category of chain complexes in a category enriched over $\mathcal{A}b$. Furthermore, for the (co)-differential to work, we have to have signs. These are exactly what is provided by the odd versions. In order to be able to define the transforms, one has to fix an odd version $\mathfrak{F}^{odd}$ of $\mathfrak{F}$, see [**KW17**, §5.2.3,§5.2.6] for full details. This is analogous to the suspension in the usual bar transforms. In fact, the following is more natural, see [**KW17, KWZ12**]. The degree is 1 for each bar and in the graph case the edges get degree 1. The basic example are graphical Feynman categories, for which the odd version is given by the twist with $\mathfrak{K}$. In general, one needs an ordered presentation.

The transforms are of interest in themselves, but one common application is that the bar-cobar transform as well as the double Feynman transform give a "free" resolution. In general, of course, "free" means co-fibrant. For this kind of statement one needs a Quillen model structure, which is provided in §D.

The Feynman transform is quasi–free, that is for $\mathcal{O} \in \mathcal{F}\text{-}\mathcal{O}ps_{\mathcal{C}}$, $F\mathcal{O} = \imath_!^{odd}(\imath^*\mathcal{O}) \in \mathcal{F}^{odd}\text{-}\mathcal{O}ps$ is free, if one ignores the differential. The dg structure is compatible with $F\mathcal{O}$ precisely if it satisfies a Master Equation, which is fixed by the choice of $\mathfrak{F}^{odd}$.

5.2.1. *Presentations.* As mentioned, in order to define the transforms, we have to give what is called an ordered presentation [**KW17**]. Rather then giving the technical conditions, we will consider the graph case and show these structures in this case.

5.2.2. *Basic example* $\mathfrak{G}$. In $\mathfrak{G}$ the presentation comes from the following set of morphisms Φ

(1) There are 4 types of basic morphisms: Isomorphisms, simple edge contractions, simple loop contractions and mergers. Call this set Φ.
(2) These morphisms generate all one–comma generators upon iteration. Furthermore, isomorphisms act transitively on the other classes. The relations on the generators are given by commutative diagrams.
(3) The relations are quadratic for edge contractions as are the relations involving isomorphisms. Finally there is a non–homogenous relation coming from a simple merger and a loop contraction being equal to an edge contraction.
(4) We can therefore assign degrees as 0 for isomorphisms and mergers, 1 for edge or loop contractions and split Φ as $\Phi^0 \amalg \Phi^1$. This gives a degree to any morphism.

Up to isomorphism any morphism of degree n can be written in $n!$ ways up to morphisms of degree 0. These are the enumerations of the edges of the ghost graph.

There is also a standard order in which isomorphisms come before mergers which come before edge contractions, cf. [**BM08, KW17**]. This gives an ordered presentation.

In general, an ordered presentation is a set of generators Φ and extra data such as the subsets Φ^0 and Φ^1; we refer to [**KW17**] for details.

5.2.3. *Differential.* Given a $d_{\Phi^1} = \sum_{[\phi_1]\in\Phi^1/\sim} \phi_1 \circ$ defines an endomorphism on the Abelian group generated by the isomorphism classes morphisms. The non–defined terms are set to zero. Φ^1 is called resolving if this is a differential.

In the graph case, this amounts to the fact that for any composition of edge contractions $\phi_e \circ \phi_{e'}$, there is precisely another pair of edge contractions $\phi_{e''} \circ \phi_{e'''}$ which contracts the edges in the opposite order.

This differential will induce differentials for the transforms, which we call by the same name. We again refer to [**KW17**] for details.

5.2.4. *Setup.* $\mathfrak{F}$ be a Feynman category enriched over $\mathcal{A}b$ and with an ordered presentation and let $\mathfrak{F}^{odd}$ be its corresponding odd version. Furthermore let Φ^1 be a resolving subset of one-comma generators and let $\mathcal{C}$ be an additive category, i.e. satisfying the analogous conditions above. In order to give the definition, we need a bit of preparation. Since $\mathcal{V}$ is a groupoid, we have that $\mathcal{V} \simeq \mathcal{V}^{op}$. Thus, given a functor $\Phi : \mathcal{V} \to \mathcal{C}$, using the equivalence we get a functor from $\mathcal{V}^{op}$ to $\mathcal{C}$ which we denote by Φ^{op}. Since the bar/cobar/Feynman transform adds a differential, the natural target category from $\mathcal{F}\text{-}\mathcal{O}ps$ is not $\mathcal{C}$, but complexes in $\mathcal{C}$, which we denote by $Kom(\mathcal{C})$. Thus any $\mathcal{O}$ may have an internal differential $d_{\mathcal{O}}$.

5.2.5. *The bar construction.* This is the functor

$$\mathsf{B}\colon \mathcal{F}\text{-}\mathcal{O}ps_{Kom(\mathcal{C})} \to \mathcal{F}^{odd}\text{-}\mathcal{O}ps_{Kom(\mathcal{C}^{op})}$$

$$\mathsf{B}(\mathcal{O}) := \imath_{\mathfrak{F}^{odd}}{}_*(\imath_{\mathfrak{F}}^*(\mathcal{O}))^{op}$$

together with the differential $d_{\mathcal{O}^{op}} + d_{\Phi^1}$.

5.2.6. *The cobar construction.* This is the functor

$$\Omega\colon \mathcal{F}^{odd}\text{-}\mathcal{O}ps_{Kom(\mathcal{C}^{op})} \to \mathcal{F}\text{-}\mathcal{O}ps_{Kom(\mathcal{C})}$$

$$\Omega(\mathcal{O}) := \imath_{\mathfrak{F}}{}_*(\imath_{\mathfrak{F}^{odd}}^*(\mathcal{O}))^{op}$$

together with the co-differential $d_{\mathcal{O}^{op}} + d_{\Phi^1}$.

5.2.7. *Feynman transform.* Assume there is a duality equivalence $\vee\colon \mathcal{C} \to \mathcal{C}^{op}$. The Feynman transform is a pair of functors, both denoted FT,

$$\mathsf{FT}\colon \mathcal{F}\text{-}\mathcal{O}ps_{Kom(\mathcal{C})} \leftrightarrows \mathcal{F}^{odd}\text{-}\mathcal{O}ps_{Kom(\mathcal{C})} : \mathsf{FT}$$

defined by

$$\mathsf{FT}(\mathcal{O}) := \begin{cases} \vee \circ \mathsf{B}(\mathcal{O}) & \text{if } \mathcal{O} \in \mathcal{F}\text{-}\mathcal{O}ps_{Kom(\mathcal{C})} \\ \vee \circ \Omega(\mathcal{O}) & \text{if } \mathcal{O} \in \mathcal{F}^{odd}\text{-}\mathcal{O}ps_{Kom(\mathcal{C})} \end{cases}$$

PROPOSITION 5.1 ([**KW17**, Lemma 7.4.2]). *The bar and cobar construction form an adjunction.*

$$\Omega\colon \mathcal{F}^{odd}\text{-}\mathcal{O}ps_{Kom(\mathcal{C}^{op})} \rightleftarrows \mathcal{F}\text{-}\mathcal{O}ps_{Kom(\mathcal{C})} : \mathsf{B}$$

The quadratic relations in the graph examples are a feature that can be generalized to the notion of *cubical* Feynman categories. The name reflects the fact that in the graph example the $n!$ ways to decompose a morphism whose ghost graph is connected and has n edges into simple edge contractions correspond to the edge paths of I^n going from $(0,\dots,0)$ to $(1,\dots,1)$. Each edge flip in the path represent one of the quadratic relations and furthermore the $\mathbb{S}_n$ action on the coordinates is transitive on the paths, with transposition acting as edge flips. We will not give the full detailed definition, but note that $\mathfrak{G}$ is cubical and refer to [**KW17**, Definition 7.2.1] for the technical details.

This is a convenient generality in which to proceed.

THEOREM 5.2 ([**KW17**, Theorem 7.4.3]). *Let $\mathfrak{F}$ be a cubical Feynman category and $\mathcal{O} \in \mathcal{F}\text{-}\mathcal{O}ps_{Kom(\mathcal{C})}$. Then the co–unit $\Omega\mathsf{B}(\mathcal{O}) \to \mathcal{O}$ of the above adjunction is a levelwise quasi-isomorphism.*

REMARK 5.3. In the case of $\mathcal{C} = dgVect$, the Feynman transform can be intertwined with the aforementioned push-forward and pull-back operations to produce new operations on the categories $\mathcal{F} - \mathcal{O}ps_{\mathcal{C}}$. A lifting (up to homotopy) of these new operations to $\mathcal{C} = Vect$ is given in [**War16**]. In particular this result shows how the Feynman transform of a push-forward (resp. pull-back) may be calculated as the push-forward (resp. pull-back) of a Feynman Transform. One could thus assert that the study of the Feynman transform belongs to the realm of Feynman categories as a whole and not just to the representations of a particular Feynman category.

5.3. Master equations. In [**KWZ12**], we identified the common background of master equations that had appeared throughout the literature for operad–like objects and extended them to all graphs examples. An even more extensive theorem for Feynman categories can also be given.

The Feynman transform is quasi–free. An algebra over $F\mathcal{O}$ is dg–if and only if it satisfies the relevant Master Equation. First, we have the tabular theorem from [**KWZ12**] for the usual suspects.

THEOREM 5.4. *([**Bar07**],[**MV09**],[**MMS09**],[**KWZ12**]) Let $\mathcal{O} \in \mathcal{F}$-$\mathcal{O}ps_{\mathcal{C}}$ and $\mathcal{P} \in \mathcal{F}^{odd}$-$\mathcal{O}ps_{\mathcal{C}}$ for an $\mathcal{F}$ represented in Table 5. Then there is a bijective correspondence:*

$$\mathrm{Hom}(\mathsf{FT}(\mathcal{P}), \mathcal{O}) \cong \mathrm{ME}(\lim_{\mathcal{V}}(\mathcal{P} \otimes \mathcal{O}))$$

Here ME is the set of solutions of the appropriate master equation set up in each instance.

Name of $\mathcal{F}$-$\mathcal{O}ps_{\mathcal{C}}$	Algebraic Structure of $F\mathcal{O}$	Master Equation (ME)
operad,[**GJ94**]	odd pre-Lie	$d(-) + - \circ - = 0$
cyclic operad [**GK95**]	odd Lie	$d(-) + \frac{1}{2}[-,-] = 0$
modular operad [**GK98**]	odd Lie + Δ	$d(-) + \frac{1}{2}[-,-] + \Delta(-) = 0$
properad [**Val07**]	odd pre-Lie	$d(-) + - \circ - = 0$
wheeled properad [**MMS09**]	odd pre-Lie + Δ	$d(-) + - \circ - + \Delta(-) = 0$
wheeled prop [**KWZ12**]	dgBV	$d(-) + \frac{1}{2}[-,-] + \Delta(-) = 0$

TABLE 5. Collection of Master Equations for operad–type examples

With Feynman categories this tabular theorem can be compactly written and generalized. The first step is the realization that the differential specifies a natural operation, in the above sense, for each arity n. Furthermore, in the Master Equation there is one term form each generator of Φ^1 up to isomorphism.

The natural operation which lives on a space associated to an $\mathcal{Q} \in \mathcal{F}$-$\mathcal{O}ps$ is denoted $\Psi_{\mathcal{Q},n}$ and is formally defined as follows:

DEFINITION 5.5. [**KW17**, §7.5] For a Feynman category $\mathfrak{F}$ admitting the Feynman transform and for $\mathcal{Q} \in \mathcal{F}$-$\mathcal{O}ps_{\mathcal{C}}$ we define the formal master equation of $\mathfrak{F}$ with respect to $\mathcal{Q}$ to be the completed co–chain $\Psi_{\mathcal{Q}} := \prod \Psi_{\mathcal{Q},n}$. If there is an N such that $\Psi_{\mathcal{Q},n} = 0$ for $n > N$, then we define the master equation of $\mathfrak{F}$ with respect to $\mathcal{Q}$ to be the finite sum:

$$d_{\mathcal{Q}} + \sum_n \Psi_{\mathcal{Q},n} = 0$$

We say $\alpha \in \lim_{\mathcal{V}}(\mathcal{Q})$ is a solution to the master equation if $d_{\mathcal{Q}}(\alpha) + \sum_n \Psi_{\mathcal{Q},n}(\alpha^{\otimes n}) = 0$, and we denote the set of such solutions as $\mathrm{ME}(\lim_{\mathcal{V}}(\mathcal{Q}))$.

Here the first term is the internal differential and the term for $n = 1$ is the differential corresponding to d_{Φ^1}, where Φ^1 is the subset of odd generators.

THEOREM 5.6 ([**KW17**, Theorem 5.7.3]). *Let $\mathcal{O} \in \mathcal{F}\text{-}\mathcal{O}ps_{\mathcal{C}}$ and $\mathcal{P} \in \mathcal{F}^{odd}\text{-}\mathcal{O}ps_{\mathcal{C}}$ for an $\mathcal{F}$ admitting a Feynman transform and master equation. Then there is a bijective correspondence:*

$$\mathrm{Hom}(\mathsf{FT}(\mathcal{P}), \mathcal{O}) \cong \mathrm{ME}(\lim_{\mathcal{V}}(\mathcal{P} \otimes \mathcal{O}))$$

6. W-construction and cubical structures

6.1. Setup. In this section we start with a cubical Feynman category $\mathfrak{F}$.

6.1.1. *The category $w(\mathfrak{F}, Y)$, for $Y \in \mathcal{F}$.* OBJECTS: The objects are the set $\coprod_n C_n(X, Y) \times [0,1]^n$, where $C_n(X,Y)$ are chains of morphisms from X to Y with n degree ≥ 1 maps modulo contraction of isomorphisms.

An object in $w(\mathfrak{F}, Y)$ will be represented (uniquely up to contraction of isomorphisms) by a diagram

$$X \xrightarrow[f_1]{t_1} X_1 \xrightarrow[f_2]{t_2} X_2 \to \cdots \to X_{n-1} \xrightarrow[f_n]{t_n} Y$$

where each morphism is of positive degree and where $t_1, \ldots, t_n$ represents a point in $[0,1]^n$. These numbers will be called weights. Note that in this labeling scheme isomorphisms are always unweighted.

MORPHISMS:

(1) Levelwise commuting isomorphisms which fix Y, i.e.:

$$\begin{array}{ccccccccc} X & \longrightarrow & X_1 & \longrightarrow & X_2 & \longrightarrow \ldots \longrightarrow & X_n & \longrightarrow & Y \\ \downarrow{\scriptstyle\cong} & & \downarrow{\scriptstyle\cong} & & \downarrow{\scriptstyle\cong} & & \downarrow{\scriptstyle\cong} & \nearrow & \\ X' & \longrightarrow & X'_1 & \longrightarrow & X'_2 & \longrightarrow \ldots \longrightarrow & X'_n & & \end{array}$$

(2) Simultaneous $\mathbb{S}_n$ action.

(3) Truncation of 0 weights: morphisms of the form $(X_1 \xrightarrow{0} X_2 \to \cdots \to Y) \mapsto (X_2 \to \cdots \to Y)$.

(4) Decomposition of identical weights: morphisms of the form $(\cdots \to X_i \xrightarrow{t} X_{i+2} \to \ldots) \mapsto (\cdots \to X_i \xrightarrow{t} X_{i+1} \xrightarrow{t} X_{i+2} \to \ldots)$ for each (composition preserving) decomposition of a morphism of degree ≥ 2 into two morphisms each of degree ≥ 1.

DEFINITION 6.1. Let $\mathcal{P} \in \mathcal{F}\text{-}\mathcal{O}ps_{\mathcal{T}op}$. For $Y \in ob(\mathcal{F})$ we define

$$W(\mathcal{P})(Y) := \mathrm{colim}_{w(\mathfrak{F}, Y)} \mathcal{P} \circ s(-)$$

THEOREM 6.2 ([**KW17**, Theorem 8.6.9]). *Let $\mathfrak{F}$ be a simple Feynman category and let $\mathcal{P} \in \mathcal{F}\text{-}\mathcal{O}ps_{\mathcal{T}op}$ be ρ-co–fibrant. Then $W(\mathcal{P})$ is a co–fibrant replacement for $\mathcal{P}$ with respect to the above model structure on $\mathcal{F}\text{-}\mathcal{O}ps_{\mathcal{T}op}$.*

Here "simple" is a technical condition satisfied by all graph examples.

6.2. W-construction yields Associahedra. To see that this indeed yields the usual W-construction, we turn to the well known case of associahedra and its cubical decomposition, see [**BV73, MSS02**]. The relevant Feynman category is the Feynman sub–category of $\mathfrak{O}$ were the vertices are restricted to be at most trivalent rooted planar corollas. This is a sub-Feynman category of $\mathfrak{O}^{pl} = \mathfrak{O}_{\mathrm{dec}\, Ass}$

Let $*_{n_+}$ be the corolla whose flags are $\{0, \ldots, n\}$ where 0 is the root with the natural linear order on the set of flags as an object of $\mathfrak{O}^{pl}$.

PROPOSITION 6.3. *Let $\mathcal{T}$ be the trivial operad to Set.* $W\mathcal{T}(*_{n_+}) = K_n$

PROOF. A basic morphism is given by an edge contraction. Any morphism can be decomposed into edge contractions $\phi = \phi_{e_1} \circ \cdots \circ \phi_{e_n}$ which up to isomorphism can be taken to be pure. That means that we get a cube of dimension n for each class of morphism of degree n. Such a class is uniquely determined by its ghost tree $\mathbb{\Gamma}(\phi)$, which is a planted planar tree. The maximal degree of a morphisms with the given target $*_{n+}$ has a source given by $n-1$ trivalent corollas. Any other morphism is obtained by shortening a maximal chain, which correspond to the boundaries $wt(e) \to 0$ by definition.

The compositions two or more morphisms of degree 1 which are identified in the relations of w with diagonals, not noted In the pictures. □

REMARK 6.4. The first two cases are given by Figure 4, which is taken from [**KS10**]. Note that the outer boundary are the pieces, where $wt(e) \to 1$ and the inner boundaries are the ones where $wt(e) \to 0$, which results in the shortening of the chain of morphisms and hence the contraction of the edge whose weight goes to 0.

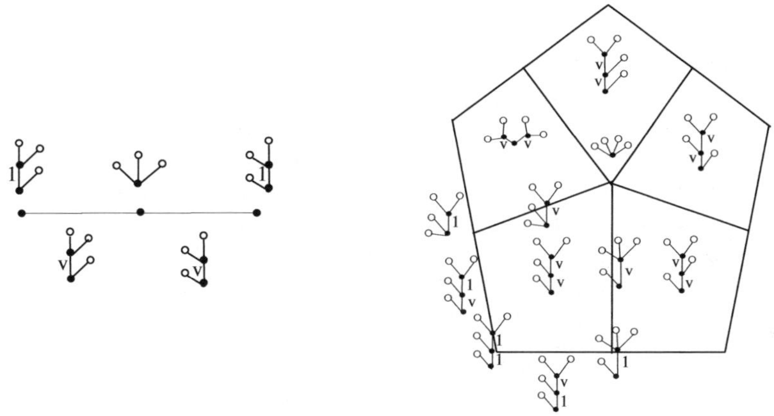

FIGURE 4. The cubical decomposition for K_3 and K_4, v indicated a variable height.

The $\mathcal{O}ps$ structure is the usual operad structure found by Stasheff [**Sta63**]. Namely, if ${}_0\circ_i : *_{n+} \amalg *_{m_+} \to *_{(m+n-1)_+}$ then $W\mathcal{T}(\phi_{e_i})$ glues the two flags 0 and i of any two points in $W\mathcal{T}(*_{n_+})$ and $W\mathcal{T}(*_{m_+})$ and marks the new edge by 1, that is we obtain the face of K_{n+m} which is $K_n \circ_i K_m = K_n \times K_m$

REMARK 6.5. It is intriguing that this cubical decomposition also turns up in the stability conditions for A_n quivers, which was pointed out to us by K. Igusa. It is furthermore worth pointing out that the central charge function results in particular in an embedding of K_4 into $\mathcal{C}$. Such embeddings were sought after since [**Sta63**], cf. e.g. [**MSS02**] for various truncations and constructions.

7. Outlook

7.1. W-construction and moduli spaces. Using the language set up in Appendix A.5.2, especially (A.4), in collaboration with C. Berger we prove [**BK20a**]:

THEOREM 7.1 ([**BK20a**]). (1) *The derived push–forward* $k_!^{\prime}W\mathcal{T}(*_{g,s,S_1,\dots,S_b})$ *is homotopy equivalent to* $M_{g,s,S_1,\dots S_b}$.
(2) *The derived push–forward* $k_!W\mathcal{O}_{cA}(*_{g,S}) = \amalg_{b,s,S=S_1\amalg\cdots\amalg S_b}$ *is homotopy equivalent to* $M_{g,s,S_1,\dots S_b}$.
(3) *The derived push–forward* $j_!W\mathcal{T}$ *is the cubical complex appearing in the Cutkosky rules in* [**BK15, Kre16**] *and Outer Space in* [**BSV18**].

This generalizes and puts the results of Igusa [**CV86, Igu02**] into this simple Feynman context. It also makes some of the constructions of Costello [**Cos07a, Cos07b**] more transparent and rigorous. The last statement is actually brand new and gives a framework for the cubical categories that have appeared in physics and topology.

THEOREM 7.2 ([**BK20a**]). $W(\mathcal{T})(*_{g,s,S_1,\dots,S_b})$ *is the cone over the combinatorial aka. Penner–Kontsevich compactification* $\overline{M}^{KP}_{g,s,S_1,\dots,S_b}$.

7.2. Further connections to representation theory. Generalizations of some of the work presented here is already in the works. The plus construction shall be generalized for arbitrary monoidal categories in [**KM20**]. The new Feynman category $\mathcal{F}in\mathcal{S}et_{<}$ can be generalized to fibers with cyclic orders. This should be related via the plus construction yield the Feynman category for cyclic operads. It is conceivable that if one bases everything on categories with a duality there is a plus construction which yields graphs.

A further point of study will be the connections to cluster algebras. Here there are already several strands especially in the form [**KKGJ15**]. The first is the appearance of the cubical decomposition of the Associahedra introduced above. Here the cubical cells are related to wall crossings. We expect a similar story for other polytopes. Especially for cyclohedra, which appear in the context of the little discs operad in the form of the cactus operad and Deligne's conjecture for A_∞ algebras. These polytopes are at the vertices of the diagrams of [**KKGJ15**, Example 4.0.2] and the cyclic versions in the Feynman theory are possibly related to [**IT15, IT19**], the appearance of cacti being a common theme.

Cluster transformations and cluster varieties yield a method to glue these local complexes together to global data. This is parallel to the results for moduli spaces or Cutkosky rules where the cubical complexes have face transitions according to shrinking edges of different graphs. The prerogative is to weave these strands together. In this realm, we also expect to encounter 2–Segal spaces of [**DK15, DK18**] which basically categorify the pentagon relation, which also appears in mathematical physics and number theory as the identity for quantum di–logarithms.

The enrichment of quivers §4.3.4 via [**KKW15**] provides a new link to mutations and is possibly related to [**HS18**].

Another intriguing aspect is given by moduli spaces and arcs, which also naturally appear in Feynman categories [**KW17, BK20b**] , operadic theory [**KLP03**] as well as in the theory of cluster algebras, see [**GS19, HKK17**].

Appendix A. Graph glossary and graphical Feynman categories

A.1. The category of graphs. Interesting examples of Feynman categories used in operad–like theories are indexed over a Feynman category built from graphs. It is important to note that although we will first introduce a category of graphs $\mathcal{G}raphs$, the relevant Feynman category is given by a full subcategory $\mathcal{A}gg$ whose objects are disjoint unions or aggregates of corollas. The corollas themselves play the role of $\mathcal{V}$.

Before giving more examples in terms of graphs it will be useful to recall some terminology. A very useful presentation is given in [**BM08**] which we follow here.

A.1.1. *Abstract graphs.* An abstract graph Γ is a quadruple $(V_\Gamma, F_\Gamma, i_\Gamma, \partial_\Gamma)$ of a finite set of vertices V_Γ, a finite set of half edges or flags F_Γ, an involution on flags $i_\Gamma\colon F_\Gamma \to F_\Gamma; i_\Gamma^2 = id$ and a map $\partial_\Gamma\colon F_\Gamma \to V_\Gamma$. We will omit the subscript Γ if no confusion arises.

Since the map i is an involution, it has orbits of order one or two. We will call the flags in an orbit of order one *tails* and denote the set of tails by T_Γ. The flags in an orbit of order two will be called internal flags and this set of tails will be denoted by F_Γ^{int}. Thus $F_\Gamma = T_g \amalg F_\Gamma^{\text{int}}$ We will call an orbit of order two an *edge* and denote the set of edges by E_Γ. The flags of an edge are its elements. The function ∂ gives the vertex a flag is incident to. It is clear that the set of vertices and edges form a 1-dimensional CW complex. The realization of a graph is the realization of this CW complex.

A graph is (simply) connected if and only if its realization is. Notice that the graphs do not need to be connected. Lone vertices, that is, vertices with no incident flags, are also possible.

We also allow the empty graph $\mathbb{1}_\emptyset$, that is, the unique graph with $V = \varnothing$. It will serve as the monoidal unit.

Example A.1. A graph with one vertex and no edges is called a *corolla*. Such a graph only has tails. For any set S the corolla $*_{p,S}$ is the unique graph with $V = \{p\}$ a singleton and $F = S$.

We fix the short hand notation $*_S$ for the corolla with $V = \{*\}$ and $F = S$.

Given a vertex v of a graph, we set $F_v = \partial^{-1}(v)$ and call it *the flags incident to* v. This set naturally gives rise to a corolla. The *tails* at v is the subset of tails of F_v.

As remarked above, F_v defines a corolla $*_v = *_{\{v\},F_v}$.

Remark A.2. The way things are set up, we are talking about (finite) sets, so changing the sets even by bijection changes the graphs.

Remark A.3. As the graphs do not need to be connected, given two graphs Γ and Γ' we can form their disjoint union:

$$\Gamma \sqcup \Gamma' = (F_\Gamma \sqcup F_{\Gamma'}, V_\Gamma \sqcup V_{\Gamma'}, i_\Gamma \sqcup i_{\Gamma'}, \partial_\Gamma \sqcup \partial_{\Gamma'})$$

One actually needs to be a bit careful about how disjoint unions are defined. Although one tends to think that the disjoint union $X \sqcup Y$ is strictly symmetric, this is not the case. This becomes apparent if $X \cap Y \neq \emptyset$. Of course there is a bijection $X \sqcup Y \overset{1-1}{\longleftrightarrow} Y \sqcup X$. Thus the categories here are symmetric monoidal, but not strict symmetric monoidal. This is important, since we consider functors into other not necessarily strict monoidal categories.

Using MacLane's theorem it is however possible to make a technical construction that makes the monoidal structure (on both sides) into a strict symmetric monoidal structure

Example A.4. An *aggregate of corollas* or aggregate for short is a finite disjoint union of corollas, that is, a graph with no edges.

Notice that if one looks at $X = \bigsqcup_{v\in I} *_{S_v}$ for some finite index set I and some finite sets of flags S_v, then the set of flags is automatically the disjoint union of the sets S_v. We will just say just say $s \in F_X$ if s is in some S_v.

A.1.2. *Grafting of graphs into vertices.* Fix a graph $\Gamma = (V, F, \partial, i)$, a collection of graphs $\Gamma_v = (F(v), V(v), \partial(v), i(v))$ indexed by $v \in V$ and a set of bijections $\beta_v : F_v \to T_{\Gamma_v}$ that is a bijection of the flags at v with the tails of Γ_v. Set $\beta = \amalg_{v\in V}\beta_v$.

We define the graph $\amalg_{v\in V}\Gamma_v \circ_\beta \Gamma$ to be the graph obtained from grafting Γ_v into the vertex v. As a graph the vertices, flags and boundary are those of $\amalg_{v\in V}\Gamma_v$, that is $\amalg_{v\in V}V(v)$ the flags are $\amalg_{v\in V}F(v)$ with the boundary map $\amalg_{v\in V}\partial(v)$. But, the involution given by $\amalg_{v\in V}i(v)$ on the internal edges of $\amalg_v\Gamma_v$ is continued to the tails $\amalg_{v\in V}T_{\Gamma_v}$ by $\beta \circ i \circ \beta^{-1}$. This equivalent to the statement that the edges of the graph $\amalg_{v\in V}\Gamma_v \circ_\beta \Gamma$ are the set $\amalg_{v\in V}E_{\Gamma_v} \amalg \beta(E_\Gamma)$.

A.1.3. *Category structure: Morphisms of graphs.*

DEFINITION A.5. [**BM08**, **KW17**] Given two graphs Γ and Γ', consider a triple (ϕ^F, ϕ_V, i_ϕ) where

(i) $\phi^F\colon F_{\Gamma'} \hookrightarrow F_\Gamma$ is an injection,
(ii) $\phi_V\colon V_\Gamma \twoheadrightarrow V_{\Gamma'}$ and i_ϕ is a surjection and
(iii) i_ϕ is a fixed point free involution on the tails of Γ not in the image of ϕ^F.

One calls the edges and flags that are not in the image of ϕ the contracted edges and flags. The orbits of i_ϕ are called *ghost edges* and denoted by $E_{ghost}(\phi)$. The ghost edges are uniquely determined by and uniquely determine i_ϕ.

Such a triple is *a morphism of graphs* $\phi\colon \Gamma \to \Gamma'$ if

(1) The involutions are compatible:
   (a) An edge of Γ is either a subset of the image of ϕ^F or not contained in it.
   (b) If an edge is in the image of ϕ^F then its pre–image is also an edge.
(2) ϕ^F and ϕ_V are compatible with the maps ∂:
   (a) Compatibility with ∂ on the image of ϕ^F:
   If $f = \phi^F(f')$ then $\phi_V(\partial f) = \partial f'$
   (b) Compatibility with ∂ on the complement of the image of ϕ^F:
   The two vertices of a ghost edge in Γ map to the same vertex in Γ' under ϕ_V.

If the image of an edge under ϕ^F is not an edge, we say that ϕ grafts the two flags.

The composition $\phi' \circ \phi\colon \Gamma \to \Gamma''$ of two morphisms $\phi\colon \Gamma \to \Gamma'$ and $\phi'\colon \Gamma' \to \Gamma''$ is defined to be $(\phi^F \circ \phi'^F, \phi'_V \circ \phi_V, i_{\phi'\circ\phi})$ where $i_{\phi'\circ\phi}$ is defined by its orbits viz. the ghost edges. This means that $E_{ghost}(\phi \circ \phi') = E_{ghost}(\phi) \amalg \phi^F(E_{ghost})(\phi')$.

More explicitly, let $F_\phi = F' \setminus \phi^F(F')$, $F'_{\phi'} = F' \setminus \phi'^F(F'')$ and $F_{\phi'\circ\phi} = F \setminus \phi^F \circ \phi'^F(F'')$, then $F_{\phi'\circ\phi} = F_\phi \amalg \phi^F(F'_{\phi'})$ and

$$i_{\phi'\circ\phi} = i_\phi \amalg \phi^F \circ i_{\phi'} \circ \phi^{F-1} : F_{\phi'\circ\phi} \to F_{\phi'\circ\phi} \tag{A.1}$$

The following definition from [**KW17**] is essential.

DEFINITION A.6. The underlying ghost graph of a morphism of graphs $\phi\colon \Gamma \to \Gamma'$ is the graph $\mathbb{\Gamma}(\phi) = (V_\Gamma, F_\Gamma, \hat{\imath}_\phi)$ where $\hat{\imath}_\phi$ is i_ϕ on the complement of $\phi^F(\Gamma')$ and identity on the image of flags of Γ' under ϕ^F. Or, alternatively, the edges of $\mathbb{\Gamma}(\phi)$ are the ghost edges of ϕ that is $E_{ghost}(\phi)$.

LEMMA A.7. *Using the usual notation, the following hold for the ghost graphs:*

(1) *Let* $\phi : \Gamma \to \Gamma'$ *then* $\mathbb{\Gamma}(\phi) = \amalg_{v' \in V'} \mathbb{\Gamma}_{v'}$ *and* $\phi^F|_{F'_{v'}} : F'_{v'} \to T_{\mathbb{\Gamma}_{v'}}$ *is a bijection.*

(2) $\mathbb{\Gamma}(\phi \amalg \psi) = \mathbb{\Gamma}(\phi) \amalg \mathbb{\Gamma}(\psi)$.

(3) *For a composition of morphisms* $\Gamma \xrightarrow{\phi} \Gamma' \xrightarrow{\phi} \Gamma''$*:* $\mathbb{\Gamma}(\phi') \circ_\beta \phi = \mathbb{\Gamma}(\phi) \circ_{\phi^F} \mathbb{\Gamma}(\phi)$

PROOF. For the first statement, we define $\mathbb{\Gamma}(v')(\phi)$ as follows: The vertices are $\phi_V^{-1}(v')$, the flags are $\amalg_{v \in \phi_V^{-1}(v')} F_v$. The morphisms ∂_v and i_v are the restriction of ∂_Γ and $\hat{i}_\phi$, which are well defined. For ∂_v this is guaranteed by condition (2) and for i_v we have to check that if a flag is in $F_{\phi_V^{-1}}(v')$ then so is $i_\phi(f)$ which is guaranteed by condition (2) (b). Now ϕ^F is a bijection onto its image, which are precisely the tails of the ghost graph. This restricts to the $\mathbb{\Gamma}_v$. The second statement follows from the first. The last statement is clear for the vertices and the flags. We have to check that the involution coincide. These are determined by the ghost edges. The ghost edges of the composition are the disjoint union of the ghost edges of ϕ and the image of those of ϕ' which is guaranteed by (A.1). □

DEFINITION A.8. We let *Graphs* be the category whose objects are abstract graphs and whose morphisms are the morphisms described in Definition A.5. We consider it to be a monoidal category with monoidal product $\sqcup$ (see Remark A.3).

A.2. The Feynman category $\mathfrak{G}$. Let *Agg* be the full subcategory of *Graphs* whose objects are aggregates of corollas. Let *Crl* be the sub–groupoid of *Agg* whose objects are corollas and whose morphisms are the isomorphisms between them and denote the inclusion by $\imath$.

LEMMA A.9. *Given a morphism* $\phi\colon X \to Y$ *where* $X = \bigsqcup_{w \in V_X} *_w$ *and* $Y = \bigsqcup_{v \in V_Y} *_v$ *are two aggregates, we can decompose* $\phi = \bigsqcup \phi_v$ *with* $\phi_v\colon X_v \to *_v$ *where* X_v *is the sub–aggregate* $\bigsqcup_{\phi_V(w)=v} *_w$*, and* $\bigsqcup_v X_v = X$.

Furthermore, $\mathbb{\Gamma}(\phi) = \amalg_{v \in V} \mathbb{\Gamma}(\phi_v)$ *and* $\phi_v'^F : F_v \to T_{\mathbb{\Gamma}(\phi_v)}$ *is a bijection.*

PROOF. Explicitly $(\phi_v)_V$ is the restriction of ϕ_V to V_{X_v} and ϕ_v^F is the restriction of ϕ^F to $(\phi^F)^{-1}(F_{X_v} \cap \phi^F(F_Y))$. This map is still injective. Finally i_{ϕ_v} is the restriction of i_ϕ to $F_{X_v} \setminus \phi^F(F_Y)$. These restrictions are possible due to the condition (2) above. The penultimate statement follows from Lemma A.7. The tails of the ghost graph are precisely the elements in the image of ϕ^F. □

PROPOSITION A.10. *The triple* $\mathfrak{G} = (\mathit{Crl}, \mathit{Agg}, \imath)$ *is a strictly strict Feynman category.*

PROOF. The proof follows from the Lemma above. Condition (i) is clear on the object level. By the lemma an isomorphism factors as a disjoint union of isomorphisms. It also shows that condition (ii) holds. Note that this condition implies that decompositions are unique up to unique isomorphism. Indeed any decomposition will be given by a permutation of the decomposition above and isomorphisms of the vertices this will uniquely determine the diagram. Finally, (iii), follows from the category of finite sets is essentially small. Strictness is clear from the definition. □

A tree	is a connected, simply connected graph.
A directed graph Γ	is a graph together with a map $F_\Gamma \to \{in, out\}$ such that the two flags of each edge are mapped to different values.
A rooted tree	is a directed tree such that each vertex has exactly one "out" flag.
A ribbon or fat graph	is a graph together with a cyclic order on each of the sets F_v.
A planar graph	is a ribbon graph that can be embedded into the plane such that the induced cyclic orders of the sets F_v from the orientation of the plane coincide with the chosen cyclic orders.
A planted planar tree	is a rooted planar tree
An oriented graph	is a graph with an orientation on the set of its edges.
An ordered graph	is a graph with an order on the set of its edges.
A γ labelled graph	is a graph together with a map $\gamma : V_\Gamma \to \mathbb{N}_0$.
A b/w graph	is a graph Γ with a map $V_\Gamma \to \{black, white\}$.
A bipartite graph	is a b/w graph whose edges connect only black to white vertices.
A c colored graph	for a set c is a graph Γ together with a map $F_\Gamma \to c$ s.t. each edge has flags of the same color.
A connected 1–PI graph	is a connected graph that stays connected, when one severs any edge.
A 1–PI graph	is a graph whose every component is 1–PI.

TABLE 6. Graph Dictionary

A.3. Extra structures.

A.3.1. *Dictionary.* This section is intended as a reference section. Recall that an order of a finite set S is a bijection $S \to \{1, \dots, |S|\}$. Thus the group $\mathbb{S}_{|S|} = Aut\{1, \dots, n\}$ acts on all orders. An orientation of a finite set S is an equivalence class of orders, where two orders are equivalent if they are obtained from each other by an even permutation. With this Table 6 provides a dictionary for standard graph terminology.

A.3.2. *Remarks and language.*

(1) In a directed graph one speaks about the "in" and the "out" edges, flags or tails at a vertex. For the edges this means the one flag of the edges is an "in" flag at the vertex. In pictorial versions the direction is indicated by an arrow. A flag is an "in" flag if the arrow points to the vertex.
(2) As usual there are edge paths on a graph and the natural notion of an oriented edge path. An edge path is a (oriented) cycle if it starts and stops at the same vertex and all the edges are pairwise distinct. It is called simple if each vertex on the cycle has exactly one incoming flag and one outgoing flag belonging to the cycle. An oriented simple cycle will be called a *wheel*. An edge whose two vertices coincide is called a *(small) loop*.

(3) There is a notion of the genus of a ribbon graph, which is the minimal dimension of the surface it can be embedded on. A ribbon graph is planar if this genus is 0.

(4) For any graph, its Euler characteristic is given by

$$\chi(\Gamma) = b_0(\Gamma) - b_1(\Gamma) = |V_\Gamma| - |E_\Gamma|;$$

where b_0, b_1 are the Betti numbers of the (realization of) Γ. Given a γ labelled graph, we define the total γ as

$$\gamma(\Gamma) = 1 - \chi(\Gamma) + \sum_{v \text{ vertex of } \Gamma} \gamma(v) \tag{A.2}$$

If Γ is *connected*, that is $b_0(\Gamma) = 1$ then a γ labeled graph is traditionally called a genus labeled graph and

$$\gamma(\Gamma) = \sum_{v \in V_\Gamma} \gamma(v) + b_1(\Gamma) \tag{A.3}$$

is called the genus of Γ. This is actually not the genus of the underlying graph, but the genus of a connected Riemann surface with possible double points whose dual graph is the genus labelled graph.

A genus labelled graph is called *stable* if each vertex with genus labeling 0 has at least 3 flags and each vertex with genus label 1 has at leas one edge.

(5) A planted planar tree induces a linear order on all sets F_v, by declaring the first flag to be the unique outgoing one. Moreover, there is a natural order on the edges, vertices and flags given by its planar embedding.

(6) A rooted tree is usually taken to be a tree with a marked vertex. Note that necessarily a rooted tree as described above has exactly one "out" tail. The unique vertex whose "out" flag is not a part of an edge is the root vertex. The usual picture is obtained by deleting this unique "out" tail.

A.3.3. *Category of directed/ordered/oriented graphs.*

(1) Define the category of directed graphs $\mathcal{G}raphs^{dir}$ to be the category whose objects are directed graphs. Morphisms are morphisms ϕ of the underlying graphs, which additionally satisfy that ϕ^F preserves orientation of the flags and the i_ϕ also only has orbits consisting of one "in" and one "out" flag, that is the ghost graph is also directed.

(2) The category of edge ordered graphs $\mathcal{G}raphs^{or}$ has as objects graphs with an order on the edges. A morphism is a morphism together with an order *ord* on all of the edges of the ghost graph.

The composition of orders on the ghost edges is as follows. $(\phi, ord) \circ \bigsqcup_{v\in V}(\phi_v, ord_v) := (\phi \circ \bigsqcup_{v\in V} \phi_v, ord \circ \bigsqcup_{v\in V} ord_v)$ where the order on the set of all ghost edges, that is $E_{ghost}(\phi) \sqcup \bigsqcup_v E_{ghost}(\phi_v)$, is given by first enumerating the elements of $E_{ghost}(\phi_v)$ in the order ord_v where the order of the sets $E(\phi_v)$ is given by the order on V, i.e. given by the explicit ordering of the tensor product in $Y = \bigsqcup_v *_v$.[6] and then enumerating the edges of $E_{ghost}(\phi)$ in their order *ord*.

[6]Now we are working with ordered tensor products. Alternatively one can just index the outer order by the set V by using [**Del90**]

(3) The oriented version $\mathcal{G}raphs^{or}$ is then obtained by passing from orders to equivalence classes.

A.3.4. *Category of planar aggregates and tree morphisms.* Although it is hard to write down a consistent theory of planar graphs with planar morphisms, if not impossible, there does exist a planar version of special subcategory of $\mathcal{G}raphs$.

We let $\mathcal{C}rl^{pl}$ have as objects planar corollas — which simply means that there is a cyclic order on the flags — and as morphisms isomorphisms of these, that is isomorphisms of graphs, which preserve the cyclic order. The automorphisms of a corolla $*_S$ are then isomorphic to $C_{|S|}$, the cyclic group of order $|S|$. Let $\mathfrak{C}^{pl}$ be the full subcategory of aggregates of planar corollas whose morphisms are morphisms of the underlying corollas, for which the ghost graphs in their planar structure induced by the source is compatible with the planar structure on the target via ϕ^F. For this we use the fact that the tails of a planar tree have a cyclic order.

Let $\mathcal{C}rl^{pl,dir}$ be directed planar corollas with one output and let $\mathfrak{O}^{pl}$ be the subcategory of $\mathcal{A}gg^{pl,dir}$ of aggregates of corollas of the type just mentioned, whose morphisms are morphisms of the underlying directed corollas such that their associated ghost graphs are compatible with the planar structures as above.

In general, one needs to use so–called almost ribbon graphs, see e.g. [**KL16**] or [**Kau09**, Appendix A1], and in §A.5.2.

A.4. Insertion. Given graphs, Γ,Γ', a vertex $v \in V_\Gamma$ and an isomorphism ϕ: $F_v \leftrightarrow T_{\Gamma'}$ we define $\Gamma \circ_v \Gamma'$ to be the graph obtained by deleting v and identifying the flags of v with the tails of Γ' via ϕ. Notice that if Γ and Γ' are ghost graphs of a morphism then it is just the composition of ghost graphs, with the morphisms at the other vertices being the identity.

A.5. Operad-types and their graphical Feynman categories. There is a substantial list of examples that are generated by decorating the graphs of $\mathfrak{G}$ and restricting to certain subcategories, see Table 7. The decorations are actually decorations in the technical sense of §3.2. Examples of the needed decorations are listed in Table 8.

A.5.1. *Flag labeling, colors, direction and roots as a decoration.* Recall that $*_S$ is the one vertex graph with flags labelled by S and these are the objects of $\mathcal{V} = \mathcal{C}rl$ for $\mathfrak{G}$. For any set X introduce the following $\mathfrak{G}$-$\mathcal{O}p$: $X(*_S) = X^S$. The compositions are simply given by restricting to the target flags.

Now let the set X have an involution $\bar{}: X \to X$. Then a natural subcategory $\mathfrak{F}^{dir}_{\text{dec } X}$ of $\mathfrak{G}_{\text{dec } X}$ is given by the wide subcategory, whose morphisms additionally satisfy that only flags marked by elements x and $\bar{x}$ are glued and then contracted; viz $\imath_\phi$ only pairs flags of marked x with edges marked by $\bar{x}$. That is the underlying ghost graph has edges whose two flags are labelled accordingly. In the notation of graphs: $X(f) = \overline{\imath_\phi(f)}$.

If X is pointed by x_0, there is the subcategory of $\mathfrak{G}_{\text{dec } X}$ whose objects are those generated by $*_S$ with exactly one flag labelled by x_0 and where the restriction on graphs is that for the underlying graph additionally, each edge has one flag labelled by x_0.

Now if $X = \mathbb{Z}/2\mathbb{Z} = \{0, 1\}$ with the involution $\bar{0} = 1$, we can call 0 "out" and 1 "in". As a result, we obtain the category of directed graphs $\mathfrak{G}_{\text{dec } \mathbb{Z}/2Z}$. Furthermore, if 0 is the distinguished element, we get the rooted version. This explains the relevant examples Table 8. More generally, in quantum field theory

$\mathfrak{F}$	Feynman category for	condition on ghost graphs $\mathbb{\Gamma}_v$ for basic morphisms and additional decoration
$\mathfrak{O}$	(pseudo)–operads	rooted trees
$\mathfrak{O}^{\neg\Sigma}$	non-Sigma operads	planar rooted trees
$\mathfrak{O}_{mult}$	operads with mult.	b/w rooted trees.
$\mathfrak{C}$	cyclic operads	trees
$\mathfrak{C}^{\neg\Sigma}$	non–Sigma cyclic operads	planar trees
$\mathfrak{G}$	unmarked nc modular operads	graphs
$\mathfrak{G}^{ctd}$	unmarked modular operads	connected graphs
$\mathfrak{M}$	modular operads	connected + genus marking
$\mathfrak{M}^{nc}$	nc modular operads	genus marking
$\mathfrak{D}$	dioperads	connected directed graphs w/o directed loops or parallel edges
$\mathfrak{P}$	PROPs	directed graphs w/o directed loops
$\mathfrak{P}^{ctd}$	properads	connected directed graphs w/o directed loops
$\mathfrak{D}^{\circlearrowleft}$	wheeled dioperads	directed graphs w/o parallel edges
$\mathfrak{P}^{\circlearrowleft,ctd}$	wheeled properads	connected directed graphs
$\mathfrak{P}^{\circlearrowleft}$	wheeled props	directed graphs
$\mathfrak{F}^{1\text{-}PI}$	1–PI algebras	1–PI connected graphs.

TABLE 7. List of Feynman categories with conditions and decorations on the graphs, yielding the zoo of examples

the involution sends a field to its anti–field and this is what decorates the lines or propagators in a Feynman graph.

Finally, if the involution is trivial, then we obtain the colored version, where ghost edges have flags of the same color.

A.5.2. *Other decorating operads and connecting the bootstraps.* The other decoration operads are

(1) $\mathbb{N} \in \mathfrak{G}^{ctd}\text{-}\mathcal{O}ps_{\mathcal{S}et}$ given by $\mathbb{N}(*_S) = \mathbb{N}$ with the composition given by addition for edge contractions and $n \mapsto n+1$ for loop contractions.
(2) $Assoc \in \mathfrak{O}\text{-}\mathcal{O}ps_{\mathcal{S}et}$, as defined in Lemma 3.37. $Assoc(*_S) = \{$orders on the set $S\}$
(3) $CycAssoc \in \mathfrak{C}\text{-}\mathcal{O}ps_{\mathcal{S}et}$: $CycAssoc(*_S) = \{$cyclic orders on the set $S\}$ here composition is given by splicing in the cyclic order, see e.g. [**Kau04**].

Using these decorations, we have a diagram of Feynman categories.

THEOREM A.11. *The following equalities can serve as a natural definition of the right hand side or as a theorem identifying the right hand side as a decorated Feynman category*

(1) $\mathfrak{O}_{\text{dec } Assoc} = opereads^{\neg\Sigma}$.
(2) *The morphism* $\mathfrak{i} : \mathfrak{O} \to \mathfrak{C}$, *given by forgetting the root, is an indexing, but neither connected nor a cover.*
(3) $i^*(CycAssoc) = Assoc$.
(4) $\mathfrak{G}^{ctd}_{dec\, j_!(\mathcal{T})} = \mathfrak{M}$. *By general theory* $\mathfrak{j}$ *factors as* $\mathfrak{j} = forget \circ \mathfrak{k}$ *with* $\mathfrak{k} : \mathfrak{C} \to \mathfrak{M}$, *where* $\mathfrak{k}$ *is the connected morphism defined by assigning the marking* $g = 0$ *to a corolla.*

$\mathfrak{F}_{\mathrm{dec}\mathcal{O}}$	Feynman category for	decorating $\mathcal{O}$	restriction
$\mathfrak{F}^{dir}$	directed version	$(\mathbb{Z}/2\mathbb{Z}, \bar{0}=1)$	
$\mathfrak{F}^{rooted}$	root	$(\mathbb{Z}/2\mathbb{Z}, \bar{0}=1)$	vertices have one output flag.
$\mathfrak{F}^{genus}$	genus marked	$\mathbb{N}$	
$\mathfrak{F}^{C\text{-}col}$	colored version	$(C, \bar{c}=c)$	
$\mathfrak{O}^{\neg\Sigma}$	non-Sigma-operads	$Assoc$	
$\mathfrak{C}^{\neg\Sigma}$	non-Sigma-cyclic operads	$CycAssoc$	
$\mathfrak{M}^{\neg\Sigma}$	non–Signa-modular	$ModAssoc$	

TABLE 8. List of decorated Feynman categories with decorating $\mathcal{O}$ and possible restriction. $\mathfrak{F}$ stands for an example based on $\mathfrak{G}$ in the list or more generally indexed over $\mathfrak{G}$ (see [**KW17**]).

(5) $\mathfrak{C}_{dec\,\mathcal{O}_{CycAssoc}} = \mathfrak{C}^{\neg\Sigma}$, *the Feynman category whose $\mathcal{O}ps$ are non–sigma cyclic operads.*

(6) $\mathfrak{M}_{dec\,k_!(\mathcal{O}_{CycAssoc})} = \mathfrak{M}^{\neg\Sigma}$, *the Feynman category whose $\mathcal{O}ps$ are non–sigma modular operads as defined in* [**Mar16, KL16**]. *Here, $k_!(CycAssoc)$ is the modular envelope of $CycAssoc$.*

These fit into the diagram (A.4), *where the upper squares are the one of the type* (3.6) *and the triangle is also such a square by using the identification $\mathfrak{F}_{dec\mathcal{T}} = \mathfrak{F}$.*

(A.4)
$$\begin{array}{ccccc}
\mathfrak{O}^{\neg\Sigma} & \xrightarrow{\ \mathfrak{i}'\ } & \mathfrak{C}^{\neg\Sigma} & \xrightarrow{\ \mathfrak{k}'\ } & \mathfrak{M}^{\neg\Sigma} \\
\downarrow{\scriptstyle p(Assoc)} & & \downarrow{\scriptstyle p(CycAssoc)} & & \downarrow{\scriptstyle p(\mathfrak{k}_!(CycAssoc))} \\
\mathfrak{O} & \xrightarrow[\ \mathfrak{i}\ ]{} & \mathfrak{C} & \xrightarrow[\ \mathfrak{k}\ ]{} & \mathfrak{M} \\
 & & & \searrow{\scriptstyle \mathfrak{j}} & \downarrow{\scriptstyle p(\mathfrak{j}_!(\mathcal{T}))} \\
 & & & & \mathfrak{G}^{ctd}
\end{array}$$

DETAILS: By calculating the push–forwards, one obtains that the basic objects of $\mathfrak{F}^{mod}$ are of genus marked corollas $*_{g,S}$ with $g \in \mathbb{N}_0$, while the basic objects of $\mathfrak{F}^{\neg\Sigma\,mod}$ are marked corollas $*_{g,s,S_1,\dots,S_b}$, where $s, g \in \mathbb{N}_0$ and $S_1, \dots, S_b$ are non–empty sets which each have a cyclic order.

As a geometric mnemonic $*_{g,s,S_1,\dots,S_b}$ represents an oriented *topological surface* of genus g with s internal punctures, b boundaries and marked points S_i on the boundary i.

These facts are now just a neat calculation using a Kan extension, with the only inputs being j and $cAss$. This is another example of a radical reduction of complicated concepts to more basic structures.

Appendix B. Graph description of $\mathfrak{F}^+$, $\mathfrak{F}^{+gcp}$ and $\mathfrak{F}^{hyp}$

In this section, we will give a graphical version for $\mathfrak{F}^+, \mathfrak{F}^{+gcp}$ and $\mathfrak{F}^{hyp}$. This is a variation of the category for operads. There is a discrete version, see [**KM20**], which uses the fact that for a strictly strict $\mathfrak{F}$, $Obj(\mathcal{F} \downarrow \mathcal{V})$ are a $Obj(\mathcal{V})$ colored operad and hence there is a decoration of the Feynman category for $Obj(V)$–colored operads.

To obtain the correct behaviour for the isomorphism, we have to consider the "$\mathcal{V}$–colored operad $Iso(\mathcal{F} \downarrow \mathcal{V})$" and regard the corresponding decorated Feynman category. This is what is captured below.

B.1. Combinatorial graph based description of $\mathfrak{F}^+$ in the strictified case. In this section, we consider $\mathfrak{F}$ to be strictly strict, that is $Iso(\mathcal{F}) = \mathcal{V}^{\otimes}$ where the latter is the strict free symmetric monoidal category.

B.1.1. *Planar planted corollas.* Recall that a corolla is a graph with one vertex and no edges. In the in the notation of [**BM08**, **KW17**], see Appendix A, a corolla with vertex v and set of flags S is given by $v_S = (\{v\}, S, \partial : S \to \{v\}, id)$. It is planted planar, if S has a linear order. We say the smallest element is the root flag, and denote is by s_0. This gives S a natural structure of pointed set (S, s_0) and we denote the planar planted corolla by $(v_S, s_0, <)$. The order is equivalent to a map $\text{lab} : S \leftrightarrow \{0, , \dots, , n\}$ with $\text{lab}(s_0) = 0$ or in other words a bijection of pointed sets.

B.1.2. *The groupoid* $Crl_{\mathfrak{F}^+}$. An $\mathcal{V}$–colored $(\mathcal{F} \downarrow \mathcal{V})$ decorated corolla is a morally rooted corolla whose lone vertex is $\phi \in Obj(\mathcal{F} \downarrow \mathcal{V})$ and whose leaves $s(\phi) = *_1 \otimes \cdots \otimes *_n$ and whose root is $t(\phi)$, where s, t are the source and target maps. This is the view taken in [**KW17**].

To use graphical notation as introduced above with all the bells and whistles, consider $(v_S, s_0, <, c_F, \text{dec}_V)$, that is a planar planted corolla together with a map c_F that decorates the flags by objects of $\mathcal{V} : c_F : S \to Obj(\mathcal{V})$, and a decoration of the vertex $\text{dec}_V : \{v\} \to Obj(\imath^{\otimes} \downarrow \imath)$ which is a compatible decoration of the lone vertex by a morphisms that is a choice $\text{dec}\,(v) = \phi_v \in Obj(\mathcal{F} \downarrow \mathcal{V})$, such that $\phi_v : c_F(\text{lab}^{-1}(1)) \otimes \cdots \otimes c_F(\text{lab}^{-1}(n)) \to c_F(\text{lab}^{-1}(0))$. Two corollas are equivalent if $(v_S, s_0, <, c_F, \text{dec}_V) \sim (v_S, s_0, <', c_F, \text{dec}_V)$ they provide the same $s(\phi)$.

$$c_F(\text{lab}^{-1}(1)) \otimes \cdots \otimes c_F(\text{lab}^{-1}(n)) = c_F(\text{lab}'^{-1}(1)) \otimes \cdots \otimes c_F(\text{lab}'^{-1}(n)) \tag{B.1}$$

An $\mathcal{V}$–colored $(\mathcal{F} \downarrow \mathcal{V})$ decorated corolla is an equivalence class of planar planted decorated corollas. $[(v_S, s_0, <, c_F, \text{dec}_V)]$.

Each morphism $\phi : *_1 \otimes \cdots \otimes *_n \to *_0$ thus naturally determines an $\mathfrak{F}$–colored corolla $[(v, \{0, 1, \to n\}, 0, c_F(n) = n, \text{dec}_V(v) = \phi)]$, which will simply be denoted by $*_\phi$. This carries over to the notation used in [**KW17**] using the formalism of [**Del90**]: $\phi : \bigotimes_{s \in S} *_s \to *_t$.

Let $Crl_{\mathfrak{F}^+}$, which is short for $\mathcal{V}$-$Crl_{(\mathcal{F}\downarrow\mathcal{V})}$, be the groupoid whose objects are $\mathcal{V}$–colored $(\mathcal{F} \downarrow \mathcal{V})$ decorated corollas.

The *morphisms* of the groupoid $Crl_{\mathfrak{F}^+}$ are isomorphisms of corollas compatible with the decorations. That is a tuple $(f, \boldsymbol{\sigma})$ where f a graph morphism $v_S \to v'_{S'}$ given by: the only possible map $f_V : \{v\} \to \{v'\}$, a bijection $f^F : S \leftrightarrow S'$ and involution $\imath_f = \emptyset$ which means no ghost edges. Note that f^F induces a permutation: $p = \text{lab} \circ (f^F) \circ \text{lab}'^{-1}$. Let $*_i = c_F(\text{lab}^{-1}(i)), *'_i = c_F(\text{lab}'^{-1}(i))$. Then $\boldsymbol{\sigma} = (\sigma_0, \sigma_1, \dots, \sigma_n)$ is an ordered tuple of isomorphisms $\sigma_i : *'_i \xrightarrow{\sim} *_{p(i)}$. Now let $C_p : *_1 \otimes \cdots \otimes *_n \to *_{p(1)} \otimes \cdots \otimes *_{p(n)}$ be the permutation commutativity constraint.

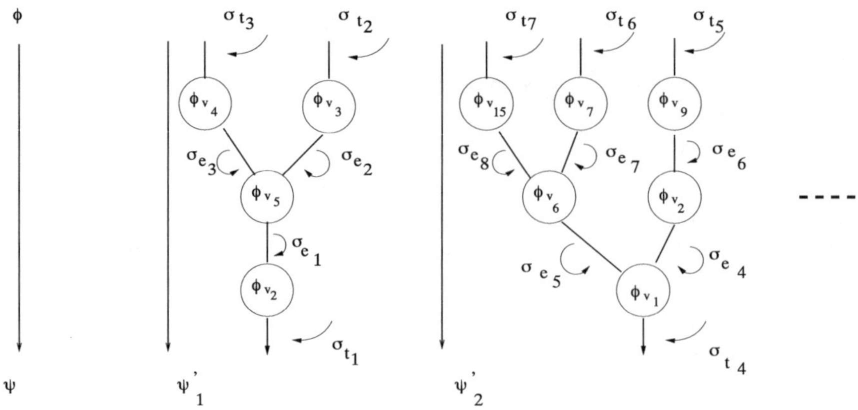

FIGURE 5. A decorated rooted forest. The $\sigma_e \in \boldsymbol{\imath}$ are written next to the internal edges, while the $\sigma_t \in \boldsymbol{\sigma}$ are written next to the tails. These trees give the flow charts to the ψ'_i and the forest is a map $\phi = \phi_1 \otimes \cdots \otimes \phi_n \to \psi = \psi_1 \otimes \cdots \otimes \psi_m$ where the ψ_j are given by (B.3).

Compatibility means that $\phi'_{v'} = \sigma_0 \circ \phi \circ C_p^{-1} \circ (\sigma_1 \otimes \cdots \otimes \sigma_n)$, i.e. the following diagram commutes.

$$
\begin{array}{ccccc}
*_1 \otimes \cdots \otimes *_n & \xrightarrow[\sim]{C_p} & *_{p(1)} \otimes \cdots \otimes *_{p(n)} & \xleftarrow[\sim]{\sigma_1 \otimes \cdots \otimes \sigma_n} & *'_1 \otimes \cdots \otimes *'_n \\
\downarrow{\scriptstyle \phi_v} & & \downarrow{\scriptstyle \phi_v \circ C_{p^{-1}}} & & \downarrow{\scriptstyle \phi_{v'}} \\
*_0 & = & *_0 & \xrightarrow[\sigma_0]{\sim} & *'_0
\end{array}
\tag{B.2}
$$

It is straightforward to check that this is compatible with the equivalence relation $\sim$ and hence independent of representative.

Note that these are precisely the isomorphisms allowed in $(\mathcal{F} \downarrow \mathcal{V})$.

B.1.3. *The monoidal category* $\mathcal{A}gg_{\mathfrak{F}^+}$. Again, $\mathcal{A}gg_{\mathfrak{F}^+}$ is short for $\mathcal{V}$-$\mathcal{A}gg_{(\mathcal{F}\downarrow\mathcal{V})}$.

Let $\mathcal{A}gg_{\mathfrak{F}^+}$ be the category whose underlying objects are $Obj(\mathcal{A}gg_{\mathfrak{F}^+}) = Obj(\mathcal{C}rl^{\otimes}_{\mathfrak{F}^+})$ and whose morphisms are given by morphisms of decorated aggregates of corollas. That is *compatible* tuples $(g, \boldsymbol{\sigma}, \boldsymbol{\imath})$, where

(1) g is an morphism of the underlying rooted corollas $g = (g_V, g^F, \imath_g)$. This means that $\mathbb{\Gamma}(g)$ is a rooted forest and every ghost edge has exactly one rooted flag that it is naturally directed $e = (f = r, \imath(f))$ with r in the set of roots.
(2) If F' is the set of flags of the target, then $\boldsymbol{\sigma} = \{\sigma'_f : f' \in F'\}$ is a collection of isomorphisms $\sigma_{f'} : \mathrm{c}(f') \xrightarrow{\sim} \mathrm{c}(g^F(f'))$ and
(3) $\boldsymbol{\imath} = \{\sigma_e\}, e \in E_{\mathbb{\Gamma}(g)}$ is a collection of isomorphisms $\sigma_e : \mathrm{c}(f) \to \mathrm{c}(\imath(f))$ for each edge directed as $e = (f, \imath_g(f))$

This again clearly compatible with $\sim$.

An example of such a decorated tree is given in Figure 5.

To explain the compatibility condition, let $\tau \subset \mathbb{\Gamma}(g)$ be a connected component, viz. a rooted tree. Each such tree corresponds to a vertex of w_j the target via the map g_v, viz. all the vertices the tree are precisely the set $f_V^{-1}(w_j)$ and thus the components are can be enumerated as τ_j. The short version of *compatibility* is that using g and $\boldsymbol{\imath}$ each decorated τ_j represents a flow chart yielding a morphism ψ_j and up to an isomorphism determined by $\boldsymbol{\sigma}$: $\psi_1 \otimes \cdots \otimes \psi_m \simeq \psi$. To define the *flow chart*, we first notice that since there is a root, each vertex has a height which is the distance from the root. Next, we notice that picking a representative of the vertices, the tree actually is planar as there is a linear order at each vertex. This gives a linear order to all the leaves. We proceed by induction. To start, assume that τ_j has only one vertex, the root vertex v_i with $g_V(v_i) = w_j$. Let $n = ar(v_i)$ be the arity, i.e. the number of incoming flags, of v_i, let the incoming flags be colored by $*_{i_1}, \ldots, *_{j_n}$, the outgoing (root) flag be colored by $*_{j_0}$ and let ϕ_i be the color of v_j that is $\phi_i : *_{j_1} \otimes \cdots \otimes *_{i_n} \to *_{j_0}$. We set $\psi'_j = \phi_j$.

We can now proceed by induction. Assume that we have defined a morphism $\psi'_j(\tau_j)$ for every decorated tree τ_j of maximal height n. Let τ_j be a tree of maximal height $n+1$. Consider the leaf vertices of τ_j. Since τ_j is planar, they are ordered. There are two types of vertices. The first is at height $n+1$ and the second is at lower height. Let v_i be such a leaf vertex then the incoming flags are colored by $*_{i_1}, \ldots, *_{i_{ar(v_i)}}$ if v_i is not at maximal height we set $\psi'_i = id_{*_{i_1}} \otimes \cdots \otimes id_{*_{i_{ar(v_i)}}}$

If v_i is of maximal height and ϕ_i, the color of v_i, maps $*_{i_1} \otimes \cdots \otimes *_{i_{ar(v_i)}}$ to $*_{i_0}$. There is a (ghost) edge $e(i_0, j_k)$ which connects v_i to some v_j. Then $\sigma_e : *_{i_0} \xrightarrow{\sim} *_{v_k}$ and we set $\psi'_i = \sigma_e \circ \phi_i$. Let $\{v_i\} i = 1 \ldots l$ be the set of leaf vertices. Let τ' be the tree obtained from τ' by cutting the outgoing edges of the leaf vertices of maximal height. Then we set $\psi'(\tau) := (\psi'_1 \otimes \cdots \otimes \psi'_l) \circ \psi'(\tau')$. This defines a morphism $\psi'(\tau_j)$. The source of the morphism is given by the color of the leaf flags. Let $*_{l_1}, \ldots, *_{l_L}$ be the color of the leaf flags and $*_{j_0}$ be the color of the root of τ_j then $\psi'(\tau_j) : *_{l_1} \otimes \cdots \otimes *_{l_L} \to *_{j_0}$.

The map g^F then identifies the flags of the vertices w_j with the tails, that is the leaves and the root, of τ_j and the decoration gives further isomorphisms. The decoration is *compatible* if using these isomorphisms $\psi'(\tau_j)$ is isomorphic to ψ_j, the color of w_j. More precisely, let $f_{j_1}, \ldots, f_{j_L}$ be the incoming flags of w_j and f_{j_0} be the root flag let $\sigma_k : c_F(f_{j_k}) \to c_F(g^F(f_{j_k}))$ be the isomorphisms provided by $\boldsymbol{\sigma}$. Just as before, g^F induces a permutation p on the image and a corresponding commutativity constraint C_p. Then the compatibility equation reads

$$\psi_j = \sigma_0^{-1} \circ \psi'_j \circ C_P^{-1} \circ (\sigma_1 \otimes \cdots \otimes \sigma_L) \tag{B.3}$$

In terms of morphisms such a forest gives a morphism of vertex colors $\phi_1 \otimes \cdots \otimes \phi_n \to \psi_1 \otimes \cdots \otimes \psi_m$. This construction is compatible with $\sim$, and thus does not depend on the choice of particular order $<$ in an equivalence class.

The monoidal structure on morphisms is given by disjoint union of decorated forests. Composition of morphisms is given by composition of the underlying graph morphisms and decorations. The composition of the graph morphisms has the effect of inserting trees τ_j into the vertices w_j of the ghost forest of the second morphism in a composition $\amalg v_k \xrightarrow{g} \amalg w_j \xrightarrow{h} \amalg u_l$. One tree per vertex, see [**KW17**], where the incoming flags of w_j are identified with the leaves of τ_j and the outgoing flag of w_j with the root of τ_j. The decoration of this "blown-up" tree is given as follows. For $\boldsymbol{\sigma}$ this is simply the concatenation of isomorphisms. I.e. if $h^F(f) = f'$ and $g^F(f') = f''$ then the isomorphism is $\sigma_{f'} \circ \sigma_{f''} : c_F(f'') \to c_F(f)$. For $\boldsymbol{\imath}$, for the edges of the τ_j, the isomorphisms remain. For the image of the edges that connect the w_j one again composes the isomorphisms. Let $e' = (r', f')$ be a directed edge of

$\mathbb{\Gamma}(h)$. And let $g^F(f') = f, g^F(r') = r$ then the edge $e = (r, f)$ is an edge of $\mathbb{\Gamma}(h \circ g)$ and the isomorphism is given by $\sigma_e = \sigma_f \circ \sigma_{e'} \circ \sigma_r : \mathrm{c}(r) \to \mathrm{c}(f)$.

It is straightforward to check that this is associative. Let $\imath$ be the natural inclusion of $\mathcal{C}rl_{\mathfrak{F}^+} \to \mathcal{A}gg_{\mathfrak{F}^+}$.

THEOREM B.1. *The triple* $\mathfrak{D}_{\mathfrak{F}^+} = (\mathcal{C}rl_{\mathfrak{F}^+}, \mathcal{A}gg_{\mathfrak{F}^+}, \imath)$ *is a Feynman category equivalent to* $\mathfrak{F}^+$.

PROOF. Condition (i) is clear on the object level by definition. For the isomorphisms, we notice that a morphism in $\mathcal{A}gg_{\mathfrak{F}^+}$ is an isomorphisms if only if the forest only has trees of height zero, that is each tree is simply a corolla. In the corolla case, the morphisms in $\mathcal{A}gg_{\mathfrak{F}^+}$ coincides with the morphism in $\mathcal{C}rl_{\mathfrak{F}^+}$ and hence $Iso(\mathcal{A}gg_{\mathfrak{F}^+}) \simeq \mathcal{C}rl^{\otimes}_{\mathfrak{F}^+}$.

For the condition (ii) we first notice that each morphism in $\mathcal{A}gg^{\mathrm{rt},<}_{\mathfrak{F}}$ is a tensor product of morphisms whose underlying graph is a tree, and a these are precisely the morphisms $Obj(\mathcal{A}gg^{\mathrm{rt},<}_{\mathfrak{F}} \downarrow \mathcal{C}rl^{\mathrm{rt},<}_{\mathfrak{F}})$. Thus the condition follows on the object level. For the morphism level, it is clear that any other decomposition up to isomorphism is given by permutation of the trees of the forest and an isomorphism of the $\psi'(\tau)_j$.

Condition (iii) holds, since it holds in $\mathfrak{F}$ and the morphisms in the slice categories are the morphisms are given by decorated trees and the trees as well as the decorations satisfy (iii), due to the said condition for $\mathfrak{D}$ and for $\mathfrak{F}$.

Thus the triple $\mathfrak{D}_{\mathfrak{F}^+}$ is a Feynman category. It remains to prove that this equivalent to $\mathfrak{F}^+$. For this it is enough to assume that $\mathfrak{F}$ is strict.

First, we show that $\mathcal{V}^+$ is equivalent to $\mathcal{C}rl_{\mathfrak{F}^+}$. The morphisms for $\mathcal{V}^+$ due to (ii) for $\mathfrak{F}$ are given two generators, permutations of the source factors and isomorphisms of the source factors and the target. The former correspond to corollas whose decoration $\boldsymbol{\sigma}$ is given by identities, but with non–trivial p. The latter correspond to non–trivial isomorphisms in $\boldsymbol{\sigma}$, but trivial p. These are also exactly the generators of $\mathcal{C}rl_{\mathfrak{F}^+}$. The relations are simply from composition of permutations and isomorphisms in both cases. Formally the functor is given by sending $\phi : *_1 \otimes \cdots \otimes *_n \to *_0$ to $*_\phi$[7]. An iso-morphism $(\sigma \Downarrow \sigma') : \phi \to \phi'$ in $\mathcal{V}^+$ to $(f, \boldsymbol{\sigma}, \sigma')$ where if $\sigma = \sigma_1 \otimes \cdots \otimes \sigma_n \circ p, p \in \mathbb{S}_n$, $f = (id, p^{-1} \amalg id, i_\emptyset)$, with $i_\emptyset$ the empty morphism, and $\boldsymbol{\sigma} = (\sigma_1, \ldots, \sigma_n)$.

To show the equivalence for $\mathcal{A}gg_{\mathfrak{F}^+}$, we proceed in a similar manner. Having already matched the isomorphisms, we define the value of a functor on an elementary morphism in $\mathfrak{F}^+$ to be to the morphism f whose ghost tree is a two level tree whose source is the aggregate corresponding to the $\phi_i, i = 0, \ldots, n$ and whose target is the decorated corolla corresponding to $\psi = \phi_0 \circ (\phi_1 \otimes \cdots \otimes \phi_n)$, where f_V is fixed by this data, f^F is the inclusion and $\imath$ is given by connecting the roots of vertices 1,..., n to the flags of the vertex 0. The decorations $\boldsymbol{\sigma}$ and $\boldsymbol{\imath}$ given by identities. Furthermore the two monoidal structures also agree in that taking tensor products in $\mathcal{F}^+$ corresponds to $\amalg$ in the $\mathcal{A}gg_{\mathfrak{F}^+}$. We extend the definition to all maps of $\mathfrak{F}^+$ by functoriality. This is possible since the relations are preserved. Composing isomorphisms composes the permutations and the isomorphisms of the sources and targets on both sides. Composing elementary morphisms corresponds to gluing in two–level trees into vertices. And composing isomorphisms with elementary morphisms is also the same on both sides. Finally, we see that there is a functor going

[7] For the fastidious reader a canonical choice of the one element set $\{v\}$, would be $\{\phi\}$, viz. using ϕ as an atom. Similarly naturally the set of flags is simply the set $\{*_1, \ldots, *_n\}$.

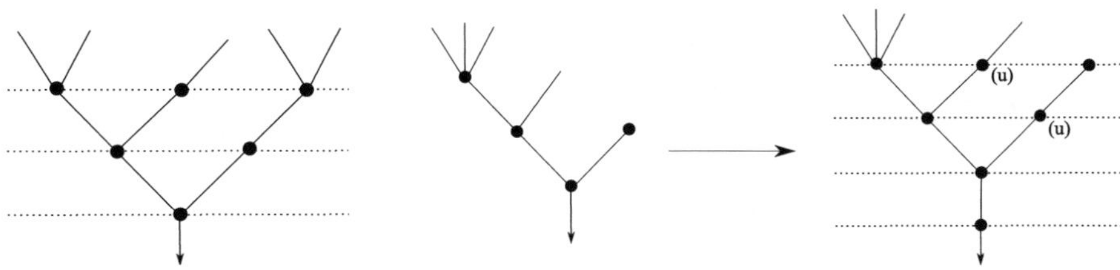

FIGURE 6. A tree with levels and the leveling of a tree. The marking (u) indicates the presence of a unit.

backwards, which is the given by the induction process above. At each step, we have a collapse of a the leaf vertices with those below them, which is a combination of the image of an elementary morphism and an isomorphism. It is straightforward to check that this allows to recover any morphism in $\mathfrak{F}^+$ by showing that every morphism of $\mathfrak{F}^+$ is encoded in a unique decorated forest. Notice that the associativity of composition manifests itself in the fact that one can "collapse" all edges in any order. Here collapse means that this is a concatenation with an elementary morphism and isomorphisms leading to a blow–up of the collapsed vertex. □

B.1.4. *Leveling.* In the description of the flow chart, we secretly leveled up the trees by introducing unit morphisms for the leaf vertices not of maximal height. This can be formalized in the tree/forest picture by introducing a new type of bivalent non-decorated vertex. We will call this a black vertex. This allows one to level–up the tree to a level tree, see Figure 6.The leveling-up replaces the flags of the vertices that are not of maximal height by a string of black vertices to get a level tree of the height of the original tree. This is what we used above to define the flow chart. This is not a unique choice. There are many choices of inserting black vertices, but these do not alter the result of the flow chart. Thus, one has to introduce the equivalence relation that two trees with black and white vertices are equivalent if they agree after deleting all black vertices. The original trees become canonical representatives as does the leveled–up tree.

For the composition of flow charts, one composes the underlying trees via insertion and then levels up again. It is in this sense that the morphisms of $\mathcal{F}^+$ are level trees/forests [**KW17**, §3.6].

B.1.5. *The decorations $(\boldsymbol{\imath}, \boldsymbol{\sigma})$ as b/w trees with marked black vertices.* One can think of $\boldsymbol{\imath}$ as a tail as follows: Using that $T(\mathbb{\Gamma}(g)) = g^G(F')$, set $c_T : T \to Mor(\mathcal{V})$ $f \mapsto \sigma_{(g^F)^{-1}(f)}$. This is used in Figure 5. The datum $\boldsymbol{\sigma}$ is directly a directed edge decoration. Replacing each edge and each tail by a black two valenced vertex, we can put the decoration on the vertex and thus encode the morphisms completely combinatorially as $\blacklozenge\,\sigma$, where the roots are now marked by $\blacklozenge\,\sigma_0^{-1}$. See Figure 7.

In particular the morphisms $(\sigma \Downarrow \sigma')$ now are encoded as marked black vertices, See Figure 8.

B.2. The graphical gcp and hyper versions $\mathfrak{F}^{gcp+}$, $\mathfrak{F}^{hyp}$. In the gcp version of the plus construction there are the extra morphisms $i_\sigma : \mathbb{1} \to *_\sigma$. We will write these morphisms as $\mathbb{1} \overset{\blacklozenge\,\sigma}{\to} *_\sigma$. Pre–composition with such maps, changes the color of the vertex $*_\sigma$ to black in the above b/w picture. The well–definedness of the flow-chart is then guaranteed by (3.19) and (3.20) as the black vertices implement the morphisms $(\sigma \Downarrow \sigma')$.

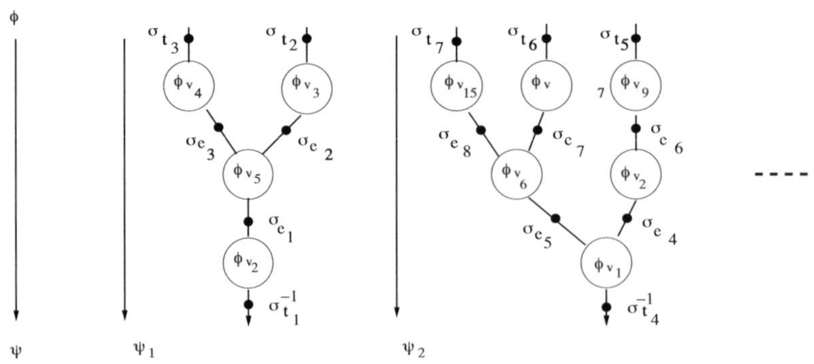

FIGURE 7. The morphism of Figure 5 as a b/w forest with isomorphism decorations on the black vertices.

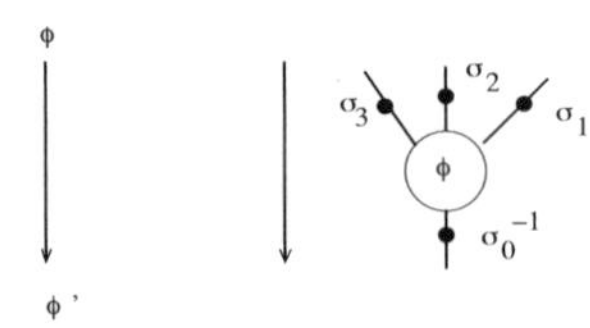

FIGURE 8. The morphism $D((\sigma \Downarrow \sigma')) : \phi \to \phi'$ with $\sigma = \sigma_1 \otimes \sigma_2 \otimes \sigma_3$ and $\sigma' = \sigma_0$ as a b/w corolla with isomorphism decorations on the black vertices.

In the case of $\mathfrak{F}^{hyp}$, one can either omit the vertices $*_\sigma$, or by abuse of notation regard black vertices $\blacklozenge\ \sigma$ in lieu of $*_\sigma$. In the first case, the morphisms are given by decorations, where none of the (non-black) vertices is decorated by an isomorphism. In the second case, one has to be careful that the $\blacklozenge\ \sigma$ are not quite vertices as they do not belong to $\mathcal{V}^{hyp}$, but rather denote an object isomorphic to $\mathbb{1}$.Making the vertex black is a useful mnemonic of this. The abuse of notation is that we identify the target of the isomorphism $\blacklozenge\ \sigma$ with the morphism itself.

Appendix C. Double categories, 2–categories and monoidal categories

In this appendix, we show that many of the constructions become natural in the language of 2–categories and double categories as founded in [**Ehr63, Bén67**] with the further developments in [**BS76, BM99**], see also [**Fio07**].

C.1. 2–categories and double categories. Before going into the definition of 2–categories, we give a natural example:

EXAMPLE C.1 (The 2-category of categories). We can set up a category of categories, whose objects are categories and whose morphisms are functors. There is another structure here though, namely there are natural transformation, which are morphisms between functors. That is morphisms of morphisms or 2–morphisms. These satisfy natural compatibility conditions which can be encoded into a 2–category.

We will need a slightly more general notion. This is based on the fact that a category can be given by the source, target and identity maps s, t, id with $(s \circ$

$id)(X) = (t \circ id)(X) = X$ and an associative unital composition $\circ$

$$\text{(C.1)} \qquad \begin{array}{c} Obj(\mathcal{C}) \\ s\!\uparrow \downarrow id \uparrow t \\ Mor(\mathcal{C}) \end{array} \qquad \circ : Mor(\mathcal{C})_s \times_t Mor(\mathcal{C}) \to Mor(\mathcal{C})$$

DEFINITION C.2. A double category is given by a diagram (C.1) in categories. That is a category of objects $\mathcal{O}bj$ a category of morphisms $\mathcal{M}or$ together with functors s, t, id and $\circ$ an associative unital composition.

The objects of $\mathcal{O}bj$ are called objects, the morphisms of $\mathcal{O}bj$ are called vertical 1–morphisms, and their composition, vertical composition, is denoted by $\circ_v$. The objects of $\mathcal{M}or$ go by the name of horizontal 1–morphisms, with horizontal composition $\circ_h$ given by the functor $\circ$. The morphisms of $\mathcal{M}or$ are referred to as 2–cells or 2–morphisms. The latter have both a horizontal and a vertical composition, the vertical composition $\circ_v$ is the composition in the category $\mathcal{M}or$, while the horizontal composition is given by the functor $\circ$. These two compositions satisfy the interchange equations.

The usual diagrams for objects, vertical, horizontal and 2-morphisms and their composition are given by

$$\text{(C.2)} \qquad \begin{array}{ccc} X & \xrightarrow{\phi} & Y \\ \sigma\downarrow & \Downarrow\alpha & \downarrow\sigma' \\ X' & \xrightarrow{\phi'} & Y' \end{array} \qquad \begin{array}{ccc} X & \xrightarrow{\phi} & Y \\ \sigma\downarrow & \Downarrow^{\alpha} & \downarrow\sigma' \\ X' & \xrightarrow{\phi'} & Y \\ \tau\downarrow & \Downarrow^{\beta} & \downarrow\tau' \\ X'' & \xrightarrow{\phi''} & Y'' \end{array} \qquad \begin{array}{ccccc} X & \xrightarrow{\phi} & Y & \xrightarrow{\psi} & Z \\ \sigma\downarrow & \Downarrow^{\alpha_1} & \downarrow\sigma' & \Downarrow^{\alpha_2} & \downarrow\sigma'' \\ X' & \xrightarrow{\phi'} & Y & \xrightarrow{\psi'} & Z \end{array}$$

The interchange relation reads:

$$\text{(C.3)} \qquad (\alpha_1 \circ_h \alpha_2) \circ_v (\beta_1 \circ_h \beta_2) = (\alpha_1 \circ_v \beta_1) \circ_h (\alpha_2 \circ_v \beta_2)$$

which means the two ways of composing the diagram (C.4), first horizontal then vertical or first vertical then horizontal, yield the same result

$$\text{(C.4)} \qquad \begin{array}{ccccc} X & \xrightarrow{\phi} & Y & \xrightarrow{\psi} & Z \\ \sigma\downarrow & \Downarrow^{\alpha_1} & \downarrow\sigma' & \Downarrow^{\alpha_2} & \downarrow\sigma'' \\ X' & \xrightarrow{\phi'} & Y & \xrightarrow{\psi'} & Z \\ \tau\downarrow & \Downarrow^{\beta_1} & \downarrow\tau' & \Downarrow^{\beta_2} & \downarrow\tau'' \\ X'' & \xrightarrow{\phi''} & Y' & \xrightarrow{\psi''} & Z'' \end{array}$$

There are also certain weakenings of the axioms for double categories and their functors, which for instance relax the condition of associativity and units either up to isomorphism or to simply having morphisms. In general, retaining an equality adds the adjective "strict", allowing for an isomorphism instead of an equality is indicated by the attribute "strong" and only postulating a morphism from one side of the equality to the other goes by the designations "lax" or "op–lax" depending on the direction.

DEFINITION C.3. A strict functor between two double categories $\mathcal{C}$ and $\mathcal{D}$ is a pair of functors (F, G): $F : \mathcal{O}bj(\mathcal{C}) \to \mathcal{O}bj(\mathcal{D})$ and $G : \mathcal{M}or(\mathcal{C}) \to \mathcal{M}or(\mathcal{D})$ which is compatible with the source, target, unit and composition functors. A horizontally lax functor means that the conditions are relaxed for the underlying categories of morphisms.

REMARK C.4. Another nice way of encoding a double category $\mathcal{D}$ is by the four sets D_0 of objects, D_1^H of horizontal 1–morphism, D_1^V of vertical 1–morphisms, and D_2 of 2–morphisms, which form four categories: the horizontal edge category $\mathcal{D}_H = \mathcal{O}bj$ given by D_0 and D_1^H, the vertical edge category $\mathcal{D}_V$ given by D_0 and D_1^V, the horizontal category of morphisms $\mathcal{D}^H_{Mor} = \mathcal{M}or$ with objects D_1^H and morphisms D_2 and finally $\mathcal{D}^V_{Mor}$ with the same morphisms, but objects D_1^V, see [**BM99**] for more details.

EXAMPLE C.5. Given a category $\mathcal{C}$, one can define a double category $\Box\mathcal{C}$ with $\mathcal{O}bj(\Box\mathcal{C}) = \mathcal{C}$ and $\mathcal{M}or(\mathcal{C})$ having objects $Mor(\mathcal{C})$ with 2–morphisms given by commutative diagrams. That is, there is precisely one 2–morphism $(\psi \Downarrow \psi') : \phi \to \phi'$ for any two morphisms ψ, ψ' with $\psi' \circ \phi = \phi' \circ \psi$.

There is a sub–double–category $\mathcal{I}_V\mathcal{C}$ of $\Box\mathcal{C}$ given by restricting the vertical morphisms to be isomorphisms. Explicitly, $\mathcal{O}bj(\mathcal{I}_V\mathcal{C}) = Iso(\mathcal{C})$ and $\mathcal{M}or(\mathcal{C}) = Iso(\mathcal{C} \downarrow \mathcal{C})$, that is $D_0 = Obj(\mathcal{C})$, $D_1^V = Mor(Iso(\mathcal{C}))$, $D_1^H = Mor(\mathcal{C})$ and $D_2(\mathcal{C}) = Mor(Iso(\mathcal{C} \downarrow \mathcal{C}))$ with horizontal composition given by $((\sigma \Downarrow \sigma'))(\phi) \circ (\sigma' \Downarrow \sigma'')(\psi) = (\sigma \Downarrow \sigma'')(\phi \circ \psi)$.

$\mathcal{I}_V\mathcal{C}$ is a double category both of whose underlying categories are groupoids, but its horizontal morphisms are not necessarily isomorphisms.

EXAMPLE C.6 (Feynman categories and double categories). A Feynman category $\mathfrak{F}$ naturally yields the double–category $\mathcal{I}_V\mathcal{F}$. The 2–morphisms are the $(\sigma \Downarrow \sigma') : \phi \to \phi'$, see (2.4).

Note that this is very natural as the condition (i) and (ii) can be rephrased as: there is an equivalence of double categories $(\mathcal{V}^{\otimes}, (\imath^{\otimes} \downarrow \imath)^{\otimes})$ and $\mathcal{I}_V\mathcal{F}$, where the horizontal composition in $(\imath^{\otimes} \downarrow \imath)^{\otimes}$ is the one naturally induced by the composition of morphisms in $\mathcal{F}$.

DEFINITION C.7. Following [**Ehr63**] we define a 2–category to be a double category whose category of objects is discrete, i.e. only has identity morphisms.

Example C.1 is the 2–category of categories whose horizontal 1–morphisms are functors and whose 2–morphisms are natural transformations.

REMARK C.8. Alternatively, one can simply omit D_1^V from the list and retain only the three sets $D_0, D_1 = D_1^H, D_2$ with their structural data. In terms of diagrams, one shrinks the vertical sides, which are by definition identity maps:

$$\text{(C.5)} \qquad \begin{array}{ccc} X & \xrightarrow{\phi} & Y \\ \| & \Downarrow\alpha & \| \\ X & \xrightarrow{\phi'} & Y \end{array} \quad \rightsquigarrow \quad X \underset{\phi'}{\overset{\phi}{\Downarrow\alpha}} Y$$

EXAMPLE C.9 (Monoidal categories as 2–categories). Just as a group G defines a category $\underline{G}$ with one object, any strict monoidal category defines a 2–category. Any strict monoidal category $(\mathcal{C}, \otimes, \mathbb{1})$ is a 2–category with one object $\underline{\mathcal{C}}$. This is $Obj(\underline{\mathcal{C}}) = \{*\}$, $1\text{-}Mor(*,*) = Obj(\mathcal{C})$ with composition $\circ$ given by $\otimes$. That is $X \circ Y := X \otimes Y$. The identity 1–morphism id_* is $\mathbb{1} \in Obj(\mathcal{C}) = 1\text{-}Mor(*,*)$, where one uses strictness. Associativity also holds due to strictness. $2\text{-}Mor(X,Y) = Mor_{\mathcal{C}}(X,Y)$ and $\circ_v = \circ$. The units in $2\text{-}Mor(X,X)$ are the id_X. The horizontal composition is given by $\otimes$ again: For $\phi \in Hom_C(X,Y), \phi' \in \mathrm{Hom}(X',Y')$: $\phi' \circ_h \psi = \phi \otimes \psi'$. Due to strictness this is associative and $id_{\mathbb{1}} = id_{id_*}$ is the unit for $\circ_h$. Finally, the interchange relation holds (2.1).

If the monoidal category is not strict, one has an example of a weak 2–category. In particular, it will be horizontally strong. The associativity for the composition $\otimes$ is not strict and given by associators, which satisfy the pentagon axiom. Likewise, the unit $\mathbb{1}_{\mathcal{E}}$ is not strict and its strong unit property for composition composition is given by the unit constraints.

In the same vein, one has lax and op–lax monoidal functors as examples of a weakening of the conditions on the functor level.

REMARK C.10. There are several relationships between double and 2–categories. Being supplied the data D_0, D_1, D_2, there are three natural double categories one can construct:

(1) The horizontal double category, given by $D_0, D_1^V = D_0, D_1^H = D_1, D_2$ with the natural structure maps. Here the elements of $D_1^V = D_0$ are viewed as identity maps, and the 2–morphisms are expanded into squares reversing the shrinking of (C.5).
(2) The vertical double category, given by $D_0, D_1^V = D_1, D_1^H = D_0, D_2$. Where the identifications are as above, just switching the roles of horizontal and vertical.
(3) The edge–symmetric double category is given by $D_0, D_1^V = D_1, D_1^H = D_1, D_2$, where the two–morphisms for a square of the type (C.2) given by a 2–morphism $\alpha : \sigma' \circ \phi \to \phi' \circ \sigma$.

Vice–versa, a double category has an underlying horizontal respectively vertical 2-category given restricting the horizontal or vertical morphisms to be identities.

EXAMPLE C.11 (A monoidal category as a double category). Given a monoidal category $\mathcal{E}$, we define $D(\mathcal{E})$ to be the horizontal realization of $\underline{\mathcal{E}}$.

DEFINITION C.12. An enrichment functor for a Feynman category $\mathfrak{F}$ with values in a monoidal category $\mathcal{E}$ is a horizontally lax functor of double categories $(F, \mathcal{D})$ from $\mathcal{I}_V\mathcal{F}$, see Example C.6 to $D(\mathcal{E})$, see Example C.11.

REMARK C.13. Note that there is only one possible component fuctor F, which is the trivial functor $F = \mathcal{T}$: $\mathcal{T}(X) = *$ and $\mathcal{T}(\sigma) = \mathbb{1}_{\mathcal{E}}$. Thus the data for an enrichment functor is:

(1) A functor

$$\mathcal{D} : Iso(\mathcal{F} \downarrow \mathcal{F}) \to \mathcal{E} \tag{C.6}$$

On objects, we have $\mathcal{D}(\phi) \in \mathcal{E}$ and for morphisms $\mathcal{D}((\sigma \Downarrow \sigma')) : \mathcal{D}(\phi) \to \mathcal{D}(\phi')$ are isomorphisms.

(2) For each pair of composable morphisms ϕ_0, ϕ_1 a natural morphism

$$\mathcal{D}(\phi_0) \otimes \mathcal{D}(\phi_1) \to \mathcal{D}(\phi_0 \circ \phi_1) \tag{C.7}$$

(3) An element in $\mathcal{D}(id_X)$ for each X. That is a morphism $id_{\mathcal{D}(X)} = id_* = \mathbb{1} \to \mathcal{D}(id_X)$ which is a unit for the maps above.

DEFINITION C.14. We say that an enrichment functor is lax–monoidal, if in addition one has the following data

(4) On horizontal 1–morphisms maps:

$$\mathcal{D}(\phi) \otimes \mathcal{D}(\psi) \to \mathcal{D}(\phi \otimes \psi) \tag{C.8}$$

(5) On 2–morphisms:

$$\mathcal{D}((\sigma \Downarrow \sigma')) \otimes \mathcal{D}((\tau \Downarrow \tau')) \to \mathcal{D}((\sigma \otimes \tau \Downarrow \sigma' \otimes \tau')) \tag{C.9}$$

(6) A unit morphism:

$$\mathbb{1} \to \mathcal{D}(id_{\mathbb{1}}) \tag{C.10}$$

Such that the constraints are associative (i.e. satisfy the pentagon identity), the interchange relation is functorially preserved and the unit constraints are transformed into each other. On vertical 1–morphisms the morphisms $\mathcal{T}(\sigma) \otimes \mathcal{T}(\sigma') = \mathbb{1}_{\mathcal{E}} \otimes \mathbb{1}_{\mathcal{E}} \to \mathcal{T}(\sigma \circ \sigma') = \mathbb{1}_{\mathcal{E}}$ are given by the unit constraints.

We say the functor is strongly monoidal if the morphisms in the last three equations are isomorphisms, and strictly monoidal if it has equalities in the last equations (C.8), (C.9) and (C.10).

PROPOSITION C.15. *For a Feynman category the data of an enrichment functor that is strict monoidal is up to equivalence, determined by*

(1) *(Groupoid data) A functor*

$$Iso(\mathcal{F} \downarrow \mathcal{V}) \to \mathcal{E} \tag{C.11}$$

(2) *(Composition data) For $\phi_0 \in (\mathcal{F} \downarrow \mathcal{V})$ and $\phi_1 = \phi_{1,1} \otimes \cdots \otimes \phi_{1,n}$, with $\phi_{1,i} \in (\mathcal{F} \downarrow \mathcal{V})$ that are composable, set $\phi := \phi_0 \circ \phi_1$. Morphisms*

$$\mathcal{D}(\phi_0) \otimes \mathcal{D}(\phi_{1,1}) \otimes \cdots \otimes \mathcal{D}(\phi_{1,n}) \to \mathcal{D}(\phi). \tag{C.12}$$

(3) *(Unit data) For each object $*_v \in \mathcal{V}$ an element*

$$\mathbb{1} \to \mathcal{D}(id_{*_v}) \tag{C.13}$$

which is a unit for the composition data.

PROOF. Due to the condition (i) for a Feynman category $Iso(\mathcal{F} \downarrow \mathcal{F}) \simeq Iso(\mathcal{F} \downarrow \mathcal{V})^{\otimes}$. Since $\mathcal{D}$ is strict monoidal $\mathcal{D}$ is fixed up to equivalence on $Iso(\mathcal{F} \downarrow \mathcal{V})$. Again, since $\mathcal{D}$ is monoidal, we can use condition (ii) to reduce to the case where $\phi_0 \in (\imath^{\otimes} \downarrow \imath)$ and $\phi_1 \in (\imath^{\otimes} \downarrow \imath^{\otimes})$. Finally, since for and X, id_X is isomorphic to $\bigotimes_{v \in V} id_{*_v}$ for some decomposition of $X \simeq \bigotimes_{v \in V} *_v$. We see that up to equivalence the unit data is fixed on the id_{*_v}. □

PROPOSITION C.16 (Ground monoid/ring). *In the lax monoidal case, we have that $R := \mathcal{D}(\mathrm{id}_{\mathbb{1}_{\mathcal{F}}})$ is a unital monoid and if $\mathcal{D}$ takes values in a linear category it is a ring. Moreover all the $Hom_{\mathcal{F}_{\mathcal{D}}}(X, Y)$ become R-R–modules, and the category is enriched in R-R–modules.*

PROOF. The multiplicative structure is given by $\mathcal{D}(id_{\mathbb{1}}) \otimes \mathcal{D}(id_{\mathbb{1}}) \to \mathcal{D}(id_{\mathbb{1}})$ corresponding to the composition $id_{\mathbb{1}} \circ id_{\mathbb{1}} = id_{\mathbb{1}}$. The R-R–module structure is given by the left and unit constraints. For the right action: $\mathcal{D}(\phi) \otimes \mathcal{D}(id_{\mathbb{1}}) \to \mathcal{D}(\phi \otimes id_{\mathbb{1}}) \to \mathcal{D}(\phi)$, where the first map is given by the lax monoidal structure and the second by the unit constraint in $\mathcal{F}$. This provides the morphisms $\mathcal{D}(\phi) \otimes R \to \mathcal{D}(\phi)$. The unit comes from the structure map $\mathbb{1} \to \mathcal{D}(id_*)$. □

C.2. Holonomy, connections, gcp and hyper functors.

C.2.1. *Holonomy and connections.* We briefly recall the pertinent elements from [**BS76, BM99, Fio07**] and sketch how to adapt and apply them to the setting of Feynman categories. A homolony for a double category $\mathcal{D}$ is a functor $^{-} : \mathcal{D}_V \to \mathcal{D}_H$ which is identity on the objects, viz. D_0. A left respectively right connection for a holonomy is assignment $\lrcorner$ respectively $\ulcorner$ from $D_1^V \to D_2$

$$\begin{array}{ccccccc} X & \xrightarrow{\bar\sigma} & Y & & X & \xrightarrow{id} & X \\ {\scriptstyle\sigma}\downarrow & \Downarrow \lrcorner(\sigma) & \downarrow{\scriptstyle id} & & {\scriptstyle id}\downarrow & \ulcorner(\sigma) \Downarrow & \downarrow{\scriptstyle\sigma} \\ Y & \xrightarrow{id} & Y & & X & \xrightarrow{\bar\sigma} & Y \end{array} \tag{C.14}$$

which satisfy a natural compatibility, see [**BS76**]. A connection pair is a holonomy together with left and a right connection for it.

A category with holonomy and a connection pair is called a *category with connection.* A functor (F, G) of double categories with connections is a functor which preserves this extra structure. It weakly preserves the holonomy if there are natural morphisms $F(\sigma) \to G(\bar\sigma)$. Such a functor is strong, if the morphisms are isomorphisms. Likewise, one can relax the condition on the connection; see §C.2.2 for a concrete application.

EXAMPLE C.17. Note that in the double category $\Box\mathcal{C}$, we can choose $^{-}$ to be the identity functor. This restricts to a functor $Iso(\mathcal{C}) \to \mathcal{C}$ which is a holonomy for $\mathcal{I}_V\mathcal{C}$, and further restricts to a holonomy for the horizontal double category of $\mathcal{C}$. Since there is precisely one two morphisms for each square in both cases, there are unique connections given by $\lrcorner(\sigma) = (\sigma \Downarrow id) : \bar\sigma = \sigma \to id$ and $\ulcorner(\sigma) = (\mathrm{id} \Downarrow \sigma) : id \to \bar\sigma = \sigma$. This related to the fact that $\Box\mathcal{C}$ has a canonical thin structure. In the horizontal double category case, σ is an identity.

C.2.2. *Applications to Feynman categories.* In particular, there is a canonical connection on the double category of a Feynman category $\mathcal{I}_V\mathcal{F}$ and the horizontal double category of $\underline{\mathcal{E}}$.

An enrichment functor weakly preserving the connection has natural morphisms $\mathcal{T}(\sigma) = \mathbb{1} \to \mathcal{D}(\bar\sigma = \sigma)$, i.e. $\mathcal{D}$ is groupoid compatibly pointed. It preserves the connection if $\mathbb{1} = \mathcal{D}(\sigma)$ and it is strong if $\mathcal{D}$ is a hyper functor. The conditions for the weakness of a gcp functor and the strongness of the hyper-functor for the connections are the diagrams (3.12) and (3.13). In particular, $\mathcal{D}(\ulcorner(\sigma)) : \mathbb{1} \to \mathcal{D}(\sigma)$ yields the groupoid pointing and $\mathcal{D}(\lrcorner(\sigma) : \mathcal{D}(\sigma) \to \mathbb{1}$ gives the splitting as in Definition 4.12.

Appendix D. Model structures

In this section we discuss Quillen model structures for $\mathcal{F}$-$\mathcal{O}ps_{\mathcal{C}}$. It turns out that these model structures can be defined if $\mathcal{C}$ satisfies certain conditions and if this is the case work for all $\mathfrak{F}$, e.g. all the previous examples.

D.1. Model structure.

THEOREM D.1 ([**KW17**, Theorem 8.2.1]). *Let $\mathfrak{F}$ be a Feynman category and let $\mathcal{C}$ be a cofibrantly generated model category and a closed symmetric monoidal category having the following additional properties:*

(1) *All objects of $\mathcal{C}$ are small.*

(2) *$\mathcal{C}$ has a symmetric monoidal fibrant replacement functor.*
(3) *$\mathcal{C}$ has $\otimes$-coherent path objects for fibrant objects.*

Then $\mathcal{F}$-$\mathcal{O}ps_{\mathcal{C}}$ is a model category where a morphism $\phi\colon \mathcal{O} \to \mathcal{Q}$ of $\mathcal{F}ops$ is a weak equivalence (resp. fibration) if and only if $\phi\colon \mathcal{O}(v) \to \mathcal{Q}(v)$ is a weak equivalence (resp. fibration) in $\mathcal{C}$ for every $v \in \mathcal{V}$.

D.1.1. *Examples.*

(1) Simplicial sets. (Straight from Theorem D.1)
(2) $dg\mathcal{V}ect_k$ for $char(k) = 0$ (Straight from Theorem D.1)
(3) $\mathcal{T}op$ (More work, see below.)

D.1.2. *Remark.* Condition (i) is not satisfied for $\mathcal{T}op$ and so we cannot directly apply the theorem. In [**KW17**] this point was first cleared up by following [**Fre10**] and using the fact that all objects in Top are small with respect to topological inclusions.

THEOREM D.2 ([**KW17**, Theorem 8.2.13]). *Let $\mathcal{C}$ be the category of topological spaces with the Quillen model structure. The category $\mathcal{F}$-$\mathcal{O}ps_{\mathcal{C}}$ has the structure of a cofibrantly generated model category in which the forgetful functor to $\mathcal{V}$-$Seq_{\mathcal{C}}$ creates fibrations and weak equivalences.*

D.2. Quillen adjunctions from morphisms of Feynman categories.

D.2.1. *Adjunction from morphisms.* We assume $\mathcal{C}$ is a closed symmetric monoidal and model category satisfying the assumptions of Theorem D.1. Let $\mathfrak{E}$ and $\mathfrak{F}$ be Feynman categories and let $\mathfrak{f}\colon \mathfrak{E} \to \mathfrak{F}$ be a morphism between them. This morphism induces an adjunction

$$f_!\colon \mathcal{E}\text{-}\mathcal{O}ps_{\mathcal{C}} \leftrightarrows \mathcal{F}\text{-}\mathcal{O}ps_{\mathcal{C}}\colon f^*$$

LEMMA D.3. *Suppose f^* restricted to $\mathcal{V}_{\mathfrak{F}}$-$\mathcal{M}ods_{\mathcal{C}} \to \mathcal{V}_{\mathfrak{E}}$-$\mathcal{M}ods_{\mathcal{C}}$ preserves fibrations and acyclic fibrations, then the adjunction $(f_!, f^*)$ is a Quillen adjunction.*

D.3. Cofibrant replacement.

THEOREM D.4. *The Feynman transform of a non-negatively graded dg $\mathcal{F}$-$\mathcal{O}p$ is co-fibrant.*

The double Feynman transform of a non-negatively graded dg-$\mathcal{F}$-$\mathcal{O}p$ in a cubical Feynman category is a co-fibrant replacement.

Acknowledgments

This work would have not been possible without the 2018 Maurice Auslander International Conference. We thank the organizers G. Todorov, A. Martsinkovsky, and K. Igusa for making this possible and creating a perfect atmosphere.

Special thanks goes to A. Martsinkovsky for his encouragement and patience during the preparation of the manuscript. We are grateful to K. Igusa for key discussions, and, as well as S. Schroll and A. Hanany for further enlightening discussions.

We also thank J. Miller and P. Patzt, for discussions about FI–algebras and the organizers and participants of the conference on "Combinatorial Categories in Algebra and topology" in Osnabrück, especially M. Spitzweck and and H. Krause, where the results on FI–algebras were first presented,

It is a pleasure to thank and acknowledge my collaborators C. Berger, M. Markl and M. Batanin for related and ongoing discussions.

References

[Aus74] Maurice Auslander, *Representation theory of Artin algebras. I, II*, Comm. Algebra **1** (1974), 177–268; ibid. 1 (1974), 269–310, DOI 10.1080/00927877408548230. MR349747

[Bar07] Serguei Barannikov, *Modular operads and Batalin-Vilkovisky geometry*, Int. Math. Res. Not. IMRN **19** (2007), Art. ID rnm075, 31, DOI 10.1093/imrn/rnm075. MR2359547

[Bat98] M. A. Batanin, *Monoidal globular categories as a natural environment for the theory of weak n-categories*, Adv. Math. **136** (1998), no. 1, 39–103, DOI 10.1006/aima.1998.1724. MR1623672

[Bat17] M. A. Batanin, *An operadic proof of Baez-Dolan stabilization hypothesis*, Proc. Amer. Math. Soc. **145** (2017), no. 7, 2785–2798, DOI 10.1090/proc/13423. MR3637930

[Bau98] Hans-Joachim Baues, *The cobar construction as a Hopf algebra*, Invent. Math. **132** (1998), no. 3, 467–489, DOI 10.1007/s002220050231. MR1625728

[BB17] M. A. Batanin and C. Berger, *Homotopy theory for algebras over polynomial monads*, Theory Appl. Categ. **32** (2017), Paper No. 6, 148–253. MR3607212

[BD98] John C. Baez and James Dolan, *Higher-dimensional algebra. III. n-categories and the algebra of opetopes*, Adv. Math. **135** (1998), no. 2, 145–206, DOI 10.1006/aima.1997.1695. MR1620826

[Bén67] Jean Bénabou, *Introduction to bicategories*, Reports of the Midwest Category Seminar, Springer, Berlin, 1967, pp. 1–77. MR0220789

[BFSV03] C. Balteanu, Z. Fiedorowicz, R. Schwänzl, and R. Vogt, *Iterated monoidal categories*, Adv. Math. **176** (2003), no. 2, 277–349, DOI 10.1016/S0001-8708(03)00065-3. MR1982884

[BK15] Spencer Bloch and Dirk Kreimer, *Cutkosky Rules and Outer Space*, `arXiv:1512.01705` preprint, 2015.

[BK17] Clemens Berger and Ralph M. Kaufmann, *Comprehensive factorisation systems*, Tbilisi Math. J. **10** (2017), no. 3, 255–277, DOI 10.1515/tmj-2017-0112. MR3742580

[BK20a] Clemens Berger and Ralph M. Kaufmann, *Derived Feynman functors and modular envelopes*, in preparation, 2020.

[BK20b] Clemens Berger and Ralph M. Kaufmann, *Derived Feynman functors and modular envelopes: chain aspects*, in preparation, 2020.

[BKM] M. A. Batanin, Ralph M. Kaufmann, and Martin Markl, *Comparing Feynman and Operadic Categories*, In preparation.

[BKW18] Michael Batanin, Joachim Kock, and Mark Weber, *Regular patterns, substitudes, Feynman categories and operads*, Theory Appl. Categ. **33** (2018), Paper No. 7, 148–192. MR3769040

[BM99] Ronald Brown and Ghafar H. Mosa, *Double categories, 2-categories, thin structures and connections*, Theory Appl. Categ. **5** (1999), No. 7, 163–175. MR1694653

[BM08] Dennis V. Borisov and Yuri I. Manin, *Generalized operads and their inner cohomomorphisms*, Geometry and dynamics of groups and spaces, Progr. Math., vol. 265, Birkhäuser, Basel, 2008, pp. 247–308, DOI 10.1007/978-3-7643-8608-5_4. MR2402406

[BM15] Michael Batanin and Martin Markl, *Operadic categories and duoidal Deligne's conjecture*, Adv. Math. **285** (2015), 1630–1687, DOI 10.1016/j.aim.2015.07.008. MR3406537

[BS76] Ronald Brown and Christopher B. Spencer, *Double groupoids and crossed modules*, Cahiers Topologie Géom. Différentielle **17** (1976), no. 4, 343–362. MR440553

[BSV18] Kai-Uwe Bux, Peter Smillie, and Karen Vogtmann, *On the bordification of outer space*, J. Lond. Math. Soc. (2) **98** (2018), no. 1, 12–34, DOI 10.1112/jlms.12124. MR3847230

[BV73] J. M. Boardman and R. M. Vogt, *Homotopy invariant algebraic structures on topological spaces*, Lecture Notes in Mathematics, Vol. 347, Springer-Verlag, Berlin-New York, 1973. MR0420609

[CK00] Alain Connes and Dirk Kreimer, *Renormalization in quantum field theory and the Riemann-Hilbert problem. I. The Hopf algebra structure of graphs and the main theorem*, Comm. Math. Phys. **210** (2000), no. 1, 249–273, DOI 10.1007/s002200050779. MR1748177

[CK01] Alain Connes and Dirk Kreimer, *Renormalization in quantum field theory and the Riemann-Hilbert problem. II. The β-function, diffeomorphisms and the renormalization group*, Comm. Math. Phys. **216** (2001), no. 1, 215–241, DOI 10.1007/PL00005547. MR1810779

[Cos07a] Kevin Costello, *A dual version of the ribbon graph decomposition of moduli space*, Geom. Topol. **11** (2007), 1637–1652, DOI 10.2140/gt.2007.11.1637. MR2350462

[Cos07b] Kevin Costello, *Topological conformal field theories and Calabi-Yau categories*, Adv. Math. **210** (2007), no. 1, 165–214, DOI 10.1016/j.aim.2006.06.004. MR2298823

[CV86] Marc Culler and Karen Vogtmann, *Moduli of graphs and automorphisms of free groups*, Invent. Math. **84** (1986), no. 1, 91–119, DOI 10.1007/BF01388734. MR830040

[Del90] P. Deligne, *Catégories tannakiennes* (French), The Grothendieck Festschrift, Vol. II, Progr. Math., vol. 87, Birkhäuser Boston, Boston, MA, 1990, pp. 111–195. MR1106898

[DK15] T. Dyckerhoff and M. Kapranov, *Crossed simplicial groups and structured surfaces*, Stacks and categories in geometry, topology, and algebra, Contemp. Math., vol. 643, Amer. Math. Soc., Providence, RI, 2015, pp. 37–110, DOI 10.1090/conm/643/12896. MR3381470

[DK18] Tobias Dyckerhoff and Mikhail Kapranov, *Triangulated surfaces in triangulated categories*, J. Eur. Math. Soc. (JEMS) **20** (2018), no. 6, 1473–1524, DOI 10.4171/JEMS/791. MR3801819

[Ehr63] Charles Ehresmann, *Catégories structurées* (French), Ann. Sci. École Norm. Sup. (3) **80** (1963), 349–426. MR0197529

[Fio07] Thomas M. Fiore, *Pseudo algebras and pseudo double categories*, J. Homotopy Relat. Struct. **2** (2007), no. 2, 119–170. MR2369164

[FL91] Zbigniew Fiedorowicz and Jean-Louis Loday, *Crossed simplicial groups and their associated homology*, Trans. Amer. Math. Soc. **326** (1991), no. 1, 57–87, DOI 10.2307/2001855. MR998125

[Fre10] Benoit Fresse, *Props in model categories and homotopy invariance of structures*, Georgian Math. J. **17** (2010), no. 1, 79–160. MR2640648

[GCKT20a] Imma Gálvez-Carrillo, Ralph M. Kaufmann, and Andrew Tonks, *Three Hopf algebras from number theory, physics & topology, and their common background I: operadic & simplicial aspects*, Commun. Number Theory Phys. **14** (2020), no. 1, 1–90, DOI 10.4310/CNTP.2020.v14.n1.a1. MR4060970

[GCKT20b] Imma Gálvez-Carrillo, Ralph M. Kaufmann, and Andrew Tonks, *Three Hopf algebras from number theory, physics & topology, and their common background II: general categorical formulation*, Commun. Number Theory Phys. **14** (2020), no. 1, 91–169, DOI 10.4310/CNTP.2020.v14.n1.a2. MR4060971

[Get09] Ezra Getzler, *Operads revisited*, Algebra, arithmetic, and geometry: in honor of Yu. I. Manin. Vol. I, Progr. Math., vol. 269, Birkhäuser Boston, Boston, MA, 2009, pp. 675–698, DOI 10.1007/978-0-8176-4745-2_16. MR2641184

[GJ94] Ezra Getzler and Jones J.D.S. *Operads, homotopy algebra and iterated integrals for double loop spaces*, arXiv:hep-th/9403055, 1994.

[GK95] E. Getzler and M. M. Kapranov, *Cyclic operads and cyclic homology*, Geometry, topology, & physics, Conf. Proc. Lecture Notes Geom. Topology, IV, Int. Press, Cambridge, MA, 1995, pp. 167–201. MR1358617

[GK98] E. Getzler and M. M. Kapranov, *Modular operads*, Compositio Math. **110** (1998), no. 1, 65–126, DOI 10.1023/A:1000245600345. MR1601666

[GS19] Edward L. Green and Sibylle Schroll, *Almost gentle algebras and their trivial extensions*, Proc. Edinb. Math. Soc. (2) **62** (2019), no. 2, 489–504, DOI 10.1017/s001309151800055x. MR3935809

[HKK17] F. Haiden, L. Katzarkov, and M. Kontsevich, *Flat surfaces and stability structures*, Publ. Math. Inst. Hautes Études Sci. **126** (2017), 247–318, DOI 10.1007/s10240-017-0095-y. MR3735868

[HS18] Amihay Hanany and Marcus Sperling, *Resolutions of nilpotent orbit closures via Coulomb branches of 3-dimensional $\mathcal{N} = 4$ theories*, J. High Energy Phys. **8** (2018), 189, front matter+35, DOI 10.1007/jhep08(2018)189. MR3861133

[Igu02] Kiyoshi Igusa, *Higher Franz-Reidemeister torsion*, AMS/IP Studies in Advanced Mathematics, vol. 31, American Mathematical Society, Providence, RI; International Press, Somerville, MA, 2002. MR1945530

[IT15] Kiyoshi Igusa and Gordana Todorov, *Cluster categories coming from cyclic posets*, Comm. Algebra **43** (2015), no. 10, 4367–4402, DOI 10.1080/00927872.2014.946138. MR3366582

[IT19] Kiyoshi Igusa and Gordana Todorov, *Cyclic posets and triangulation clusters*, Sci. China Math. **62** (2019), no. 7, 1289–1316, DOI 10.1007/s11425-018-9507-3. MR3974094

[JR79] S. A. Joni and G.-C. Rota, *Coalgebras and bialgebras in combinatorics*, Stud. Appl. Math. **61** (1979), no. 2, 93–139, DOI 10.1002/sapm197961293. MR544721

[Kas95] Christian Kassel, *Quantum groups*, Graduate Texts in Mathematics, vol. 155, Springer-Verlag, New York, 1995. MR1321145

[Kau04] Ralph M. Kaufmann, *Operads, moduli of surfaces and quantum algebras*, Woods Hole mathematics, Ser. Knots Everything, vol. 34, World Sci. Publ., Hackensack, NJ, 2004, pp. 133–224, DOI 10.1142/9789812701398_0004. MR2123369

[Kau09] Ralph M. Kaufmann, *Dimension vs. genus: a surface realization of the little k-cubes and an E_∞ operad*, Algebraic topology—old and new, Banach Center Publ., vol. 85, Polish Acad. Sci. Inst. Math., Warsaw, 2009, pp. 241–274, DOI 10.4064/bc85-0-17. MR2503531

[Kau17] Ralph M. Kaufmann, *Lectures on Feynman categories*, 2016 MATRIX annals, MATRIX Book Ser., vol. 1, Springer, Cham, 2018, pp. 375–438. MR3792531

[Kel82] Gregory Maxwell Kelly, *Basic concepts of enriched category theory*, London Mathematical Society Lecture Note Series, vol. 64, Cambridge University Press, Cambridge-New York, 1982. MR651714

[KKGJ15] Igusa Kiyoshi, Orr Kent, Todorov Gordana, and Weyman Jerzy, *Modulated semi-invariants,* `arXiv:1507.03051v2`, 2015.

[KKW15] R M Kaufmann, S Khlebnikov, and B Wehefritz—Kaufmann, *Projective representations from quantum enhanced graph symmetries*, Journal of Physics: Conference Series, 597:012048, apr 2015.

[KKWK16] Ralph M. Kaufmann, Sergei Khlebnikov, and Birgit Wehefritz-Kaufmann, *Re-gauging groupoid, symmetries and degeneracies for graph Hamiltonians and applications to the gyroid wire network*, Ann. Henri Poincaré **17** (2016), no. 6, 1383–1414, DOI 10.1007/s00023-015-0443-8. MR3500218

[KL16] Ralph Kaufmann and Jason Lucas, *Decorated Feynman categories*, J. Noncommut. Geom. **11** (2017), no. 4, 1437–1464, DOI 10.4171/JNCG/11-4-8. MR3743229

[KLP03] Ralph M. Kaufmann, Muriel Livernet, and R. C. Penner, *Arc operads and arc algebras*, Geom. Topol. **7** (2003), 511–568, DOI 10.2140/gt.2003.7.511. MR2026541

[KM20] Ralph M. Kaufmann and Michael Monaco, *The plus construction*, Preprint, 2020.

[Kre16] Dirk Kreimer, *Cutkosky Rules from Outer Space*, `arXiv:1607.04861` preprint, 2016.

[KS10] Ralph M. Kaufmann and R. Schwell, *Associahedra, cyclohedra and a topological solution to the A_∞ Deligne conjecture*, Adv. Math. **223** (2010), no. 6, 2166–2199, DOI 10.1016/j.aim.2009.11.001. MR2601011

[KW13] Ralph M. Kaufmann and Benjamin C. Ward, *Feynman categories* (English, with English and French summaries), Astérisque **387** (2017), vii+161. MR3636409

[KW17] Ralph M. Kaufmann and Benjamin C. Ward, *Feynman categories* (English, with English and French summaries), Astérisque **387** (2017), vii+161. MR3636409

[KWK18] Ralph M. Kaufmann and Birgit Wehefritz-Kaufmann, *Theoretical properties of materials formed as wire network graphs from triply periodic CMC surfaces, especially the gyroid*, The role of topology in materials, Springer Ser. Solid-State Sci., vol. 189, Springer, Cham, 2018, pp. 173–200. MR3931091

[KWZ12] Ralph M. Kaufmann, Benjamin C. Ward, and J. Javier Zúñiga, *The odd origin of Gerstenhaber brackets, Batalin-Vilkovisky operators, and master equations*, J. Math. Phys. **56** (2015), no. 10, 103504, 40, DOI 10.1063/1.4932962. MR3412702

[Law70] F. William Lawvere, *Equality in hyperdoctrines and comprehension schema as an adjoint functor*, Applications of Categorical Algebra (Proc. Sympos. Pure Math., Vol. XVII, New York, 1968), Amer. Math. Soc., Providence, R.I., 1970, pp. 1–14. MR0257175

[Lei04] Tom Leinster, *Higher operads, higher categories*, London Mathematical Society Lecture Note Series, vol. 298, Cambridge University Press, Cambridge, 2004. MR2094071

[Lod98] Jean-Louis Loday, *Cyclic homology*, 2nd ed., Grundlehren der Mathematischen Wissenschaften [Fundamental Principles of Mathematical Sciences], vol. 301, Springer-Verlag, Berlin, 1998. Appendix E by María O. Ronco; Chapter 13 by the author in collaboration with Teimuraz Pirashvili. MR1600246

[Mar16] Martin Markl, *Modular envelopes, OSFT and nonsymmetric (non-Σ) modular operads*, J. Noncommut. Geom. **10** (2016), no. 2, 775–809, DOI 10.4171/JNCG/248. MR3519052

[ML98] Saunders Mac Lane, *Categories for the working mathematician*, 2nd ed., Graduate Texts in Mathematics, vol. 5, Springer-Verlag, New York, 1998. MR1712872

[MMS09] M. Markl, S. Merkulov, and S. Shadrin, *Wheeled PROPs, graph complexes and the master equation*, J. Pure Appl. Algebra **213** (2009), no. 4, 496–535, DOI 10.1016/j.jpaa.2008.08.007. MR2483835

[MSS02] Martin Markl, Steve Shnider, and Jim Stasheff, *Operads in algebra, topology and physics*, Mathematical Surveys and Monographs, vol. 96, American Mathematical Society, Providence, RI, 2002. MR1898414

[MV09] Sergei Merkulov and Bruno Vallette, *Deformation theory of representations of prop(erad)s. I*, J. Reine Angew. Math. **634** (2009), 51–106, DOI 10.1515/CRELLE.2009.069. MR2560406

[PR02] T. Pirashvili and B. Richter, *Hochschild and cyclic homology via functor homology*, K-Theory **25** (2002), no. 1, 39–49, DOI 10.1023/A:1015064621329. MR1899698

[So03] Jolanta Słomińska, *Decompositions of the category of noncommutative sets and Hochschild and cyclic homology*, Cent. Eur. J. Math. **1** (2003), no. 3, 327–331, DOI 10.2478/BF02475213. MR1992895

[SS17] Steven V. Sam and Andrew Snowden, *Gröbner methods for representations of combinatorial categories*, J. Amer. Math. Soc. **30** (2017), no. 1, 159–203, DOI 10.1090/jams/859. MR3556290

[SS19] Steven V. Sam and Andrew Snowden, *Representations of categories of G-maps*, J. Reine Angew. Math. **750** (2019), 197–226, DOI 10.1515/crelle-2016-0045. MR3943321

[Sta63] James Dillon Stasheff, *Homotopy associativity of H-spaces. I, II*, Trans. Amer. Math. Soc. 108 (1963), 275-292; ibid. **108** (1963), 293–312, DOI 10.1090/s0002-9947-1963-0158400-5. MR0158400

[Val07] Bruno Vallette, *A Koszul duality for PROPs*, Trans. Amer. Math. Soc. **359** (2007), no. 10, 4865–4943, DOI 10.1090/S0002-9947-07-04182-7. MR2320654

[War16] Benjamin C. Ward, *Maurer-Cartan elements and cyclic operads*, J. Noncommut. Geom. **10** (2016), no. 4, 1403–1464, DOI 10.4171/JNCG/263. MR3597148

[War19] Benjamin C. Ward, *Six operations formalism for generalized operads*, Theory Appl. Categ. **34** (2019), Paper No. 6, 121–169, DOI 10.1515/eqc-2018-0029. MR3921622

Purdue University Department of Mathematics, West Lafayette, Indiana 47907
Email address: rkaufman@math.purdue.edu

Contemporary Mathematics
Volume **769**, 2021
https://doi.org/10.1090/conm/769/15420

Preprojective roots and graph monoids of Coxeter groups

Mark Kleiner

Abstract. Certain results on representations of quivers have analogs in the structure theory of general Coxeter groups. A fixed Coxeter element turns the Coxeter graph into an acyclic quiver, allowing for the definition of a preprojective root. A positive root is an analog of an indecomposable representation of the quiver. The Coxeter group is finite if and only if every positive root is preprojective, which is analogous to the well-known result that a quiver is of finite representation type if and only if every indecomposable representation is preprojective. Combinatorics of orientation-admissible words in the graph monoid of the Coxeter graph relates strongly to reduced words and the weak order of the group.

Introduction

Coxeter groups, in the crystallographic case, have been used in several recent papers on representations of quivers or more general finite dimensional algebras, see for example [**IS10**, **IR11**, **AIRT12**, **ORT15**]. On the other hand, Pelley and the author used representations of quivers to prove that the powers of a Coxeter element in an infinite irreducible crystallographic group are reduced [**KP07**], and then Speyer proved the result for a general Coxeter group [**Spe09**], using combinatorics of [**KP07**] and stripping out the quiver theory. The current paper shows that certain results on representations of quivers have analogs in the theory of Coxeter groups. The road from quivers to Coxeter groups goes through the notion of root in view of the results of Kac [**Kac80**], for a root of a Coxeter group is an analog of a real root of a quiver, and each positive real root of a quiver is the dimension vector of a unique up to isomorphism indecomposable representation. Therefore we view a positive root of a Coxeter group $\mathcal{W}$ as an analog of an indecomposable representation, and view a finite set of positive roots as an analog of a representation that need not be indecomposable but has no isomorphic direct summands.

Among the indecomposable representations of an acyclic quiver that correspond to real roots, the most important are preprojective and preinjective representations introduced by Bernstein, Gelfand, and Ponomarev [**BGP73**] as those annihilated by a power of the Coxeter functor. The analog of the Coxeter functor is a Coxeter element, so Igusa and Schiffler [**IS10**] fix a Coxeter element $c \in \mathcal{W}$ and define a c-preprojective (resp. c-projective) root as a positive root sent to a negative root by a positive power of c (resp. by c). The element c determines a unique acyclic orientation of the Coxeter graph Γ of $\mathcal{W}$ and thus turns it into a quiver. The inverse

2020 *Mathematics Subject Classification.* Primary 20F55; Secondary 16G20.

Coxeter element, c^{-1}, yields the opposite quiver, so the c^{-1}-preprojective roots are analogs of preinjective representations.

To study c-preprojective or c-projective roots, we give a different, but equivalent, definition. Following the suggestion of [**BGP73**, Note 2, p. 25], we say that a positive root is c-preprojective if there exists a c-admissible sequence of vertices of Γ, called $(+)$-admissible in [**BGP73**], for which the associated product of simple reflections sends the root to a negative root. In this context a simple reflection is the analog of a reflection functor, and the advantage is that the collection of c-admissible sequences has a rich combinatorial structure. The collection has a natural equivalence relation and a preorder structure that induce on the set of equivalence classes a partial order closely related to the weak order; thus obtained partially ordered set is a distributive lattice; there is a canonical form for each equivalence class; etc. These and other results of Pelley, Tyler, and the author [**KP07, KT05, KT08**] hold for an arbitrary acyclic quiver. Based on these results and using the work of Howlett [**How82**] and Speyer [**Spe09**], we show that properties of c-preprojective or c-projective roots are similar to well-known properties of preprojective or projective representations of a quiver.

We prove that the Coxeter group $\mathcal{W}$ is finite if and only if each positive root is c-preprojective (Theorem 3.1), and $\mathcal{W}$ is an elementary abelian 2-group if and only if each positive root is c-projective (Proposition 3.2). These statements are analogs of the following well-known results. A quiver is of finite representation type if and only if each indecomposable representation is preprojective [**ARS97**, Section VIII.1], and a quiver consists of isolated vertices if and only if each indecomposable representation is projective. We obtain (Theorem 2.17) an explicit description of the c-projective roots similar to that of the indecomposable projective representations of a quiver [**ARS97**, Section III.1], and show that the linear operators $-c$ and $-c^{-1}$ establish a bijection between the c-projective and c^{-1}-projective roots.

To better handle the combinatorics of c-admissible sequences, we use an equivalent but more convenient language of graph monoids introduced by Cartier and Foata [**CF69**]. For any finite undirected graph, the *graph monoid* $\mathfrak{M}$ is the quotient of the free monoid on the set of vertices modulo the congruence generated by the binary relation $vwRwv$, for all pairs $\{v, w\}$ of distinct vertices not connected by an edge. When Γ is the graph, a subset (not a submonoid) $\mathfrak{M}(c)$ of $\mathfrak{M}$ corresponds to the equivalence classes of c-admissible sequences and, thus, is a distributive lattice having the properties mentioned above. We say that the elements of $\mathfrak{M}(c)$ are the *c-admissible words* of $\mathfrak{M}$. The surjective monoid homomorphism $\rho : \mathfrak{M} \to \mathcal{W}$ sending each vertex to the associated simple reflection relates combinatorics of $\mathfrak{M}$ to that of $\mathcal{W}$. An element $w \in \mathcal{W}$ is *c-admissible* if it has a c-admissible preimage under ρ. If $X \in \mathfrak{M}$, we say that the word $\rho(X)$ in $\mathcal{W}$ is *reduced* if the length of the element $\rho(X)$ of $\mathcal{W}$ equals the length of X.

Throughout the paper we assume $\mathcal{W}$ irreducible. It is straightforward to extend our results to the case when $\mathcal{W}$ is a finite direct product of irreducible Coxeter groups.

In Section 1 we recall definitions and results about c-admissible words. The notion of *principal* c-admissible word is important here. Section 2 deals with c-preprojective or c-projective roots. If α is a c-preprojective root, we consider the set of elements $X \in \mathfrak{M}(c)$ for which $\rho(X)\alpha$ is a negative root, and show that the set is a sublattice of $\mathfrak{M}(c)$ with a unique least element W_α, which must be a principal word. Likewise, the set of elements of $\mathfrak{M}(c)$ sending to a negative root each element of a finite set Ψ of c-preprojective roots is a sublattice with a unique least element W_Ψ, which must be a join of principal words. The latter two statements are parts of Theorem 2.12, which plays a major role in the paper. The section also contains various properties and characterizations of c-preprojective or c-projective roots that are analogs of well-known results on representations of quivers. Section 3 presents the main result, a characterization of finite Coxeter groups in terms of c-preprojective roots. In Section 4 we relate the partial order on $\mathfrak{M}(c)$ to the left weak order on the set of c-admissible elements of $\mathcal{W}$, and show among other things that if $X \in \mathfrak{M}(c)$, then the word $\rho(X)$ is reduced if and only if $X = W_\Psi$, where Ψ is a finite *independent* set of c-preprojective roots; here Ψ is independent if the decomposition of W_Ψ as a join of principal words has the smallest possible number of terms. Combining these results with a simple inductive construction that produces the canonical form of each principal word in $\mathfrak{M}(c)$ [**KT05**], we hope to continue our study of reduced c-admissible words in $\mathcal{W}$.

1. Graph monoid

We begin by recalling some facts, definitions, and notation, using freely [**BGP73, Hum90, KP07, KT05**]. Several results presented here were obtained in [**KP07, KT05**] using the language of (+)-admissible sequences. Our consistent use of the algebraic language of graph monoids allows us to streamline the presentation and to strengthen some of the aforementioned results. Denote by $|S|$ the cardinality of a set S.

Given a finite undirected graph $\Gamma = (\Gamma_0, \Gamma_1)$ with the set of vertices Γ_0, the set of edges Γ_1, and no loops or multiple edges, consider a set $\hat{\Gamma}_0$ for which a bijection $\hat{\Gamma}_0 \to \Gamma_0$ is given and, for each $v \in \Gamma_0$, $\hat{v}$ is the preimage of v. We view the set $\hat{\Gamma}_0$ as an alphabet and its elements as the letters, and denote by Γ_0^* the *free monoid* on $\hat{\Gamma}_0$. The elements of Γ_0^* are all *words* $\hat{X} = \hat{x}_l \dots \hat{x}_1$, $l \geq 0$, with $x_j \in \Gamma_0$ for all j. When $l = 0$ we get the empty word $\hat{1}$ with no letters. The binary operation (multiplication) in Γ_0^* is concatenation. If $\hat{X}, \hat{Y} \in \Gamma_0^*$, we write $\hat{X}\hat{Y}$ for their product. The *transpose* of $\hat{X}$ is $\hat{X}^T = \hat{x}_1 \dots \hat{x}_l$.

Consider a (symmetric) binary relation R on Γ_0^* where $\hat{X} R \hat{Y}$ if and only if $\hat{X} = \hat{v}\hat{w}$ and $\hat{Y} = \hat{w}\hat{v}$, for some distinct vertices v and w of Γ not joined by an edge. Denote by $\approx$ the *congruence* (equivalence relation compatible with multiplication) on Γ_0^* generated by R. The *graph monoid* $\mathfrak{M} = \mathfrak{M}_\Gamma$ of Γ [**CF69**] is the quotient set of $\approx$, the elements of $\mathfrak{M}$ are the congruence classes. We denote by 1 the congruence class of the empty word $\hat{1}$ and denote by v the congruence class of the one-letter word $\hat{v}$, $v \in \Gamma_0$. Since $\hat{v} \approx \hat{w}$ if and only if $\hat{v} = \hat{w}$, if and only if $v = w$, $v, w \in \Gamma_0$, we view the vertices of the graph as the letters of the alphabet Γ_0. Denote by X the congruence class of each word $\hat{X} \in \Gamma_0^*$. If $\hat{X} = \hat{x}_l \dots \hat{x}_1$ then $X = x_l \dots x_1$ and we call the latter expression a *word* in $\mathfrak{M}$. If X and Y are words in $\mathfrak{M}$, then $X = Y$ if and only if $\hat{X} \approx \hat{Y}$, if and only if Y can be obtained from X by a finite number of interchanges of adjacent letters that are vertices not joined by an edge. The *transpose* of X is $X^T = x_1 \dots x_l$.

The following notions are well defined for the elements of $\mathfrak{M}$. The *length* of $X = x_l \dots x_1$, $\ell(X)$, is l; the *support* of X, $\operatorname{Supp} X$, is the set of distinct vertices among x_j, $1 \leq j \leq l$; and the *multiplicity* of $v \in \Gamma_0$ in X, $m_X(v)$, is the (nonnegative) number of times v appears among the x_j. The element X is *multiplicity-free* if $m_X(v) \leq 1$, $v \in \Gamma_0$. Denote by $\mathbb{Z}_{\geq 0}$ the additive monoid of nonnegative integers and denote by $\mathbb{Z}_{\geq 0}^{|\Gamma_0|}$ the direct sum of $|\Gamma_0|$ copies of $\mathbb{Z}_{\geq 0}$. Then the *multiplicity function* $m : \mathfrak{M} \to \mathbb{Z}_{\geq 0}^{|\Gamma_0|}$ given by $m(X) = (m_X(v))_{v \in \Gamma_0}$ is a surjective homomorphism of monoids [**CF69**, p. 8, bottom].

We relate the words in $\mathfrak{M}$ to the sequences of vertices of Γ by recalling the equivalence relation $\sim$ on the set of all sequences [**KT05**, Definition 1.2], which is the reflexive and transitive closure of a (symmetric) binary relation r where, for sequences U and V, UrV if and only if $U = x_1, \ldots, x_i, x_{i+1}, \ldots, x_l$, $V = x_1, \ldots, x_{i+1}, x_i, \ldots, x_l$, $x_i \neq x_{i+1}$, and no edge of Γ joins x_i and x_{i+1}. Every sequence of vertices $x_1, \ldots, x_l$ gives rise to the word $X = x_l \ldots x_1$ in $\mathfrak{M}$. If $Y = y_m \ldots y_1$, then $X = Y$ if and only if the sequences $x_1, \ldots, x_l$ and $y_1, \ldots, y_m$ are equivalent under $\sim$. Hence every statement from [**KT05, KP07, KT08**] about the equivalence classes of $\sim$ translates verbatim into a statement about the elements of $\mathfrak{M}$.

Given $X, Y \in \mathfrak{M}$, one sets $X \preceq Y$ if $Y = ZX$ for some $Z \in \mathfrak{M}$ [**KT05**, Definition 2.1] and writes $X \prec Y$ if $X \preceq Y$ and $X \neq Y$. The binary relation $\preceq$ is a partial order on $\mathfrak{M}$, and 1 is the least element of the *partially ordered set (poset)* $\mathfrak{M} = (\mathfrak{M}, \preceq)$. The poset satisfies the *descending chain condition.* Extending the natural total order on $\mathbb{Z}_{\geq 0}$ to the *product (componentwise) order* on $\mathbb{Z}_{\geq 0}^{|\Gamma_0|}$, which is given by $(a_v)_{v \in \Gamma_0} \leq (b_v)_{v \in \Gamma_0}$ if and only if $a_v \leq b_v$, $v \in \Gamma_0$, we note that the multiplicity function $m : \mathfrak{M} \to \mathbb{Z}_{\geq 0}^{|\Gamma_0|}$ is a homomorphism of posets and, thus, a surjective homomorphism of ordered monoids.

An *orientation* Λ of Γ consists of two functions, $s : \Gamma_1 \to \Gamma_0$ and $e : \Gamma_1 \to \Gamma_0$, assigning to each edge $a \in \Gamma_1$ its starting point $s(a)$ and endpoint $e(a)$. The pair (Γ, Λ) is a *quiver* (directed graph). In the quiver, each edge $a \in \Gamma_1$ becomes an *arrow* $a : s(a) \to e(a)$ *from* $s(a)$ *to* $e(a)$. Denote by Λ^{op} the orientation obtained by reversing the direction of each arrow of (Γ, Λ). There results the *opposite quiver* of (Γ, Λ), which we denote by $(\Gamma, \Lambda)^{op} = (\Gamma, \Lambda^{op})$.

For each $x \in \Gamma_0$ and each orientation Λ, denote by $x \cdot \Lambda$ the orientation obtained from Λ by reversing the direction of each arrow incident to x and preserving the remaining arrows. This extends uniquely to a left action of the monoid $\mathfrak{M}$ on the (finite) set of all orientations: if $X = x_l \ldots x_1$ then $X \cdot \Lambda = x_l \cdot (\ldots (x_1 \cdot \Lambda) \ldots)$.

A *path* in the quiver (Γ, Λ) is either a *nontrivial path* or a *trivial path.* A nontrivial path is a word $p = a_t \ldots a_1$, $t > 0$, in the alphabet Γ_1 for which $e(a_j) = s(a_{j+1})$, $1 \leq j < t$; here $t = \ell(p)$ is the *length* of p. By definition, $s(a_1)$ is the starting point, and $e(a_t)$ is the endpoint, of p. One writes $p : s(p) \to e(p)$ and says that p is a *path from* $s(p)$ *to* $e(p)$. There are precisely $|\Gamma_0|$ trivial paths, one for each vertex. The trivial path e_x at $x \in \Gamma_0$ by definition satisfies $s(e_x) = e(e_x) = x$ and $\ell(e_x) = 0$. The paths compose as follows. For any path r one sets $e_{e(r)} r = r e_{s(r)} = r$. If $p = a_t \ldots a_1$ and $q = b_u \ldots b_1$ are nontrivial paths satisfying $s(q) = e(p)$, then $qp = b_u \ldots b_1 a_t \ldots a_1$. For all composable pairs of paths p and q, we have $\ell(qp) = \ell(q) + \ell(p)$. A path p is an *oriented cycle* if $\ell(p) > 0$ and $s(p) = e(p)$. We consider only those orientations that yield an *acyclic quiver*, i.e., a quiver without oriented cycles. In an acyclic quiver, setting $x \leq y$ if and only if there exists a path from the vertex x to the vertex y yields a partial order on Γ_0, so that $\ell(p) \leq |\Gamma_0| - 1$ for all paths p. We refer to this partial order as the *reachability order* and denote the obtained poset by (Γ_0, Λ). For each $X \in \mathfrak{M}$, the subset $\operatorname{Supp} X$ of Γ_0 becomes a poset with respect to the induced order. A vertex x of the quiver (Γ, Λ) is a *sink* (resp. *source*) if it is a maximal (resp. minimal) element of (Γ_0, Λ).

For the remainder of this section we fix a connected acyclic quiver (Γ, Λ).

Recall that a subset Q of a poset P is a *filter* if $x \in Q$ and $y \geq x$ imply $y \in Q$. If S is any subset of P, denote by $\langle S \rangle$ the filter of P *generated* by S, that is the smallest filter of P containing S, which is equal to the intersection of all filters containing S. The *principal filter* $\langle x \rangle$ generated by an element $x \in P$ is the filter generated by the subset $\{x\}$. We have $\langle x \rangle = \{y \in P \,|\, y \geq x\}$.

REMARK 1.1. Let S be a subset of P.

(a) It is straightforward that $\langle S \rangle = \bigcup_{x \in S} \langle x \rangle$. If S is finite, for instance if P is finite, then $\langle S \rangle = \bigcup_{x \in \min S} \langle x \rangle$ where $\min S$ is the set of minimal elements of S in the order induced by P.

(b) If $T \subset P$ then $\langle S \cup T \rangle = \bigcup_{x \in S \cup T} \langle x \rangle = \left(\bigcup_{x \in S} \langle x \rangle \right) \cup \left(\bigcup_{x \in T} \langle x \rangle \right) = \langle S \rangle \cup \langle T \rangle$.

(c) If T is a filter of P then $S \subset T$ if and only if $\langle S \rangle \subset T$.

A subset $Q \subset P$ is a *chain* if any two elements of Q are comparable, and it is an *antichain* if no two distinct elements of Q are comparable. The set $\min P$ (resp. $\max P$) of minimal (resp. maximal) elements of P is an example of antichain.

REMARK 1.2. If Q is a filter of a poset P then $(\min P) \setminus Q = \min(P \setminus Q)$.

If $P = (\Gamma_0, \Lambda)$ and $Q \subset \Gamma_0$, there are two natural partial orders on Q: the induced order and the reachability order on the full subquiver of (Γ_0, Λ) determined by the vertices in Q. In general, the latter is finer than the former, but the two orders coincide if Q is a filter. If $x \in \Gamma_0$ then $\langle x \rangle$ is the set of endpoints of all paths in (Γ, Λ) starting at x.

DEFINITION 1.3. A word $X = x_l \dots x_1$, $l \geq 0$, in $\mathfrak{M}$ is Λ-*admissible* ((+)-admissible in the terminology of [**BGP73**]) if x_1 is a sink in (Γ, Λ), x_2 is a sink in $(\Gamma, x_1 \cdot \Lambda)$, x_3 is a sink in $(\Gamma, x_2 x_1 \cdot \Lambda)$, and so on. An element $X \in \mathfrak{M}$ is Λ-admissible if there exists a Λ-admissible word $X = x_l \dots x_1$. Denote by $\mathfrak{M}(\Lambda)$ the set of all Λ-admissible elements of $\mathfrak{M}$.

Part (a) of the following statement shows that the notion of Λ-admissible element is well-defined.

LEMMA 1.4. *Let $X = x_l \dots x_1$ and $Y = y_m \dots y_1$ be words in $\mathfrak{M}$.*

(a) *If $X = Y$ as elements of $\mathfrak{M}$, then the word X is Λ-admissible if and only if so is the word Y.*

(b) *Let $Y = ZX$. Then $Y \in \mathfrak{M}(\Lambda)$ if and only if $X \in \mathfrak{M}(\Lambda)$ and $Z \in \mathfrak{M}(X \cdot \Lambda)$; in particular, $\mathfrak{M}(\Lambda)$ is an ideal of the poset $(\mathfrak{M}, \preceq)$.*

(c) *Let y_j be a sink, $0 < j \leq m$, and let i be the smallest index satisfying $y_j = y_i$, $0 < i \leq j$. Then, for all $h < i$, no edge of Γ joins y_i and y_h. It follows that $Y = y_m \dots y_{i+1} y_{i-1} \dots y_1 y_i$.*

PROOF. (a) See [**BGP73**, proof of Lemma 1.2, p. 24].

(b) This is clear.

(c) If an edge of Γ joins y_i and y_h, the quiver (Γ, Λ) contains an arrow $a : y_h \to y_i$. Suppose $h < i$ and denote by f the smallest index satisfying $y_f = y_h$.

Then each of the vertices $y_1, \dots, y_{f-1}$ is different from y_f, so that the successive reversings of the directions of the arrows incident to $y_1, \dots, y_{f-1}$ do not affect the arrow a. Hence y_f is not a sink in the quiver $(\Gamma, y_{f-1} \dots y_1 \cdot \Lambda)$, which contradicts the assumption that Y is Λ-admissible. □

If x is a sink, the quiver $(\Gamma, x \cdot \Lambda)$ is acyclic. Therefore, the quiver $(\Gamma, X \cdot \Lambda)$ is acyclic for all $X \in \mathfrak{M}(\Lambda)$.

Replacing sinks with sources, one can also consider (−)-admissible words, but we do not use them in this paper.

LEMMA 1.5. *Let Λ and Λ' be two (acyclic) orientations of the graph Γ and let Θ be a subset of Γ_0.*

(a) *Assume that, for all $z \in \Theta$, each edge of Γ incident to z has the same direction in both of the quivers (Γ, Λ) and (Γ, Λ'). Suppose $X \in \mathfrak{M}$ and* $\operatorname{Supp} X \subset \Theta$. *Then $X \in \mathfrak{M}(\Lambda)$ if and only if $X \in \mathfrak{M}(\Lambda')$.*

(b) *Let $x \notin \Theta$ be a sink of (Γ, Λ) and suppose that no vertex in Θ is joined to x by an edge. If $X \in \mathfrak{M}$ and* $\operatorname{Supp} X \subset \Theta$, *then $X \in \mathfrak{M}(\Lambda)$ if and only if $X \in \mathfrak{M}(x \cdot \Lambda)$.*

PROOF. (a) For $X = x_l \dots x_1$ we show that if $X \in \mathfrak{M}(\Lambda)$ then $X \in \mathfrak{M}(\Lambda')$.

Proceed by induction on $l \geq 0$. If $l = 0$ then $X = 1$ and the statement holds. Let $l > 0$ and suppose that, for all acyclic orientations Δ and Δ' of Γ with the property that, for every $z \in \Theta$, each edge of Γ incident to z has the same direction in both of the quivers (Γ, Δ) and (Γ, Δ'), if $Z \in \mathfrak{M}$ satisfies $\operatorname{Supp} Z \subset \Theta$ and $\ell(Z) = l - 1$ then $Z \in \mathfrak{M}(\Delta)$ implies $Z \in \mathfrak{M}(\Delta')$.

By Lemma 1.4(a), x_1 is a sink in the quiver (Γ, Λ). Hence the direction of each edge $a : x_1 - v$ of Γ is $a : v \to x_1$ in both of the quivers (Γ, Λ) and (Γ, Λ'), and we conclude that x_1 is a sink in the quiver (Γ, Λ'). In both of the quivers $(\Gamma, x_1 \cdot \Lambda)$ and $(\Gamma, x_1 \cdot \Lambda')$, the direction of each edge $a : x_1 - v$ incident to x_1 is $a : x_1 \to v$, while every edge not incident to x_1 but incident to some $z \in \Theta$ has the same direction as that in the quivers (Γ, Λ) and (Γ, Λ'). Hence for all $z \in \Theta$, each edge of Γ incident to z has the same direction in both of the quivers $(\Gamma, x_1 \cdot \Lambda)$ and $(\Gamma, x_1 \cdot \Lambda')$. By the necessity of Lemma 1.4(b), $Z = x_l \dots x_2$ belongs to $\mathfrak{M}(x_1 \cdot \Lambda)$. Since $\operatorname{Supp} Z \subset \Theta$ and $\ell(Z) = l - 1$, the induction hypothesis says that $Z \in \mathfrak{M}(x_1 \cdot \Lambda')$. By the sufficiency of Lemma 1.4(b), $X = Zx_1$ belongs to both $\mathfrak{M}(\Lambda)$ and $\mathfrak{M}(\Lambda')$.

(b) In view of (a), we show that, for all $z \in \Theta$, each edge of Γ incident to z has the same direction in both of the quivers (Γ, Λ) and $(\Gamma, x \cdot \Lambda)$. When passing from Λ to $x \cdot \Lambda$, one preserves the direction of each arrow of (Γ, Λ) except those of the form $v \to x$, for some $v \in \Gamma_0$, whose direction one reverses. By assumption, $v \notin \Theta$, so one preserves the direction of each arrow incident to a vertex in Θ. □

The partial order $\preceq$ on $\mathfrak{M}$ induces a partial order on $\mathfrak{M}(\Lambda)$, which we denote by the same symbol. The poset $(\mathfrak{M}(\Lambda), \preceq)$ has nice properties.

It is easy to see that the multiplicity function $m : \mathfrak{M} \to \mathbb{Z}_{\geq 0}^{|\Gamma_0|}$ preserves, but generally does not reflect, partial order. However, the restriction of m to $\mathfrak{M}(\Lambda)$ does reflect partial order and, hence, is injective.

PROPOSITION 1.6. *For all $X, Y \in \mathfrak{M}(\Lambda)$, $X \preceq Y$ if and only if $m(X) \leq m(Y)$. It follows that $X = Y$ if and only if $m(X) = m(Y)$.*

PROOF. The necessity is clear.

For the sufficiency, assume that $m_X(v) \leq m_Y(v)$ for all $v \in \Gamma_0$. If $\ell(X) = 0$ then $X = 1$, so the statement holds. Suppose $\ell(X) = l > 0$ and the statement holds for all pairs $U \preceq V$ of Δ-admissible elements with $\ell(U) < l$, where Δ runs through all acyclic orientations of Γ. By Definition 1.3, there is a sink $x \in \operatorname{Supp} X$. By assumption, $x \in \operatorname{Supp} Y$. Using parts (c) and (b) of Lemma 1.4, we obtain $X = X'x$ and $Y = Y'x$ for some $X', Y' \in \mathfrak{M}(x \cdot \Lambda)$. Since the quiver $(\Gamma, x \cdot \Lambda)$ is acyclic, $\ell(X') = l - 1$, and $m_{X'}(v) \leq m_{Y'}(v)$ for all $v \in \Gamma_0$, the induction hypothesis gives $X' \preceq Y'$ whence $Y' = ZX'$. Then $Y = Y'x = ZX'x = ZX$ so that $X \preceq Y$. □

We obtain the following cancellation property.

COROLLARY 1.7. *Let $X, Y \in \mathfrak{M}(\Lambda)$.*

(a) *For all $U \in \mathfrak{M}$, if $UX \preceq UY$ then $X \preceq Y$. Hence $UX = UY$ if and only if $X = Y$.*

(b) *For all $V \in \mathfrak{M}$, $XV \preceq YV$ if and only if $X \preceq Y$. Hence $XV = YV$ if and only if $X = Y$.*

PROOF. (a) We have

$$m(U) + m(X) = m(UX) \leq m(UY) = m(U) + m(Y)$$

whence $m(X) \leq m(Y)$. By Proposition 1.6, $X \preceq Y$.

(b) The proof of the necessity is similar to that of (a), the sufficiency is clear. □

PROPOSITION 1.8. (a) *If $X \in \mathfrak{M}(\Lambda)$ then* $\operatorname{Supp} X$ *is a filter of the poset (Γ_0, Λ).*

(b) *Let $\Theta = \{v_1, \dots, v_t\}$ be a filter of (Γ_0, Λ). If each arrow $v_j \to v_i$ satisfies $j > i$ (such a numbering of the vertices exists), then $X = v_t \dots v_1$ is a unique multiplicity-free element of $\mathfrak{M}(\Lambda)$ satisfying $\Theta = \operatorname{Supp} X$.*

PROOF. (a) Given an element $X = x_l \dots x_1$ of $\mathfrak{M}(\Lambda)$, let $u \in \operatorname{Supp} X$ and $u < w$. To show $w \in \operatorname{Supp} X$, it suffices to consider an arrow $a : u \to v$ of (Γ, Λ) and show $v \in \operatorname{Supp} X$. By assumption, $u = x_i$ where $1 \leq i \leq l$. Since u is a sink in the quiver $(\Gamma, x_{i-1} \dots x_1 \cdot \Lambda)$, v must be one of the vertices $x_1, \dots, x_{i-1}$, for otherwise the arrow a will be preserved in the latter quiver.

(b) The uniqueness in the second assertion follows from Proposition 1.6, we prove the existence. If $t = 0$ then $\Theta = \emptyset$ and $X = 1$. If $t > 0$, we show that v_j is a sink in $(\Gamma, v_{j-1} \dots v_1 \cdot \Lambda)$, $1 \leq j \leq t$. If $j = 1$ then v_1 is a sink in (Γ, Λ) by the assumption on the numbering of vertices in Θ. Suppose $j > 1$ and $a : v_j - x$ is an edge of Γ. If $x \notin \Theta$ or $x = v_i$ with $i > j$, then the direction of a in (Γ, Λ) is $a : x \to v_j$, the same as the direction of a in $(\Gamma, v_{j-1} \dots v_1 \cdot \Lambda)$, because $x \neq v_h$, $1 \leq h < j$. If $x = v_i$ with $i < j$, the direction of a in (Γ, Λ) is $a : v_j \to x$. When passing from Λ to $v_{j-1} \dots v_1 \cdot \Lambda$, one reverses the direction of a one, and only one, time, when passing from $v_{i-1} \dots v_1 \cdot \Lambda$ to $v_i v_{i-1} \dots v_1 \cdot \Lambda$. Hence the direction of a in $(\Gamma, v_{j-1} \dots v_1 \cdot \Lambda)$ is $a : x \to v_j$. We have shown that v_j is a sink in the latter quiver. □

By Proposition 1.8, there exists a unique multiplicity-free element $K \in \mathfrak{M}(\Lambda)$ satisfying $\operatorname{Supp} K = \Gamma_0$. We call the element *complete* and always denote it by K. For all integers $t \geq 0$, $K^t \in \mathfrak{M}(\Lambda)$ and $K^t \cdot \Lambda = \Lambda$.

LEMMA 1.9. *Let $X, Y \in \mathfrak{M}(\Lambda)$ and* $\operatorname{Supp} X \cap \operatorname{Supp} Y = \emptyset$, *Then:*

(a) *If $x \in \operatorname{Supp} X$ and $y \in \operatorname{Supp} Y$, then no edge of Γ joins x and y.*

(b) $XY = YX \in \mathfrak{M}(\Lambda)$.

PROOF. (a) If there is an arrow $x \to y$ in (Γ, Λ), then $y \in \operatorname{Supp} X$ by Proposition 1.8(a), a contradiction. Likewise, (Γ, Λ) has no arrow $y \to x$.

(b) By (a), $XY = YX$. To show $XY \in \mathfrak{M}(\Lambda)$, proceed by induction on $\ell(X) + \ell(Y) \geq 0$. Note that if either X or Y equals 1, the statement is obvious. If $\ell(X) + \ell(Y) = 0$ then $X = Y = 1$, so the statement holds. Suppose now that $\ell(X) + \ell(Y) = l > 0$ and that, for all acyclic orientations Δ of Γ and all $U, V \in \mathfrak{M}(\Delta)$, if $\ell(U) + \ell(V) < l$ and $\operatorname{Supp} U \cap \operatorname{Supp} V = \emptyset$ then $UV \in \mathfrak{M}(\Delta)$.

Assuming $X \neq 1, Y \neq 1$, we have $X = X'x_1$, $Y = Y'y_1$ where x_1 and y_1 are distinct sinks of (Γ, Λ). By Lemma 1.4(b), $X' \in \mathfrak{M}(x_1 \cdot \Lambda)$. Since $\operatorname{Supp} X' \subset \operatorname{Supp} X$, no vertex in $\operatorname{Supp} X'$ is joined by an edge to y_1. Since y_1 is a sink in $(\Gamma, x_1 \cdot \Lambda)$, Lemma 1.5(b) says that $X' \in \mathfrak{M}(y_1 x_1 \cdot \Lambda)$. Likewise $Y' \in \mathfrak{M}(y_1 x_1 \cdot \Lambda)$ (remember, $x_1 y_1 = y_1 x_1$). By the induction hypothesis, $X'Y' \in \mathfrak{M}(y_1 x_1 \cdot \Lambda)$. Since $y_1 x_1 \in \mathfrak{M}(\Lambda)$, then $XY = X'Y'(y_1 x_1) \in \mathfrak{M}(\Lambda)$ by the sufficiency of Lemma 1.4(b). □

DEFINITION 1.10. Two vertices of a graph are *adjacent* if they are joined by an edge. For a subset $\Theta \subset \Gamma_0$, a vertex is adjacent to Θ if the vertex is not in Θ but is adjacent to a vertex in Θ. Denote by $adj(\Theta)$ the set of vertices adjacent to Θ.

LEMMA 1.11. *Let S and T be subsets of Γ_0. Then:*

(a) $adj(S \cup T) \subset adj(S) \cup adj(T)$.
(b) $adj(S) \subset T \cup adj(S \cup T)$.
(c) $(S \cup T) \cup adj(S \cup T) = (S \cup adj(S)) \cup (T \cup adj(T))$.

For the remaining assertions, suppose $S \cup adj(S) \subset S'$ and $T \cup adj(T) \subset T'$ where S' and T' are subsets of Γ_0.

(d) $(S \cup T) \cup adj(S \cup T) \subset S' \cup T'$.
(e) $(S \cap T) \cup adj(S \cap T) \subset S' \cap T'$.

PROOF. (a) If $x \in adj(S \cup T)$ then $x \notin S \cup T$ and there is an edge $x - v$ where $v \in S \cup T$. If $v \in S$ then $v \in adj(S)$. If $v \in T$ then $v \in adj(T)$.

(b) If $x \in adj(S)$ then $x \notin S$ and there is an edge $x - s$ with $s \in S$. If $x \notin T$ then $x \in adj(S \cup T)$.

(c) Using (a) and then (b), we get

$$(S \cup T) \cup adj(S \cup T) \subset S \cup T \cup adj(S) \cup adj(T) = (S \cup adj(S)) \cup (T \cup adj(T)) \subset$$

$$(S \cup (T \cup adj(S \cup T))) \cup (T \cup (S \cup adj(S \cup T))) = (S \cup T) \cup adj(S \cup T).$$

Hence all the above inclusions are equalities.

(d) This is an immediate consequence of (c).

(e) To show that $adj(S \cap T) \subset S' \cap T'$, let $x \in adj(S \cap T)$. Then $x \notin S \cap T$ and there is an edge $x - v$ in Γ where $v \in S \cap T$. If $x \in S$ then $x \in S'$. If $x \notin S$ then $x \in adj(S)$. In both cases $x \in S'$. Likewise $x \in T'$. □

PROPOSITION 1.12. *Let $Z \in \mathfrak{M}$, let $X \in \mathfrak{M}(\Lambda)$ be multiplicity-free, and assume that $\langle \operatorname{Supp} Z \cup adj(\operatorname{Supp} Z) \rangle \subset \operatorname{Supp} X$.*

(a) *For all edges $a : z - v$ of Γ with $z \in \operatorname{Supp} Z$, the direction of a in the quivers (Γ, Λ) and $(\Gamma, X \cdot \Lambda)$ is the same.*
(b) *$Z \in \mathfrak{M}(\Lambda)$ if and only if $Z \in \mathfrak{M}(X \cdot \Lambda)$.*
(c) *If $\operatorname{Supp} Z = \operatorname{Supp} X$ then $\operatorname{Supp} X = \Gamma_0$.*

PROOF. (a) Let $X = x_l \dots x_1$. By assumption, there exist unique integers i and j for which $1 \le i < j \le k$ and $\{v, z\} = \{x_i, x_j\}$. Since X is multiplicity-free, when computing the orientation $X \cdot \Lambda$, one reverses the direction of the edge a twice, first at x_i and then at x_j. Therefore, the direction of a in $(\Gamma, X \cdot \Lambda)$ is the same as in (Γ, Λ).

(b) This follows immediately from (a) and Lemma 1.5(a).

(c) By assumption, $adj(\operatorname{Supp} Z) = \emptyset$ whence $\operatorname{Supp} Z = \Gamma_0$ because the graph Γ is connected. □

The following statement establishes a bijection between the elements of $\mathfrak{M}(\Lambda)$ and certain descending sequences of filters of the poset (Γ_0, Λ).

PROPOSITION 1.13. (a) *If $1 \neq X \in \mathfrak{M}(\Lambda)$, there exists a unique decomposition $X = X_r \dots X_1$ where $1 \neq X_i \in \mathfrak{M}(\Lambda)$ is multiplicity-free for all i and the nonempty filters $F_i = \operatorname{Supp} X_i$ of (Γ_0, Λ) satisfy $F_{i+1} \cup adj(F_{i+1}) \subset F_i$, $1 \le i < r$,*

(b) *If $F_1, \dots, F_r$, $r > 0$, are nonempty filters of (Γ_0, Λ) satisfying $F_{i+1} \cup adj(F_{i+1}) \subset F_i$, $1 \le i < r$, denote by $X_i \in \mathfrak{M}(\Lambda)$ the unique multiplicity-free element for which $F_i = \operatorname{Supp} X_i$, $i = 1, \dots, r$ (Proposition 1.8(b)). Then $1 \neq X_r \dots X_1 \in \mathfrak{M}(\Lambda)$.*

PROOF. (a) By Proposition 1.8(a), $\operatorname{Supp} X$ is a filter of (Γ_0, Λ), so we denote by X_1 a unique multiplicity-free element of $\mathfrak{M}(\Lambda)$ satisfying $\operatorname{Supp} X_1 = \operatorname{Supp} X$. By Proposition 1.6, Corollary 1.7(b), and Lemma 1.4(b), there exists a unique $Z_1 \in \mathfrak{M}$ satisfying $X = Z_1 X_1$ and we must have $Z_1 \in \mathfrak{M}(X_1 \cdot \Lambda)$. We now show that $\langle \operatorname{Supp} Z_1 \cup adj(\operatorname{Supp} Z_1) \rangle \subset F_1 = \operatorname{Supp} X_1$. Since $\operatorname{Supp} Z_1 \subset F_1$, we only have to show that $adj(\operatorname{Supp} Z_1) \subset F_1$.

Let $v \in adj(\operatorname{Supp} Z_1)$ and let $a : z - v$ be an edge of Γ with $z \in \operatorname{Supp} Z_1$. If $a : z \to v$ is the direction of a in (Γ, Λ), then $v \in F_1$ because F_1 is a filter of (Γ_0, Λ). Suppose the direction is $a : v \to z$ and assume, to the contrary, that $v \notin F_1$. Write $X_1 = x_l \dots x_1$ and $Z_1 = z_m \dots z_1$. Since X_1 is multipilcity-free, $z = x_i$ for a unique i, $1 \le i \le l$. Hence $a : z \to v$ is the direction of a in the quiver $(\Gamma, X_1 \cdot \Lambda)$. Let j be the smallest index for which $z = z_j$, $1 \le j \le m$. Since $v \neq z_h$, $1 \le h < j$, then $a : z \to v$ is the direction of a in the quiver $(\Gamma, z_{j-1} \dots z_1 X \cdot \Lambda)$, in contradiction with z being a sink in the latter quiver. Thus $v \in F_1$.

By Proposition 1.12(b), $Z_1 \in \mathfrak{M}(\Lambda)$. If $Z_1 = 1$ then $X = X_1$ is the desired decomposition and $r = 1$. If $Z_1 \neq 1$ then $F_2 = \operatorname{Supp} Z_1$ is a nonempty filter of (Γ_0, Λ), and we denote by $X_2 \in \mathfrak{M}(\Lambda)$ a unique multiplicity-free element satisfying $F_2 = \operatorname{Supp} X_2$. We have $Z_1 = Z_2 X_2$ where $Z_2 \in \mathfrak{M}$ is uniquely determined and, moreover, $Z_2 \in \mathfrak{M}(X_2 \cdot \Lambda)$. We get $X = Z_2 X_2 X_1$. If $Z_2 = 1$ then $X = X_2 X_1$ is the desired decomposition and $r = 2$. If $Z_2 \neq 1$, Proposition 1.12(b) says that $Z_2 \in \mathfrak{M}(\Lambda)$ and we continue in the same way. The process must stop after finitely many steps because $\ell(X)$ is finite. We have shown the existence of the decomposition.

The decomposition is unique because the inclusion condition satisfied by the filters F_i says that $F_1 = \operatorname{Supp} X$ and $F_i = \operatorname{Supp} Z_{i-1}$ for $i > 1$, where the elements Z_j, $j > 0$, are uniquely determined by X.

(b) By Proposition 1.12(b), $X_r \in \mathfrak{M}(X_{r-1} \cdot \Lambda)$ so that $X_r X_{r-1} \in \mathfrak{M}(\Lambda)$ by the sufficiency of Lemma 1.4(b). Since $F_r \subset F_{r-1}$ then $F_{r-1} = \operatorname{Supp} X_r X_{r-1}$. Then Proposition 1.12(b) and Lemma 1.4(b) say that $X_r X_{r-1} \in \mathfrak{M}(X_{r-2} \cdot \Lambda)$ so that $X_r X_{r-1} X_{r-2} \in \mathfrak{M}(\Lambda)$, and so on. □

DEFINITION 1.14. A unique decomposition $X = X_r \dots X_1$ of a nonidentity element $X \in \mathfrak{M}(\Lambda)$ given by Proposition 1.13(a) is the *canonical form* of X, and we also say that $X = X_r \dots X_1$ is a Λ-admissible element in the canonical form. The positive integer r is the *size* of X. The sequence $F_1 \supset \dots \supset F_r$ where $F_i = \operatorname{Supp} X_i$ is the *filter sequence* of X. By definition, the size of 1 is zero and the filter sequence of 1 is empty.

Note that the size of a Λ-admissible element is the largest multiplicity of a vertex in that element. If $X \in \mathfrak{M}(\Lambda)$ is of size r then $X \preceq K^r$.

We now use the canonical form to describe the poset $(\mathfrak{M}(\Lambda), \preceq)$ and to show it is a lattice.

THEOREM 1.15. *Let $X = X_r \dots X_1$ and $Y = Y_s \dots Y_1$ be the canonical forms of the elements $X, Y \in \mathfrak{M}(\Lambda)$, $r, s \geq 0$. Let $F_i = \operatorname{Supp} X_i$, $1 \leq i \leq r$, $G_j = \operatorname{Supp} Y_j$, $1 \leq j \leq s$, be the filter sequences. Then:*

(a) *$X \preceq Y$ if and only if $r \leq s$ and one of the following two equivalent conditions holds.*
  (i) *$X_i \preceq Y_i$ for all $i \leq r$.*
  (ii) *$F_i \subset G_i$ for all $i \leq r$.*

Without loss of generality, assume $r \leq s$ and set $F_{r+1} = \dots = F_s = \emptyset$.

(b) *Denote by $U_i \in \mathfrak{M}(\Lambda)$ a unique multiplicity-free element for which $E_i = F_i \cap G_i = \operatorname{Supp} U_i$, $1 \leq i \leq s$. Suppose $E_q \neq \emptyset$ and $E_{q+1} = \dots E_s = \emptyset$ where $0 \leq q \leq r \leq s$. Then $U = U_q \dots U_1$ is the* greatest lower bound (meet) *of X and Y in the poset $(\mathfrak{M}(\Lambda), \preceq)$ (we write $U = X \wedge Y$), and $E_1 \supset \dots \supset E_q$ is the filter sequence of U.*
(c) *Denote by $Z_j \in \mathfrak{M}(\Lambda)$ a unique multiplicity-free element for which $H_j = F_j \cup G_j = \operatorname{Supp} Z_j$. Then $Z = Z_s \dots Z_1$ is the* least upper bound (join) *of X and Y in $(\mathfrak{M}(\Lambda), \preceq)$ (we write $Z = X \vee Y$), and $H_1 \supset \dots \supset H_s$ is the filter sequence of Z.*
(d) *$\mathfrak{M}(\Lambda)$ is a distributive lattice.*

PROOF. (a) Since each X_i and Y_j is multiplicity-free, conditions (i) and (ii) are equivalent according to Proposition 1.6.

Suppose $X \preceq Y$. Then $m_X(v) \leq m_Y(v)$ for all $v \in \Gamma_0$ so that the largest multiplicity of a vertex in X does not exceed that in Y. Hence $r \leq s$. Let $v \in F_i$, $1 \leq i \leq r$. Since the filter sequence is descending, then $v \in F_h$, $1 \leq h \leq i$, whence $m_X(v) \geq i$. If $v \notin G_i$ then $v \notin G_j$, $i \leq j \leq s$, whence $m_Y(v) < i$. We obtained a contradiction, so $v \in G_i$. Thus (ii) holds.

If $r \leq s$ and (ii) holds, then $m_X(v) \leq m_Y(v)$ for all $v \in \Gamma_0$. Hence $X \preceq Y$.

(b) It is straightforward that the intersection of two filters is a filter. Proposition 1.13(a) and Lemma 1.11(a) say that $\langle E_{i+1} \cup adj(E_{i+1}) \rangle \subset E_i$, $1 \leq i \leq q$. By Proposition 1.13(b), $U \in \mathfrak{M}(\Lambda)$. Since the decomposition $U = U_q \dots U_1$ satisfies the conditions of Proposition 1.13(a), it is the canonical form of U, and $E_1 \supset \dots \supset E_q$ is the filter sequence of U. By (a), U is the meet of X and Y.

(c) The proof is similar to that of (b), using part (b) of Lemma 1.11 instead of part (a).

(d) This follows from (b) and (c). □

We record some properties of the lattice $(\mathfrak{M}(\Lambda), \preceq)$.

PROPOSITION 1.16. *Let $X, Y \in \mathfrak{M}(\Lambda)$.*

(a) $X = V(X \wedge Y)$ *and* $Y = W(X \wedge Y)$ *where* $V, W \in \mathfrak{M}((X \wedge Y) \cdot \Lambda)$ *and* $\operatorname{Supp} V \cap \operatorname{Supp} W = \emptyset$.
(b) *If* $x \in \operatorname{Supp} V$ *and* $y \in \operatorname{Supp} W$ *then no edge of* Γ *joins* x *and* y.
(c) $VW = WV \in \mathfrak{M}((X \wedge Y) \cdot \Lambda)$.
(d) $X \vee Y = VW(X \wedge Y) = WX = VY$.

Proof. (a) We have $V, W \in \mathfrak{M}((X \wedge Y) \cdot \Lambda)$ by Lemma 1.4(b), and Theorem 1.15(b) says that $V = S(V \wedge W)$ and $W = T(V \wedge W)$. Substituting, we get $X = S(V \wedge W)(X \wedge Y)$ and $Y = T(V \wedge W)(X \wedge Y)$. By Lemma 1.4(b), $(V \wedge W)(X \wedge Y) \in \mathfrak{M}(\Lambda)$ whence $(V \wedge W)(X \wedge Y) \preceq X$ and $(V \wedge W)(X \wedge Y) \preceq Y$ imply $(V \wedge W)(X \wedge Y) \preceq X \wedge Y$. By Corollary 1.7(b), $V \wedge W \preceq 1$ so that $V \wedge W = 1$ and $\operatorname{Supp}(V \wedge W) = \emptyset$. Now Theorem 1.15(b) says that $\operatorname{Supp} V \cap \operatorname{Supp} W = \emptyset$.

(b) This is an immediate consequence of (a) and Lemma 1.9(a).

(c) This follows directly from (a) and Lemma 1.9(b).

(d) In view of (a) and Theorem 1.15(c) we have

$$X \vee Y = W'X = V'Y = W'V(X \wedge Y) = V'W(X \wedge Y)$$

so that, according to Corollary 1.7(b),

$$W'V = V'W. \tag{1.1}$$

In view of (a), (c), and sufficiency of Lemma 1.4(b),

$$WX = VY = VW(X \wedge Y) \in \mathfrak{M}(\Lambda).$$

Since $X, Y \preceq WX = VY$ then $X \vee Y = W'X = V'Y \preceq WX = VY$, so that, by Corollary 1.7(b), $V' \preceq V$ and $W' \preceq W$. Therefore, $V = PV'$ and $W = QW'$ so that $\operatorname{Supp} V' \subset \operatorname{Supp} V$, $\operatorname{Supp} P \subset \operatorname{Supp} V$ and $\operatorname{Supp} W' \subset \operatorname{Supp} W$, $\operatorname{Supp} Q \subset \operatorname{Supp} W$. Substituting into (1.1) and using (b), we get

$$W'V = VW' = PV'W' = V'W = V'QW' = QV'W'.$$

By Corollary 1.7(b), $P = Q$. By (a), $\operatorname{Supp} P \cap \operatorname{Supp} Q = \emptyset$ whence $P = Q = 1$ so that $V' = V$ and $W' = W$. □

Definition 1.17. A Λ-admissible element $X \in \mathfrak{M}$ of size $r > 0$ is *Λ-principal* if its filter sequence $F_1 \supset \cdots \supset F_r$ satisfies the following two conditions:
(i) F_r is a principal filter of (Γ_0, Λ),
(ii) $\langle F_{j+1} \cup adj(F_{j+1})\rangle = F_j$, $1 \le j < r$.

If (i) and (ii) hold and $F_r = \langle x \rangle$, $x \in \Gamma_0$, we write $X = W_{r,x}$.

By definition, 1 is Λ-principal of size 0. Denote by $\mathfrak{P}(\Lambda)$ the set of Λ-principal elements of $\mathfrak{M}$.

The canonical form of $X = W_{r,x}$, $r > 0$, is $X = X_r \dots X_1$ where (i) each X_j is Λ-admissible and multiplicity-free and (ii) $\operatorname{Supp} X_r = \langle x \rangle$ and $\langle \operatorname{Supp} X_{j+1} \cup adj(\operatorname{Supp} X_{j+1})\rangle = \operatorname{Supp} X_j$, $1 \le j < r$. By Proposition 1.13, the element $W_{r,x}$ is uniquely determined by x and r.

Clearly, $\mathfrak{P}(\Lambda) \subset \mathfrak{M}(\Lambda) \subset \mathfrak{M}$. If $X \in \mathfrak{P}(\Lambda)$, the full subgraph of Γ determined by $\operatorname{Supp} X$ is connected.

Proposition 1.18. *Let* $r > 0$ *and* $x \in \Gamma_0$.

(a) *Let* $Y = Y_q \dots Y_1$ *be a Λ-admissible element of* $\mathfrak{M}$ *in the canonical form. Then* $W_{r,x} \preceq Y$ *if and only if* $r \le q$ *and* $x \in \operatorname{Supp} Y_r$. *In particular,* $W_{r,x} \preceq W_{r,y}$ *if and only if* $y \le x$.
(b) *If* $W_{r,x} = x_l \dots x_1$ *then* $x = x_l$.

PROOF. (a) The first assertion follows from Proposition 1.15(a) and Definition 1.17 by an easy induction argument. For the second assertion, the first assertion says that $W_{r,x} \preceq W_{r,y}$ if and only if $x \in \langle y \rangle$, if and only if $y \leq x$.

(b) Let $W_{r,x} = X = X_r \dots X_1$ be the canonical form. Since $\operatorname{Supp} X_r = \langle x \rangle$ then $x \in \operatorname{Supp} X_i$ for all i, whence $r = m_X(x)$ is the largest multiplicity of a vertex in X. Let j be the largest index for which $x = x_j$, $1 \leq j \leq l$, and set $Y = x_j \dots x_1$. By Proposition 1.6, $Y \preceq X$. Then $Y \in \mathfrak{M}(\Lambda)$ by Lemma 1.4(b) and $r = m_Y(x) = m_X(x)$ is the largest multiplicity of a vertex in Y. Therefore Y is an element of size r, its canonical form is $Y = Y_r \dots Y_1$, and $x \in \operatorname{Supp} Y_i$ for all i. Since $x \in \operatorname{Supp} Y_r$, (a) says that $X \preceq Y$. Therefore $X = Y$ and $j = \ell(Y) = \ell(X) = l$. $\square$

PROPOSITION 1.19. *Let $1 \neq X \in \mathfrak{M}(\Lambda)$ be a (unique) element with the filter sequence $F_1 \supset \cdots \supset F_r$. Denote by $\bar{X} \in \mathfrak{M}(\Lambda)$ the element with the filter sequence $\langle F_1 \cup adj(F_1) \rangle \supset F_1 \supset \cdots \supset F_r$ (see Proposition 1.13).*

(a) *Let G be a filter of (Γ_0, Λ) satisfying $G \supset F_1 \cup adj(F_1)$ and let $Y \in \mathfrak{M}(\Lambda)$ be a unique multiplicity-free element for which $G = \operatorname{Supp} Y$. Then $XY = \bar{X} \vee Y$.*

(b) *If $X = W_{r,x}$ then $\bar{X} = W_{r+1,x}$*

(c) *If $X = Y_1 \vee \cdots \vee Y_m$ where $1 \neq Y_j \in \mathfrak{M}(\Lambda)$ for all j, then $\bar{X} = \bar{Y}_1 \vee \cdots \vee \bar{Y}_m$.*

PROOF. (a) Denote by $X_i \in \mathfrak{M}(\Lambda)$ the unique element for which $F_i = \operatorname{Supp} X_i$. Then $XY = X_r \dots X_1 Y$ is an element of $\mathfrak{M}(\Lambda)$ in the canonical form whose filter sequence is $G \supset F_1 \supset \cdots \supset F_r$. Since $G = G \cup \langle F_1 \cup adj(F_1) \rangle$ by Remark 1.1(c), Theorem 1.15(c) says that $XY = \bar{X} \vee Y$.

(b) This follows directly from Definition 1.17.

(c) Denote by $G_1^{(j)} \supset \cdots \supset G_{s_j}^{(j)}$ the filter sequence of Y_j. By Theorem 1.15(c), $r = \max_j s_j$ and $F_i = G_i^{(1)} \cup \cdots \cup G_i^{(m)}$ where we set $G_i^{(j)} = \emptyset$ whenever $s_j < i$. By Lemma 1.11(c),

$$F_1 \cup adj(F_1) = \left(G_1^{(1)} \cup \cdots \cup G_1^{(m)}\right) \cup adj\left(G_1^{(1)} \cup \cdots \cup G_1^{(m)}\right) =$$

$$\left(G_1^{(1)} \cup adj\left(G_1^{(1)}\right)\right) \cup \cdots \cup \left(G_1^{(m)} \cup adj\left(G_1^{(m)}\right)\right)$$

so that Remark 1.1(b) says that

$$\langle F_1 \cup adj(F_1) \rangle = \left\langle G_1^{(1)} \cup adj\left(G_1^{(1)}\right)\right\rangle \cup \cdots \cup \left\langle G_1^{(m)} \cup adj\left(G_1^{(m)}\right)\right\rangle.$$

By Theorem 1.15(c), $\bar{X} = \bar{Y}_1 \vee \cdots \vee \bar{Y}_m$. $\square$

DEFINITION 1.20. A finite subset $\{S_1, \dots, S_m\}$ of $\mathfrak{P}(\Lambda)$ is *independent* if whenever $S_1 \vee \cdots \vee S_m = T_1 \vee \cdots \vee T_q$ with $T_j \in \mathfrak{P}(\Lambda)$, $1 \leq j \leq q$, then $m \leq q$.

One obtains a bijection between the set of antichains in $\mathfrak{P}(\Lambda)$ and the set $\mathfrak{M}(\Lambda)$ by sending each antichain to its join.

PROPOSITION 1.21. *Let $X \in \mathfrak{M}(\Lambda)$ be an element of size $r > 0$ and let $F_1 \supset \cdots \supset F_r$ be the filter sequence of X. Set $F_{r+1} = \emptyset$.*

(a) *$X = \bigvee_{(h,v)} W_{h,v}$ where $1 \leq h \leq r$ and, for each h, $v \in (\min F_h) \setminus \langle F_{h+1} \cup adj(F_{h+1}) \rangle$.*

(b) *Let* $X = S_1 \vee \cdots \vee S_m$ *where* $S_i \in \mathfrak{P}(\Lambda)$ *for all* i *and let* (h, v) *be a pair satisfying the assumptions of (a). Then* $W_{h,v} = S_i$ *for some* i, *so that the subset* $\{W_{h,v}\}$ *of* $\mathfrak{P}(\Lambda)$ *is independent.*

(c) *There exists an independent subset* $\{S_1, \ldots, S_m\}$ *of* $\mathfrak{P}(\Lambda)$ *satisfying* $X = S_1 \vee \cdots \vee S_m$. *If* $\{T_1, \ldots, T_q\}$ *is an independent subset of* $\mathfrak{P}(\Lambda)$ *satisfying* $X = T_1 \vee \cdots \vee T_q$, *then* $q = m$ *and, after reindexing,* $T_i = S_i$ *for all* i.

(d) *A finite subset* $S = \{S_1, \ldots, S_m\}$ *of* $\mathfrak{P}(\Lambda)$ *is independent if and only if it is an antichain of the poset* $(\mathfrak{M}, \preceq)$.

PROOF. Denote by $X_h \in \mathfrak{M}(\Lambda)$ a unique multiplicity-free element for which $F_h = \operatorname{Supp} X_h$. Then $X = X_r \ldots X_1$ is the canonical form of X.

(a) Proceed by induction on r. If $r = 1$ then $F_2 = \emptyset$ and Remark 1.1(a) says that $F_1 = \bigcup_{v \in \min F_1} \langle v \rangle$. Since $\langle v \rangle = \operatorname{Supp} W_{1,v}$, Theorem 1.15(c) says that $X = X_1 = \bigvee_{v \in \min F_1} W_{1,v}$. Thus the statement holds.

Suppose $r > 1$ and the statement holds for all elements of $\mathfrak{M}(\Lambda)$ of size $r - 1$. Since $Y = X_r \ldots X_2$ is an element of $\mathfrak{M}(\Lambda)$ of size $r - 1$ with the filter sequence $F_2 \supset \cdots \supset F_r$, the induction hypothesis says that $Y = \bigvee_{(h,v)} W_{h-1,v}$ where $2 \le h \le r$ and, for each h, $v \in (\min F_h) \setminus \langle F_{h+1} \cup adj(F_{h+1}) \rangle$. Parts (c) and (b) of Proposition 1.19 say that $\bar{Y} = \bigvee_{(h,v)} \bar{W}_{h-1,v} = \bigvee_{(h,v)} W_{h,v}$; here $\bar{Y} \in \mathfrak{M}(\Lambda)$ is the element with the filter sequence $\langle F_2 \cup adj(F_2) \rangle \supset F_2 \supset \cdots \supset F_r$, and $\bar{W}_{h-1,v}$ is similarly defined. Since $F_1 \supset F_2 \cup adj(F_2)$, Proposition 1.19(a) and the base of induction say that

$$X = YX_1 = \bar{Y} \vee X_1 = \left(\bigvee_{(h,v)} W_{h,v} \right) \vee \left(\bigvee_{v \in \min F_1} W_{1,v} \right) \tag{1.2}$$

where $2 \le h \le r$ and, for each h, $v \in (\min F_h) \setminus \langle F_{h+1} \cup adj(F_{h+1}) \rangle$. If $v \in (\min F_1) \cap \langle F_2 \cup adj(F_2) \rangle$ then $W_{1,v} \preceq \bar{Y}$ by Proposition 1.18(a). Hence, for such a v, the term $W_{1,v}$ can be omitted from the second join in formula (1.2) and we can rewrite the formula as

$$X = \left(\bigvee_{(h,v)} W_{h,v} \right) \bigvee \left(\bigvee_{v \in (\min F_1) \setminus \langle F_2 \cup adj(F_2) \rangle} W_{1,v} \right).$$

The statement holds.

(b) Without loss of generality we may assume that $S_i \neq 1$ for all i. Denote by $G_1^{(i)} \supset \cdots \supset G_{s_i}^{(i)}$ the filter sequence of S_i. Since $S_i \preceq X$, Proposition 1.18(a) says that $s_i \le r$. We set $G_j^{(i)} = \emptyset$ if $s_i < j \le r$. By Theorem 1.15(c), $F_j = \bigcup_{i=1}^{m} G_j^{(i)}$ so that $v \in G_h^{(i)}$ for some i.

CLAIM. $G_{h+1}^{(i)} = \emptyset$.

If the claim is correct then Definition 1.17 says that $G_h^{(i)} = \langle x \rangle$, for some $x \in F_h$, so that $S_i = W_{h,x}$. Since $v \in \langle x \rangle$ implies $x \le v$, we must have $x = v$ because $v \in \min F_h$.

Proof of Claim. We have $F_{h+1} = G_{h+1}^{(i)} \cup H$. By Lemma 1.11(b),

$$adj\left(G_{h+1}^{(i)}\right) \subset H \cup adj\left(G_{h+1}^{(i)} \cup H\right)$$

so that

$$\left\langle G_{h+1}^{(i)} \cup adj\left(G_{h+1}^{(i)}\right)\right\rangle \subset \left\langle G_{h+1}^{(i)} \cup H \cup adj\left(G_{h+1}^{(i)} \cup H\right)\right\rangle = \langle F_{h+1} \cup adj\,(F_{h+1})\rangle.$$

If $G_{h+1}^{(i)} \neq \emptyset$ then $G_h^{(i)} = \left\langle G_{h+1}^{(i)} \cup adj\left(G_{h+1}^{(i)}\right)\right\rangle$ by Definition 1.17, whence $v \in \langle F_{h+1} \cup adj\,(F_{h+1})\rangle$, a contradiction.

(c) This is an immediate consequence of (a) and (b).

(d) The case $m = 0$ is trivial. Suppose $m > 0$ so that $S_i \neq 1$ for all i.

If S is independent and $S_i \preceq S_j$ where $i \neq j$, then $\overset{m}{\underset{h=1}{\vee}} S_h = \underset{h\neq j}{\vee} S_h$, a contradiction. Hence S is an antichain.

Suppose S is an antichain. Then $X = S_1 \vee \cdots \vee S_m$ is an element of $\mathfrak{M}(\Lambda)$ of size $r > 0$. By (a) and (b), $\{W_{h,v}\} \subset S$ and we claim that $\{W_{h,v}\} = S$. If not, then without loss of generality we may assume that $\{W_{h,v}\} = \{S_1, \ldots, S_k\}$ where $1 \leq k < m$ and $S_l \notin \{W_{h,v}\}$, $k < l \leq m$. By (a), $X = S_1 \vee \cdots \vee S_k$.

Let $F_1 \supset \cdots \supset F_r$ be the filter sequence of X and let $G_1^{(i)} \supset \cdots \supset G_{s_i}^{(i)}$ be the filter sequence of S_i, $1 \leq i \leq k$. Since $S_i \preceq X$, we must have $1 \leq s_i \leq r$. Setting $G_j^{(i)} = \emptyset$ if $s_i < j \leq r$, by Theorem 1.15(c) we get $F_j = \overset{k}{\underset{i=1}{\cup}} G_j^{(i)}$.

Let $k < l \leq m$. Since $S_l = W_{q,u} \preceq X$, Proposition 1.18(a) says that $q \leq r$ and $u \in F_q = \overset{k}{\underset{i=1}{\cup}} G_q^{(i)}$ whence $u \in G_q^{(i)}$ for some i with $1 \leq i \leq k$. Applying Proposition 1.18(a) again, we obtain $S_l \preceq S_i$ in contradiction with S being an antichain. □

Note that [**KP07**, Proposition 4.2(3)] replaces the assumption that $v \in (\min F_h) \setminus \langle F_{h+1} \cup adj(F_{h+1})\rangle$ of Proposition 1.21(a) by $v \in \min\,(F_{h+1} \setminus \langle F_{h+2} \cup adj(F_{h+2})\rangle)$. The two conditions are equivalent according to Remark 1.2,

2. Preprojective roots

We now consider quivers arising from Coxeter groups, using freely the terminology of [**Hum90**].

Let $\mathcal{W} = (\mathcal{W}, S)$ be a *Coxeter system*, where $\mathcal{W}$ is a group with a finite set of generators S, and let $n = |S|$. The defining relations are $(ss')^{m(s,s')} = 1$, with $s, s' \in S$ and $m(s, s') < \infty$, where the $m(s, s')$ are the entries of a *Coxeter matrix* $M = (m(s, s'))_{s,s' \in S}$. Here M is a symmetric $n \times n$ matrix with $m(s, s') \in \mathbb{Z} \cup \{\infty\}$; $m(s, s) = 1$ for all $s \in S$; and $m(s, s') > 1$ whenever $s \neq s'$. Denote by $\Gamma = (\Gamma_0, \Gamma_1)$ the *Coxeter graph* of $\mathcal{W}$. The vertices of Γ are defined by a bijection $\rho : \Gamma_0 \to S$. There exists one, and only one, edge joining vertices x and y if and only if $2 < m(\rho(x), \rho(y)) \leq \infty$. In this paper we assume that $\mathcal{W}$ is *irreducible*, i.e., that the graph Γ is *connected*. We denote by the same letter ρ a unique (surjective) monoid homomorphism $\mathfrak{M} \to \mathcal{W}$ induced by the bijection ρ; here $\mathfrak{M}$ is the graph monoid of the Coxeter graph Γ. The *length* of $w \in \mathcal{W}$ is $\ell(w) = \ell(X)$ if $w = \rho(X)$ and $\ell(X) \leq \ell(Y)$ for all $Y \in \mathfrak{M}$ satisfying $w = \rho(Y)$. If $\ell(w) = \ell(X)$ and $X = x_l \ldots x_1$, we say that the word $\rho(X) = \rho(x_l) \ldots \rho(x_1)$ in $\mathcal{W}$ is *reduced*, and we also say that $\rho(X)$ is a *reduced expression for* w.

REMARK 2.1. No edge of Γ joins vertices x and y if and only if $\rho(x)\rho(y) = \rho(y)\rho(x)$. Hence $X = Y$ implies $\rho(X) = \rho(Y)$, i.e., the map $\rho : \mathfrak{M} \to \mathcal{W}$ is well defined.

DEFINITION 2.2. Let $s_1, s_2, \dots, s_n$ be the elements of S in some order, and let $\Gamma_0 = \{v_1, \dots, v_n\}$ where $s_j = \rho(v_j)$ for all j. The element $c = s_n \dots s_1$ is a *Coxeter element* of $\mathcal{W}$. The *c-orientation* of Γ is the orientation for which every arrow $v_j \to v_i$ satisfies $j > i$. There result an acyclic quiver (Γ, c), where we denote the orientation by the same letter c; a poset (Γ_0, c); the subset $\mathfrak{M}(c)$ of $\mathfrak{M}$ consisting of all *c-admissible elements*; and the subset $\mathfrak{P}(c)$ of $\mathfrak{M}(c)$ consisting of all *c-principal elements*. An element $w \in \mathcal{W}$ is *c-admissible* if one of its preimages under ρ is c-admissible.

REMARK 2.3. If Γ is the Coxeter graph, each acyclic orientation Λ of Γ is the c-orientation for some Coxeter element c. If c and d are Coxeter elements, one can verify that $c = d$ if and only if the c-orientation and the d-orientation of Γ coincide, that is, if and only if $(\Gamma, c) = (\Gamma, d)$ as quivers.

For the rest of the paper we fix an irreducible Coxeter system $\mathcal{W} = (\mathcal{W}, S)$, a Coxeter element $c = s_n \dots s_1$, and a bijection $\rho : \Gamma_0 \to S$ given by $\rho(v_j) = s_j$, $j = 1, \dots, n$.

REMARK 2.4. Let $X = x_l \dots x_1$ be in $\mathfrak{M}(c)$.

(a) The element $K = v_n \dots v_1$ is complete c-admissible, $\rho(K) = c$, and $XK \in \mathfrak{M}(c)$. The element $K^T = v_1 \dots v_n$ is complete c^{-1}-admissible with $\rho(K^T) = c^{-1}$. We have $(\Gamma, c)^{op} = (\Gamma, c^{-1})$.

(b) If $x_1 = v_j$, for some j, then $K = v_n \dots v_{j+1} v_{j-1} \dots v_1 v_j$ by Lemma 1.4(c), and $c = s_n \dots s_{j+1} s_{j-1} \dots s_1 s_j$ by (a). Therefore $s_j c s_j = s_j s_n \dots s_{j+1} s_{j-1} \dots s_1$ is a Coxeter element and $x_1 \cdot c = s_j c s_j$. By induction, $\rho(X) c \rho(X^T)$ is a Coxeter element and $X \cdot c = \rho(X) c \rho(X^T)$. In other words, the action of a c-admissible element X of the graph monoid $\mathfrak{M}$ on the c-orientation yields the d-orientation, where the Coxeter element d is the conjugate of c by the image of X under the canonical homomorphism of monoids ρ.

Thus the fact that all Coxeter elements are conjugate if Γ is a tree, is a consequence of [**BGP73**, Theorem 1.2, part 1)], saying that if Λ and Λ' are orientations of a tree Γ, then $\Lambda' = Y \cdot \Lambda$, for some $Y \in \mathfrak{M}(\Lambda)$.

Let $\mathbb{V}$ be a real vector space of dimension n with a formal basis $\{\alpha_s \mid s \in S\}$ and symmetric bilinear form B given by $B(\alpha_s, \alpha_{s'}) = -2\cos\dfrac{\pi}{m(s, s')}$, $s, s' \in S$, where $\dfrac{\pi}{\infty} = 0$ by convention. The values of B are twice those of the bilinear form defined in [**Hum90**, p. 109]. We need the modification in order to simplify the forthcoming explicit description of c-projective roots in Theorem 2.17(b). For any $\alpha = \sum\limits_{s \in S} c_s \alpha_s$ in $\mathbb{V}$, $c_s \in \mathbb{R}$ is the *s-coordinate* of α, and the *support* of α is $\operatorname{Supp} \alpha = \{x \in \Gamma_0 \mid c_{\rho(x)} \neq 0\}$.

The group $\mathcal{W}$ acts on $\mathbb{V}$ by $s\lambda = \lambda - B(\alpha_s, \lambda)\alpha_s$, $s \in S, \lambda \in \mathbb{V}$, and by extending the action from the generators to the whole group. The action preserves the bilinear form B. A vector $\alpha = w(\alpha_s)$, for some $w \in \mathcal{W}, s \in S$, is a *root*. The element $s_\alpha = w s w^{-1}$ of $\mathcal{W}$, which does not depend on the choice of either w or s, is the *reflection associated with* α. The basis vectors α_s are the *simple roots*. The elements of S are the *simple reflections*. A root α is *positive*, $\alpha > 0$ (resp. *negative*, $\alpha < 0$) if all of its coordinates are nonnegative (resp. nonpositive). Every root is either positive or negative. Denote by Φ the *root system* of $\mathcal{W}$, which is the set of all roots; denote by Φ^+ the set of all positive roots; and denote by T the set of all reflections in $\mathcal{W}$.

We quote [**Bri98**, p. 372, bottom].

PROPOSITION 2.5. *For all $\alpha \in \Phi$ the full subgraph of Γ determined by* $\operatorname{Supp}\alpha$ *is connected.*

DEFINITION 2.6. An element $X \in \mathfrak{M}$ *negates* a root $\alpha > 0$ if $\rho(X)\alpha < 0$, and X negates a finite subset Ψ of Φ^+ if X negates each element of Ψ. Denote by $\mathfrak{N}(\alpha)$ (resp. $\mathfrak{N}(\Psi)$) the set of all elements of $\mathfrak{M}$ that negate α (resp. Ψ). We call X a *minimal element* of $\mathfrak{N}(\alpha)$ (resp. $\mathfrak{N}(\Psi)$) if X is minimal with respect to the induced partial order $\preceq$ on $\mathfrak{N}(\alpha)$ (resp. $\mathfrak{N}(\Psi)$).

REMARK 2.7. Let $X \in \mathfrak{N}(\alpha)$ for some $\alpha \in \Phi^+$.

(a) $\operatorname{Supp}\alpha \subset \operatorname{Supp} X$.

(b) Suppose X is a minimal element of $\mathfrak{N}(\alpha)$. If $X = ZY$ and $Z \neq 1$, then $\rho(Y)\alpha > 0$ and Z is a minimal element of $\mathfrak{N}(\rho(Y)\alpha)$.

LEMMA 2.8. *For $V, W \in \mathfrak{M}$, suppose that* $\operatorname{Supp} V \cap \operatorname{Supp} W = \emptyset$ *and no edge of Γ joins a vertex from* $\operatorname{Supp} V$ *and a vertex from* $\operatorname{Supp} W$. *If $\alpha \in \Phi^+$, then $\rho(VW)\alpha < 0$ if and only if either $\rho(V)\alpha < 0$ or $\rho(W)\alpha < 0$.*

PROOF. Remark 2.1 says that

$$\rho(VW)\alpha = \rho(WV)\alpha = \rho(V)[\rho(W)\alpha] = \rho(W)[\rho(V)\alpha].$$

If $\rho(VW)\alpha < 0$ then Remark 2.7(a) says that $\operatorname{Supp}\alpha \subset \operatorname{Supp} VW = \operatorname{Supp} V \cup \operatorname{Supp} W$, whence $\operatorname{Supp}\alpha = (\operatorname{Supp}\alpha \cap \operatorname{Supp} V) \cup (\operatorname{Supp}\alpha \cap \operatorname{Supp} W)$. Since no edge of Γ joins a vertex from $\operatorname{Supp}\alpha \cap \operatorname{Supp} V$ and a vertex from $\operatorname{Supp}\alpha \cap \operatorname{Supp} W$, then Proposition 2.5 says that either $\operatorname{Supp}\alpha \cap \operatorname{Supp} V = \emptyset$, or $\operatorname{Supp}\alpha \cap \operatorname{Supp} W = \emptyset$. Say, the latter holds. Then $\operatorname{Supp}\alpha \subset \operatorname{Supp} V$ so that the action of $\rho(W)$ does not change the nonzero coordinates of $\rho(V)\alpha$. Since $\rho(W)[\rho(V)\alpha] < 0$ we must have $\rho(V)\alpha < 0$.

If $\rho(V)\alpha < 0$ then $\operatorname{Supp}\alpha \subset \operatorname{Supp} V$ by Remark 2.7(a). Hence $\operatorname{Supp}\alpha \cap \operatorname{Supp} W = \emptyset$, so that the action of $\rho(W)$ does not change the negative coordinates of $\rho(V)\alpha$. Therefore $\rho(VW)\alpha < 0$. The case $\rho(W)\alpha < 0$ is similar. $\square$

DEFINITION 2.9. A root $\alpha > 0$ is *c-preprojective* (resp. *c-projective*) if some $X \in \mathfrak{M}(c)$ (resp. multiplicity-free $X \in \mathfrak{M}(c)$) negates α. Denote by $\mathcal{P}(c)$ the subset of Φ^+ consisting of all c-preprojective roots.

REMARK 2.10. A root $\alpha > 0$ is c-preprojective if and only if $w(\alpha) < 0$ for some c-admissible $w \in \mathcal{W}$.

THEOREM 2.11. *Let $\alpha \in \mathcal{P}(c)$ and let Ψ be a finite nonempty subset of $\mathcal{P}(c)$. Then $\mathfrak{N}(\alpha) \cap \mathfrak{M}(c)$ and $\mathfrak{N}(\Psi) \cap \mathfrak{M}(c)$ are nonempty sublattices of $\mathfrak{M}(c)$.*

PROOF. Let $X, Y \in \mathfrak{N}(\alpha) \cap \mathfrak{M}(c)$. By parts (a) and (b) of Proposition 1.16, $X = V(X \wedge Y)$ and $Y = W(X \wedge Y)$ where $\operatorname{Supp} V \cap \operatorname{Supp} W = \emptyset$ and no edge of Γ joins a vertex from $\operatorname{Supp} V$ to a vertex from $\operatorname{Supp} W$. Since $\operatorname{Supp}\alpha \subset \operatorname{Supp} X \cap \operatorname{Supp} Y$ by Remark 2.7(a), Theorem 1.15(b) implies $X \wedge Y \neq 1$.

If $\rho(X \wedge Y)\alpha > 0$, then $V, W \in \mathfrak{N}(\rho(X \wedge Y)\alpha)$. By Remark 2.7(a), $\operatorname{Supp}\rho(X \wedge Y)\alpha \subset \operatorname{Supp} V$ and $\operatorname{Supp}\rho(X \wedge Y)\alpha \subset \operatorname{Supp} W$, whence $\operatorname{Supp} V \cap \operatorname{Supp} W \neq \emptyset$, a contradiction. Thus $\rho(X \wedge Y)\alpha < 0$, that is $X \wedge Y \in \mathfrak{N}(\alpha) \cap \mathfrak{M}(c)$.

By Proposition 1.16(d), $X \vee Y = VW(X \wedge Y)$. Since $-\rho(X \wedge Y)\alpha > 0$ and $\rho(V)[-\rho(X \wedge Y)\alpha] > 0$, $\rho(W)[-\rho(X \wedge Y)\alpha] > 0$ by assumption, the contrapositive of Lemma 2.8 says that

$$\rho(X \vee Y)\alpha = \rho(VW)[\rho(X \wedge Y)\alpha] = -\rho(VW)[-\rho(X \wedge Y)\alpha] < 0.$$

Thus $X \vee Y \in \mathfrak{N}(\alpha) \cap \mathfrak{M}(c)$. We have proved that $\mathfrak{N}(\alpha) \cap \mathfrak{M}(c)$ is a sublattice of $\mathfrak{M}(c)$. By Definition 2.9, $\mathfrak{N}(\alpha) \cap \mathfrak{M}(c) \neq \emptyset$.

If $\Psi = \{\alpha_1 \ldots, \alpha_m\}$ then $\mathfrak{N}(\Psi) = \bigcap_{i=1}^{m} \mathfrak{N}(\alpha_i)$. Hence

$$\mathfrak{N}(\Psi) \cap \mathfrak{M}(c) = \bigcap_{i=1}^{m} [\mathfrak{N}(\alpha_i) \cap \mathfrak{M}(c)]$$

where each $\mathfrak{N}(\alpha_i) \cap \mathfrak{M}(c)$ is a nonempty sublattice of $\mathfrak{M}(c)$ by what we have just proved. Since the set of all sublattices of a lattice is closed under intersections, $\mathfrak{N}(\Psi) \cap \mathfrak{M}(c)$ is a sublattice too. Since the poset $\mathfrak{M} = (\mathfrak{M}, \preceq)$ satisfies the descending chain condition, each nonempty subposet of $\mathfrak{M}$ that is a lattice must have a unique least element. If X_i is the unique least element of $\mathfrak{N}(\alpha_i) \cap \mathfrak{M}(c)$, then $X_1 \wedge \cdots \wedge X_m$ is the unique least element of $\mathfrak{N}(\Psi) \cap \mathfrak{M}(c)$; in particular, $\mathfrak{N}(\Psi) \cap \mathfrak{M}(c) \neq \emptyset$. □

Denote by $\mathfrak{f}(c)$ the set of finite subsets of $\mathcal{P}(c)$.

THEOREM 2.12. *In the setting of Theorem 2.11:*

(a) *The nonempty lattice $\mathfrak{N}(\alpha) \cap \mathfrak{M}(c)$ (resp. $\mathfrak{N}(\Psi) \cap \mathfrak{M}(c)$) has a unique least element W_α (resp. W_Ψ), which is a minimal element of $\mathfrak{N}(\alpha)$ (resp. $\mathfrak{N}(\Psi)$). We obtain a map $\nu : \mathcal{P}(c) \to \mathfrak{M}(c)$ given by $\nu(\alpha) = W_\alpha$, which extends to the map $\xi : \mathfrak{f}(c) \to \mathfrak{M}(c)$ given by $\xi(\Psi) = W_\Psi$.*
(b) $\operatorname{Im} \nu \subset \mathfrak{P}(c)$. *Hence $W_\alpha = W_{r,x}$ for uniquely determined $r > 0$, $x \in \Gamma_0$.*
(c) $\alpha = \rho\left(W_{r,x}^T\right)\left(-\alpha_{\rho(x)}\right)$, *and* $\rho\left(W_{i,x}^T\right)\left(-\alpha_{\rho(x)}\right) > 0$ *if* $0 < i < r$.
(d) *The map $\nu : \mathcal{P}(c) \to \mathfrak{M}(c)$ is injective.*

PROOF. (a) Since the poset $\mathfrak{M}$ satisfies the descending chain condition, every nonempty subposet of $\mathfrak{M}$ that is a lattice has a unique least element. To show, say, that W_Ψ is a minimal element of $\mathfrak{N}(\Psi)$, suppose $Y \preceq W_\Psi$ and $Y \in \mathfrak{N}(\Psi)$. Then Lemma 1.4(b) says that $Y \in \mathfrak{M}(c)$, whence $Y \in \mathfrak{N}(\Psi) \cap \mathfrak{M}(c)$ and we must have $W_\Psi \preceq Y$. Thus $Y = W_\Psi$.

(b) Let $W_\alpha = x_l \ldots x_1$. To prove $W_\alpha \in \mathfrak{P}(c)$, proceed by induction on $l = \ell(W_\alpha)$. If $l = 1$, then [**Hum90**, Proposition 5.6(a)] says that $\alpha = \alpha_{\rho(x_1)}$ is a simple root. Since x_1 is a sink of (Γ, c), the statement holds.

If $l > 1$, suppose that, for all Coxeter elements $d \in \mathcal{W}$, the statement holds for all $\beta \in \mathcal{P}(d)$ satisfying $\ell(W_\beta) < l$. Since $l > 1$, Remark 2.4(b) says that $\rho(x_1)\alpha \in \mathcal{P}(\rho(x_1)c\rho(x_1))$, and Remark 2.7(b) says that $W_{\rho(x_1)\alpha} = x_l \ldots x_2$. By the inductive hypothesis, $W_{\rho(x_1)\alpha} \in \mathfrak{P}(\rho(x_1)c\rho(x_1))$. Since $W_{\rho(x_1)\alpha}$ is $\rho(x_1)c\rho(x_1)$-principal, the full subgraph Θ of Γ determined by $\operatorname{Supp} W_{\rho(x_1)\alpha}$ is connected. Now [**KT08**, Proposition 2.5] says that $W_\alpha \in \mathfrak{P}(c)$ if the full subgraph Ω of Γ determined by $\operatorname{Supp} W_\alpha$ is connected.

Assume, to the contrary, that Ω is disconnected. Then $x_1 \notin \operatorname{Supp} W_{\rho(x_1)\alpha}$ and no edge of Γ joins x_1 and a vertex from $\operatorname{Supp} W_{\rho(x_1)}$. Therefore $W_\alpha = W_{\rho(x_1)\alpha}x_1 = x_1 W_{\rho(x_1)\alpha}$ whence $W_{\rho(x_1)\alpha} \in \mathfrak{M}(c)$ by Lemma 1.4(b). Since $\rho(W_\alpha)\alpha < 0$, Lemma 2.8 says that either $x_1 \in \mathfrak{N}(\alpha)$ or $W_{\rho(x_1)\alpha} \in \mathfrak{N}(\alpha)$. The former contradicts $\ell(W_\alpha) >$

1. The latter contradicts (a), for $W_{\rho(x_1)\alpha} \prec W_\alpha$. Thus Ω is connected and $W_\alpha \in \mathfrak{P}(c)$.

By Definition 1.17, $W_\alpha = W_{r,x}$ for some r and x. If $W_\alpha = W_{q,y}$ then $W_{r,x} = W_{q,y}$, so that $r = q$ and $x = y$ by Proposition 1.18(a).

(c) Since $W_\alpha = x_l \dots x_1 = W_{r,x}$ by (b), then $x_l = x$ by Proposition 1.18(b). By (a), if $0 < j < l$ then $\rho(x_j)\dots\rho(x_1)\alpha > 0$, and $\rho(x)\rho(x_{l-1})\dots\rho(x_1)\alpha < 0$. By [**Hum90**, Proposition 5.6(a)], $\alpha_{\rho(x)}$ is the only positive root negated by $\rho(x)$. Hence $\rho(x_{l-1})\dots\rho(x_1)\alpha = \alpha_{\rho(x)}$ and $\rho(W_{r,x})\alpha = -\alpha_{\rho(x)}$. Since $\rho(X)^{-1} = \rho\left(X^T\right)$ for any X, then $\alpha = \rho\left(W_{r,x}^T\right)\left(-\alpha_{\rho(x)}\right)$. If $0 < i < r$, then $W_{i,x} \prec W_{r,x}$ by Proposition 1.18(a). Now the remaining inequality follows from (a).

(d) If $\beta \in \mathcal{P}(c)$ and $\nu(\alpha) = \nu(\beta)$, then (b) says that $W_\alpha = W_\beta = W_{r,x}$ for uniquely determined $r > 0$, $x \in \Gamma_0$. By (c), $\alpha = \beta = \rho\left(W_{r,x}^T\right)\left(-\alpha_{\rho(x)}\right)$. □

The statement of Theorem 2.12(d) is not true for the map $\xi : \mathfrak{f}(c) \to \mathfrak{M}(c)$.

DEFINITION 2.13. If $\alpha \in \mathcal{P}(c)$ satisfies $W_\alpha = W_{r,x}$, for some $r > 0$, $x \in \Gamma_0$, we say that α is a c-preprojective root of *size* r. Denote by $\mathcal{P}(c, r)$ the set of c-preprojective roots of size r.

Note that $\mathcal{P}(c, 1)$ is the set of c-projective roots, for if a multiplicity-free word $X \in \mathfrak{M}(c)$ negates a root $\alpha > 0$, then W_α must be multiplicity-free because $W_\alpha \preceq X$ by Theorem 2.12(a). By Theorem 2.12(b), $\mathcal{P}(c, q) \cap \mathcal{P}(c, r) = \emptyset$ if $q \neq r$.

REMARK 2.14. Let $\alpha \in \mathcal{P}(c, r)$ so that $W_\alpha = W_{r,x}$, $x \in \Gamma_0$, and set $s = \rho(x)$. If $X = UW_\alpha$ and $x \notin \operatorname{Supp} U$, then $X \in \mathfrak{N}(\alpha)$.

Indeed, Theorem 2.12(c) says that $\rho(X)\alpha = \rho(U)\left[\rho(W_\alpha)\alpha\right] = \rho(U)(-\alpha_s) < 0$, for the s-coordinate of $\rho(X)\alpha$ is -1.

The following definition makes use of Definition 1.20 and of Theorem 2.12(b) saying that $W_\alpha \in \mathfrak{P}(c)$ for all $\alpha \in \mathcal{P}(c)$.

DEFINITION 2.15. A finite subset $\{\alpha_1, \dots, \alpha_m\}$ of $\mathcal{P}(c)$ is *independent* if $\{W_{\alpha_1}, \dots, W_{\alpha_m}\}$ is an independent subset of $\mathfrak{P}(c)$. Denote by $\mathfrak{i}(c)$ the subset of $\mathfrak{f}(c)$ consisting of all independent subsets of $\mathcal{P}(c)$, and denote by $\xi : \mathfrak{i}(c) \to \mathfrak{M}(c)$ the restriction of the map $\xi : \mathfrak{f}(c) \to \mathfrak{M}(c)$.

PROPOSITION 2.16. (a) *If $\Psi = \{\alpha_1, \dots, \alpha_m\}$, $m > 0$, is an independent subset of $\mathcal{P}(c)$, then $W_\Psi = W_{\alpha_1} \vee \dots \vee W_{\alpha_m}$.*

(b) *The map $\xi : \mathfrak{i}(c) \to \mathfrak{M}(c)$ given by $\xi(\Psi) = W_\Psi$ is injective.*

PROOF. (a) We may assume $m > 1$. Since $\mathfrak{N}(\Psi) \cap \mathfrak{M}(c) = \bigcap_{i=1}^{m} \left[\mathfrak{N}(\alpha_i) \cap \mathfrak{M}(c)\right]$, Theorem 2.12(a) says that $W_{\alpha_i} \preceq W_\Psi$ for all i. It suffices to show that $X = W_{\alpha_1} \vee \dots \vee W_{\alpha_m}$ belongs to $\mathfrak{N}(\alpha_i)$ for all i, for then $X \in \mathfrak{N}(\Psi) \cap \mathfrak{M}(c)$ and $X \preceq W_\Psi$, which forces $X = W_\Psi$. Set $Y = W_{\alpha_1} \vee \dots \vee W_{\alpha_{i-1}} \vee W_{\alpha_{i+1}} \vee \dots \vee W_{\alpha_m}$.

Using Theorem 1.16, we get $W_{\alpha_i} = V(W_{\alpha_i} \wedge Y)$, $Y = W(W_{\alpha_i} \wedge Y)$, and $X = VW\left(W_{\alpha_i} \wedge Y\right)$, where $\operatorname{Supp} V \cap \operatorname{Supp} W = \emptyset$ and no edge of Γ joins a vertex from $\operatorname{Supp} V$ and a vertex from $\operatorname{Supp} W$. Since Ψ is independent, then $W_{\alpha_i} \npreceq Y$ whence $W_{\alpha_i} \npreceq W_{\alpha_i} \wedge Y$ so that $\rho(W_{\alpha_i} \wedge Y)\alpha_i > 0$. Since $\rho(V)\left[\rho(W_{\alpha_i} \wedge Y)\alpha_i\right] < 0$, Lemma 2.8 says that $\rho(X)\alpha_i < 0$. Hence $X \in \mathfrak{N}(\alpha_i)$.

(b) Suppose $\Theta \in \mathfrak{i}(c)$ and $W_\Psi = W_\Theta$, where Ψ is from (a) and $\Theta = \{\beta_1, \dots, \beta_q\}$. Then $q > 0$ and $W_\Theta = W_{\beta_1} \vee \dots \vee W_{\beta_q}$ by (a). By Proposition 1.21(c), $m = q$ and there exists a reindexing so that $W_{\alpha_i} = W_{\beta_i}$ for all i. By Theorem 2.12(d), $\alpha_i = \beta_i$ for all i, that is $\Psi = \Theta$. □

We define the values of the bilinear form B on the paths in (Γ, c) by setting $B(e_x) = 1$ if e_x is the trivial path at $x \in \Gamma_0$, and

$$B(p) = (-1)^t B\left(\alpha_{\rho(e(a_t))}, \alpha_{\rho(s(a_t))}\right) \dots B\left(\alpha_{\rho(e(a_1))}, \alpha_{\rho(s(a_1))}\right)$$

if $p = a_t \dots a_1$ is a nontrivial path. It is straightforward that $B(p) > 0$ for all paths p, and that $B(qp) = B(q)B(p)$ whenever p and q are paths satisfying $s(q) = e(p)$.

The next statement gives a description of the c-projective roots.

THEOREM 2.17. *Let $s \in S$ and denote by x the unique vertex of Γ satisfying $s = \rho(x)$.*

(a) *The root $\pi_s(c) = \rho\left(W_{1,x}^T\right)(-\alpha_s)$ is c-projective with $W_{\pi_s(c)} = W_{1,x}$. The map $S \to \mathcal{P}(c, 1)$ given by $s \mapsto \pi_s(c)$ is bijective, so that $\mathcal{P}(c, 1) = \{\pi_s(c) \mid s \in S\}$.*

(b) *Let $t = \rho(y) \in S$, $y \in \Gamma_0$. If $x \not\leq y$, the t-coordinate of $\pi_s(c)$ is 0. If $x \leq y$, the t-coordinate of $\pi_s(c)$ is $\sum_p B(p)$, where p runs through all paths from x to y in (Γ, c). Hence*

$$\pi_s(c) = \sum_{y \in \Gamma_0,\, x \leq y} \left(\sum_{p: x \to y} B(p) \right) \alpha_{\rho(y)}.$$

(c) *$\pi_s(c) = -c^{-1}\pi_s(c^{-1})$ and $\pi_s(c^{-1}) = -c\pi_s(c)$. Therefore, the linear operators $-c$ and $-c^{-1}$ on $\mathbb{V}$ induce mutually inverse bijections $-c : \mathcal{P}(c, 1) \to \mathcal{P}(c^{-1}, 1)$ and $-c^{-1} : \mathcal{P}(c^{-1}, 1) \to \mathcal{P}(c, 1)$.*

PROOF. Let $W_{1,x} = x_l \dots x_1$.

(a) We have $\rho(W_{1,x})\,\pi_s(c) = \rho(W_{1,x})\left[\rho\left(W_{1,x}^T\right)(-\alpha_s)\right] = -\alpha_s < 0$. Since $x_l = x$ by Proposition 1.18(b), the s-coordinate of $\pi_s(c)$ is 1. Therefore $\pi_s(c) \in \mathcal{P}(c, 1)$ and $x \in \operatorname{Supp} W_{\pi_s(c)}$ so that $\operatorname{Supp} W_{1,x} = \langle x \rangle \subset \operatorname{Supp} W_{\pi_s(c)}$ because $\operatorname{Supp} W_{\pi_s(c)}$ is a filter of (Γ_0, c) by Proposition 1.8(a). By Theorem 2.12(a), $W_{1,x} = V W_{\pi_s(c)}$ whence $V = 1$ because $W_{1,x}$ is multiplicity-free. Thus $W_{1,x} = W_{\pi_s(c)}$.

To show the map $s \mapsto \pi_s(c)$ is injective, let $t \in S$ satisfy $t = \rho(y)$ and $\pi_s(c) = \pi_t(c)$. Then $W_{\pi_s(c)} = W_{\pi_t(c)}$ by Theorem 2.12(a) so that $W_{1,x} = W_{1,y}$ by what we have just proved. By Proposition 1.18(a), $x = y$ so that $s = t$.

To show the map is surjective, let $\alpha \in \mathcal{P}(c, 1)$. By Theorem 2.12(a), W_α is multiplicity-free. Then Theorem 2.12(b) says that $W_\alpha = W_{1,x}$, for some $x \in \Gamma_0$, and Theorem 2.12(c) gives $\alpha = \rho(W_{1,x}^T)\left(-\alpha_{\rho(x)}\right) = \pi_{\rho(x)}(c)$.

(b) By (a), $W_{\pi_s(c)} = W_{1,x}$, and we proceed by induction on l.

If $l = 1$, then $\pi_s(c) = \alpha_s$, so that the t-coordinate is 1 if $t = s$, and it is 0 if $t \neq s$. Since x is a sink, $x \leq y$ implies $x = y$. Since e_x is the only path from x to x, and $B(e_x) = 1$ by definition, the statement holds.

If $l > 1$, suppose that, for all Coxeter elements $d \in \mathcal{W}$, the statement holds for all $u \in S$ satisfying $\ell\left(W_{\pi_u(d)}\right) < l$. Set $Y = x_l \dots x_2$ and $\beta = \rho(x_1)\pi_s(c)$. Remark 2.4(b) says that $d = x_1 \cdot c = \rho(x_1)c\rho(x_1)$, and Lemma 1.4(b) says that $Y \in \mathfrak{M}(d)$. By Remark 2.7(b), $\beta > 0$ and Y is a minimal element of $\mathfrak{N}(\beta)$. Therefore $\beta \in \mathcal{P}(d, 1)$ and $Y = W_\beta$ according to Theorem 2.12(a). By (a), $\beta = \pi_u(d)$ for some $u \in S$, and $Y = W_{1,z} \in \mathfrak{P}(d)$ where $z \in \Gamma_0$ satisfies $u = \rho(z)$. By Proposition 1.18(b), $z = x_l = x$ so that $\beta = \pi_s(d)$ and $Y = W_{1,x}$. In particular, $\{x_2, \dots, x_l\} = \operatorname{Supp} Y$ is the principal filter of (Γ_0, d) generated by x. Since $\ell(Y) = l - 1$, the statement holds for $\pi_s(d)$, so

$$\rho(x_1)\pi_s(c) = \pi_s(d) = \sum_{j=2}^{l} \left(\sum_{q:x \to x_j} B(q) \right) \alpha_{\rho(x_j)}$$

where, for each j, q runs through all paths from x to x_j in the quiver (Γ, d). Applying $\rho(x_1)$ to both sides of the above equality, we get

$$\pi_s(c) = \sum_{j=2}^{l} \left(\sum_{q:x \to x_j} B(q) \right) \left[\alpha_{\rho(x_j)} - B\left(\alpha_{\rho(x_1)}, \alpha_{\rho(x_j)}\right) \alpha_{\rho(x_1)}\right]$$

$$= \sum_{j=2}^{l} \left(\sum_{q:x \to x_j} B(q) \right) \alpha_{\rho(x_j)} + \sum_{j=2}^{l} \left(\sum_{q:x \to x_j} - B\left(\alpha_{\rho(x_1)}, \alpha_{\rho(x_j)}\right) B(q) \right) \alpha_{\rho(x_1)}$$

$$= \sum_{j=1}^{l} \left(\sum_{p:x \to x_j} B(p) \right) \alpha_{\rho(x_j)}.$$

Note that since x_1 is a sink in (Γ, c) and a source in (Γ, d), the paths $x \to x_j$ are the same in both quivers for $j > 1$. And each path $p : x \to x_1$ in (Γ, c) satisfies $p = aq$ for a unique path $q : x \to x_j$ with $j > 1$ and a unique arrow $a : x_j \to x_1$ in (Γ, c) (remember, $x \neq x_1$ because $l > 1$). Here

$$B(p) = B(a)B(q) = -B\left(\alpha_{\rho(x_1)}, \alpha_{\rho(x_j)}\right) B(q).$$

(c) The ideal (x) of (Γ_0, c) generated by x is the filter of the poset $\left(\Gamma_0, c^{-1}\right)$ generated by x. By Proposition 1.8(b), there exists a unique multiplicity-free $Z \in \mathfrak{M}(c^{-1})$ for which $(x) = \operatorname{Supp} Z$. Applying (a) to c^{-1} instead of c, in view of Definition 1.17 we get that $\pi_s(c^{-1}) = \rho(Z^T)(-\alpha_s)$ is a c^{-1}-projective root. Using Remark 2.4(a), we obtain $K^T = VZ$, for some V. Then $K = Z^T V^T$ and Lemma 1.4(b) says that $V^T \in \mathfrak{M}(c)$. Setting $U = x_{l-1} \dots x_1$, we note that $U \in \mathfrak{M}(c)$ and $\operatorname{Supp} U \subset \operatorname{Supp} V^T$, for $\operatorname{Supp} Z \cap \operatorname{Supp} W_{1,x} = (x) \cap \langle x \rangle = \{x\}$. Since U and V^T are multiplicity-free, Proposition 1.6 says that $U \preceq V^T$ so that $V^T = Y^T U$, for some Y. Then $V = U^T Y$ and $K^T = U^T Y Z$ where $\operatorname{Supp} Y = \Gamma_0 \setminus [\langle x \rangle \cup (x)]$.

Let $Y = y_m \dots y_1$ and $0 < i \leq m$. Then no edge of Γ joins x and y_i, for there is no arrow $x \to y_i$ in (Γ, c) because $y_i \notin \langle x \rangle$, and no arrow $y_i \to x$ in (Γ, c) because $y_i \notin (x)$. Hence $\rho(y_i)(\alpha_s) = \alpha_s$ because $B\left(\alpha_s, \alpha_{\rho(y_i)}\right) = 0$. It follows that $\rho(Y)(\alpha_s) = \alpha_s$.

In view of Remark 2.4(a), we have

$$-c^{-1}\pi_s\left(c^{-1}\right) = -\rho(K^T)\rho(Z^T)(-\alpha_s) = -\rho(U^T)\rho(Y)\rho(Z)\rho(Z^T)(-\alpha_s) =$$

$$-\rho(U^T)\rho(Y)(-\alpha_s) = -\rho(U^T)(-\alpha_s) = \rho(U^T)s(-\alpha_s) =$$

$$\rho(U^T)\rho(x)(-\alpha_s) = \rho(W_{1,x}^T)(-\alpha_s) = \pi_s(c)$$

because $x_l = x$. The rest is clear. □

By Theorem 2.17(a), there is a bijection between the vertices of Γ and the c-projective roots, which is similar to the well-known bijection between the vertices of a quiver and the nonisomorphic indecomposable projective representations of the quiver. The bijection of Theorem 2.17(c) is the analog of the bijection between the indecomposable projective and indecomposable injective representations.

DEFINITION 2.18. The root $\pi_s(c)$ of Theorem 2.17(a) is the c-projective root *associated with* $s \in S$.

Parts (a) and (c) of the following statement are analogs of the well-known properties of indecomposable preprojective representations of a quiver.

PROPOSITION 2.19. *Let* $\alpha \in \mathcal{P}(c,r)$ *and let* $X_r \dots X_1$ *be the canonical form of* $W_\alpha = W_{r,x}$, $r > 0$, $x \in \Gamma_0$. *Set* $s = \rho(x)$. *Let* i *be an integer satisfying* $0 < i < r$.

(a) $c^i\alpha > 0$ *and* $c^r\alpha < 0$. *Hence* $c^i\alpha \in \mathcal{P}(c)$.

(b) $c^i\alpha = \rho(X_i \dots X_1)\alpha$ *and* $X_r \dots X_{i+1}$ *is the canonical form of* $W_{c^i\alpha}$.

(c) $c^{-i}\pi_s(c) > 0$ *and* $\alpha = c^{-r+1}\pi_s(c)$, *where* $\pi_s(c)$ *is the* c*-projective root associated with* s.

PROOF. Set $K = v_n \dots v_1$ so that $c = \rho(K)$.

(a) If $j > 0$ and $c^j\alpha < 0$, then Remark 2.4(a) and Theorem 2.12(a) say that $W_{r,x} \preceq K^j$. By Proposition 1.18(a), $r \leq j$ and $K^r = UW_\alpha$ where $x \notin \operatorname{Supp} U$. Hence $c^i\alpha > 0$ and Remark 2.14 says that $K^r \in \mathfrak{N}(\alpha)$, whence $c^r\alpha = \rho(K^r)\alpha < 0$.

(b) There is nothing to prove if $r = 1$, so let $r > 1$ and suppose the statement holds for all $\beta \in \mathcal{P}(c, r-1)$. We show first that $c\alpha = \rho(X_1)\alpha$ and $X_r \dots X_2$ is the canonical form of $W_{c\alpha}$, i.e., that the statement holds for $i = 1$.

By (a), $c^{r-1}(c\alpha) < 0$ and $c^i(c\alpha) > 0$ if $0 \leq i < r-1$. Hence $c\alpha \in \mathcal{P}(c, r-1)$ and $W_{c\alpha} = W_{r-1,y}$ for some $y \in \Gamma_0$. Clearly, $W_{r-1,x}K$ and $W_{r-1,y}K$ belong to $\mathfrak{M}(c)$. Since $\rho(W_{c\alpha})c\alpha = \rho(W_{r-1,y}K)\alpha < 0$, Theorem 2.12(a) says that $W_{r,x} \preceq W_{r-1,y}K$. By Proposition 1.18(a), $x \in \langle y \rangle$ and $W_{r,x} \preceq W_{r-1,x}K$. Therefore $y \leq x$ and $W_{r-1,x}K = UW_{r,x}$ where $x \notin \operatorname{Supp} U$. By Remark 2.14, $W_{r-1,x}K \in \mathfrak{N}(\alpha)$ whence $W_{r-1,x} \in \mathfrak{N}(c\alpha)$ so that $W_{r-1,y} \preceq W_{r-1,x}$. By Proposition 1.18(a), $x \leq y$. Thus $y = x$ and $W_{c\alpha} = W_{r-1,x} = X_r \dots X_2 = W_{\rho(X_1)\alpha}$ in view of Remark 2.7(b). Since $W_{c\alpha} = W_{\rho(X_1)\alpha}$, then $c\alpha = \rho(X_1)\alpha$ by Theorem 2.12(d). We have proved that the statement holds for $i = 1$.

The inductive hypothesis says that if $0 < j < r-1$, then

$$c^j(c\alpha) = c^j\left[\rho(X_1)\alpha\right] = \rho\left(X_{j+1} \dots X_2\right)\left[\rho(X_1)\alpha\right]$$

and $X_r \dots X_{j+2}$ is the canonical form of $W_{c^j[\rho(X_1)\alpha]} = W_{c^{j+1}\alpha}$ Setting $i = j+1$, we see that the statement holds if $1 < i < r$.

(c) By (a) and (b), $c^{r-1}\alpha > 0$ and $W_{c^{r-1}\alpha} = X_r = W_{1,x}$. Since $W_{\pi_s(c)} = W_{1,x}$ by Theorem 2.17(a), then $W_{c^{r-1}\alpha} = W_{\pi_s(c)}$ whence $c^{r-1}\alpha = \pi_s(c)$ by Theorem 2.12(d). Therefore $\alpha = c^{-r+1}\pi_s(c)$. If $0 < i < r$, then $c^{-i}\pi_s(c) = c^{-i+r-1}\alpha > 0$ by (a). □

We show how to construct the c-preprojective roots of size r.

PROPOSITION 2.20. *Let* $r > 0$ *be an integer and let* $s = \rho(x)$, $x \in \Gamma_0$. *If* $\rho\left(W_{i,x}^T\right)(-\alpha_s) > 0$ *when* $1 < i \leq r$, *then* $\alpha = \rho\left(W_{r,x}^T\right)(-\alpha_s) \in \mathcal{P}(c,r)$ *and* $W_\alpha = W_{r,x}$.

PROOF. It is straightforward that $\rho(W_{r,x})\alpha = -\alpha_s$, so $\alpha \in \mathcal{P}(c)$. We have to prove that $W_\alpha = W_{r,x}$. Let $X_r \dots X_1$ be the canonical form of $W_{r,x}$.

By Theorem 2.12(b), $W_\alpha = W_{q,y}$ for some $q > 0$, $y \in \Gamma_0$. Setting $t = \rho(y)$, we note that $W_{q,y} \preceq W_{r,x}$ by Theorem 2.12(a), so Proposition 1.18(a) says that $q \leq r$ and $y \in \operatorname{Supp} X_q$. Set $Y = X_q \dots X_1$. Then $W_{q,y} \preceq Y$ so that $Y = UW_{q,y}$, where $y \notin \operatorname{Supp} U$ because each X_j is multiplicity-free. By Remark 2.14,

$$\rho(Y)\alpha = \left[\rho(X_q \dots X_1)\rho(X_1^T \dots X_r^T)\right](-\alpha_s) < 0$$

whence $q \geq r$ because, by assumption,

$$\rho(X^T_{q+1} \ldots X^T_r)(-\alpha_s) = \rho\left(W^T_{r-q,x}\right)(-\alpha_s) > 0$$

if $0 < q < r$. Therefore $q = r$ so that $Y = W_{r,x}$. In view of Theorem 2.12(c),

$$-\alpha_s = \rho\left(W_{r,x}\right)\alpha = \rho(U)\left[\rho\left(W_{r,y}\right)\alpha\right] = \rho(U)(-\alpha_t).$$

Since $y \notin \operatorname{Supp} U$, we must have $t = s$ whence $y = x$. Thus $W_\alpha = W_{r,x}$. □

THEOREM 2.21. *The following are equivalent for a root $\alpha > 0$ and an integer $r > 0$.*

(a) $\alpha \in \mathcal{P}(c, r)$.
(b) $c^i\alpha > 0$ *whenever* $0 < i < r$, *and* $c^r\alpha < 0$.
(c) *There is an $s \in S$ for which $c^{-i}\pi_s(c) > 0$ if $0 < i < r$, and $\alpha = c^{-r+1}\pi_s(c)$; here $\pi_s(c)$ is the c-projective root associated with $s \in S$.*
(d) *There is an $x \in \Gamma_0$ for which $\rho\left(W^T_{i,x}\right)(-\alpha_s) > 0$ if $1 < i \leq r$, and $\alpha = \rho\left(W^T_{r,x}\right)(-\alpha_s)$, where $s = \rho(x)$.*

If α satisfies any of the conditions (a)-(d), the elements s in (c) and x in (d) are uniquely determined.

PROOF. (a) $\Longrightarrow$ (b) and (a) $\Longrightarrow$ (c) These are parts (a) and (c) of Proposition 2.19.

(d) $\Longrightarrow$ (a) This is Proposition 2.20.

(c) $\Longrightarrow$ (b) This is an immediate consequence of Theorem 2.17(c).

(b) $\Longrightarrow$ (d) By definition, $\alpha \in \mathcal{P}(c)$ so that Theorem 2.12(b) says that $W_\alpha = W_{q,x}$, for some $q > 0, x \in \Gamma_0$. By Theorem 2.12(c), $\rho(W^T_{i,x})(-\alpha_s) > 0$ if $0 < i \leq q$, and $\alpha = \rho(W^T_{q,x})(-\alpha_s)$. By Proposition 2.19(a), $c^i\alpha > 0$ if $0 < i < q$, and $c^q\alpha < 0$. Comparing the latter with the assumption, we get $q = r$.

The uniqueness is an immediate consequence of our prior results. □

The final statement of this section follows immediately from Theorems 2.21 and 2.17(c).

COROLLARY 2.22. (a) *A root $\alpha > 0$ is c-preprojective if and only if $c^r\alpha < 0$, for some integer $r > 0$. The root α is c-projective if and only if $c\alpha < 0$.*
(b) *For all $r > 0$,*

$$\begin{aligned}\mathcal{P}(c, r) &= \{\rho\left(W^T_{r,x}\right)(-\alpha_{\rho(x)}) \mid x \in \Gamma_0,\ \rho\left(W^T_{i,x}\right)(-\alpha_{\rho(x)}) > 0 \text{ if } 1 < i \leq r\} \\ &= \{c^{-r+1}\pi_{\rho(x)} \mid x \in \Gamma_0,\ c^{-i}\pi_{\rho(x)} > 0 \text{ if } 0 < i < r\}.\end{aligned}$$

3. Preprojective roots and finite Coxeter groups

THEOREM 3.1. *For a Coxeter element $c \in \mathcal{W}$, the following are equivalent.*

(a) *The group $\mathcal{W}$ is finite.*
(b) *The set $\mathcal{P}(c)$ is finite.*
(c) *All positive roots are c-preprojective.*
(d) *All simple roots are c-preprojective.*

PROOF. (a) $\Longrightarrow$ (b) and (c) $\Longrightarrow$ (d) are trivial.

(b) $\Longrightarrow$ (a) By [**Hum90**, Proposition 5.6(b)], the length of the element c^t, for some $t > 0$, is the number of positive roots α satisfying $c^t\alpha < 0$. Since a power of c is c-admissible by Remark 2.4(a), each such α belongs to $\mathcal{P}(c)$ by Remark 2.10. Since $\mathcal{P}(c)$ is a finite set, the lengths of the powers c^t, $t > 0$, are bounded by $|\mathcal{P}(c)|$.

Hence there are only finitely many distinct elements among such powers, so that the order of c is finite. By [**How82**, Theorem 4.1], $\mathcal{W}$ is finite.

(a) $\Longrightarrow$ (c) The unique element $w_0 \in \mathcal{W}$ of maximal length satisfies $w_0\alpha < 0$, for all roots $\alpha > 0$, according to [**Hum90**, Proposition 5.6 and Exercises 1, 2, p. 115]. By [**Spe09**, Theorem 3], w_0 is c-admissible. Hence all positive roots are c-preprojective.

(d) $\Longrightarrow$ (a) By assumption, $\Psi = \{\alpha_s \mid s \in S\} \subset \mathcal{P}(c)$. By Theorem 2.12(a), the word $W_\Psi \in \mathfrak{M}(c)$ satisfies $\rho(W_\Psi)\alpha_s < 0$, for all $s \in S$, whence $\rho(W_\Psi)\alpha < 0$ for all $\alpha \in \Phi^+$. Therefore the group $\mathcal{W}$ must be finite. □

PROPOSITION 3.2. *For a Coxeter element $c \in \mathcal{W}$, the following are equivalent.*

(a) $|\mathcal{W}| = 2$.
(b) *All positive roots are c-projective.*
(c) *All simple roots are c-projective.*

PROOF. (a) $\Longrightarrow$ (b) and (b) $\Longrightarrow$ (c) are clear.

(c) $\Longrightarrow$ (a) For each $x \in \Gamma_0$, set $s = \rho(x)$. Since α_s is c-projective, Theorem 2.17(b) says that $\pi_s(c) = \alpha_s$, whence e_x is the only path in (Γ, c) starting at x; in particular, no arrow starts at x. Hence $\Gamma_1 = \emptyset$. Since we assume $\mathcal{W}$ irreducible, Γ is connected and must consist of the single vertex x. Thus $S = \{s\}$ and $|\mathcal{W}| = 2$. □

4. Reduced c-admissible words

We quote [**BB05**, formula (1.19), p. 26, and Definition 3.1.1(ii)].

DEFINITION 4.1. The set of *right associated reflections* for $w \in \mathcal{W}$ is $T_R(w) = \{t \in T \mid \ell(wt) < \ell(w)\}$. For $u, v \in \mathcal{W}$, set $u \leq_L v$ if and only if $v = s_k s_{k-1} \dots s_1 u$, for some $s_i \in S$ satisfying $\ell(s_i s_{i-1} \dots s_1 u) = \ell(u) + i$, $0 \leq i \leq k$. The binary relation $\leq_L$ is the *left weak order* on $\mathcal{W}$.

PROPOSITION 4.2. *A root $\alpha > 0$ is c-preprojective (resp. c-projective) if and only if $s_\alpha \in T_R(w)$, where $w = \rho(X)$ for some $X \in \mathfrak{M}(c)$ (resp. multiplicity-free $X \in \mathfrak{M}(c)$).*

PROOF. By [**Hum90**, Proposition 5.7] $\ell(ws_\alpha) < \ell(w)$ if and only if $w(\alpha) < 0$. □

We relate the partial order $\preceq$ on $\mathfrak{M}(c)$ to the left weak order on the set of c-admissible elements of $\mathcal{W}$.

LEMMA 4.3. *Let $X, Y \in \mathfrak{M}$.*

(a) *If $X \preceq Y$ and the word $\rho(Y)$ is reduced, then the word $\rho(X)$ is reduced and $\rho(X) \underset{L}{\leq} \rho(Y)$.*
(b) *For any $\alpha \in \mathcal{P}(c)$ and $Y \in \mathfrak{M}(c)$, if $\rho(W_\alpha) \underset{L}{\leq} \rho(Y)$ then $W_\alpha \preceq Y$.*

PROOF. (a) The proof is straightforward.

(b) Since $\rho(W_\alpha) \underset{L}{\leq} \rho(Y)$, [**BB05**, Proposition 3.1.3] says that $T_R(\rho(W_\alpha)) \subset T_R(\rho(Y))$. Since $\rho(W_\alpha)\alpha < 0$, [**Hum90**, Proposition 5.7] says that $\ell(\rho(W_\alpha)s_\alpha) < \ell(\rho(W_\alpha))$ whence $s_\alpha \in T_R(\rho(W_\alpha))$. Then $s_\alpha \in T_R(\rho(Y))$ whence $\ell(\rho(Y)s_\alpha) < \ell(\rho(Y))$ so that $\rho(Y)\alpha < 0$. By Theorem 2.12(a), $W_\alpha \preceq Y$. □

THEOREM 4.4. *Let $\Psi = \{\alpha_1, \dots, \alpha_m\}$ be a finite subset of $\mathcal{P}(c)$.*

(a) *The word* $\rho(W_\Psi)$ *is reduced. In particular, for all* $\alpha \in \mathcal{P}(c)$, *the word* $\rho(W_\alpha)$ *is reduced.*

(b) *If* $\alpha, \beta \in \mathcal{P}(c)$, *then* $W_\alpha \preceq W_\beta$ *if and only if* $\rho(W_\alpha) \underset{L}{\leq} \rho(W_\beta)$.

(c) *Assume* $W_\Psi = W_{\alpha_1} \vee \cdots \vee W_{\alpha_m}$. *For all* $Y \in \mathfrak{M}(c)$, *if* $\rho(W_\Psi) \underset{L}{\leq} \rho(Y)$ *then* $W_\Psi \preceq Y$.

(d) *If* $\Psi, \Theta \in \mathfrak{i}(c)$, *then* $W_\Psi \preceq W_\Theta$ *if and only if* $\rho(W_\Psi) \underset{L}{\leq} \rho(W_\Theta)$.

PROOF. (a) If $\mathcal{W}$ is infinite, the word $\rho(X)$ is reduced for all $X \in \mathfrak{M}(c)$ by [**Spe09**, Theorem 2]. Since $W_\Psi \in \mathfrak{M}(c)$, the statement holds.

Suppose $\mathcal{W}$ is finite. By [**Spe09**, Theorem 3], the element w_0 of maximal length satisfies $w_0 = \rho(U)$, where $U \in \mathfrak{M}(c)$ and the word $\rho(U)$ is reduced. Since $U \in \mathfrak{N}(\Psi)$ then $W_\Psi \preceq U$. By Lemma 4.3(a), the word $\rho(W_\Psi)$ is reduced.

(b) This follows from (a) and Lemma 4.3.

(c) For each i we have $W_{\alpha_i} \preceq W_\Psi$, so Lemma 4.3(a) gives $\rho(W_{\alpha_i}) \underset{L}{\leq} \rho(W_\Psi)$ in view of (a). Then $\rho(W_{\alpha_i}) \underset{L}{\leq} \rho(Y)$ by transitivity, so Lemma 4.3(b) gives $W_{\alpha_i} \preceq Y$. Since i is arbitrary, $W_\Psi \preceq Y$ by the property of join.

(d) The necessity follows from Lemma 4.3(a) because $\rho(W_\Theta)$ is reduced according to (a). The sufficiency is a special case of (c). □

LEMMA 4.5. *Let* $X = x_l \dots x_1$ *belong to* $\mathfrak{P}(c)$, $l > 1$, *and set* $Y = x_l \dots x_2$. *If* $Y = W_\beta$ *for some* $\beta \in \mathcal{P}(\rho(x_1)c\rho(x_1))$ *where* $\beta \neq \alpha_{\rho(x_1)}$, *then* $\alpha = \rho(x_1)\beta \in \mathcal{P}(c)$ *and* $X = W_\alpha$.

PROOF. Lemma 1.4(b) and Remark 2.4(b) say that $Y \in \mathfrak{M}(\rho(x_1)c\rho(x_1))$. Since $\beta > 0$ and $\beta \neq \alpha_{\rho(x_1)}$, [**Hum90**, Proposition 5.6(a)] says that $\alpha > 0$. By Theorem 2.12(c), $\beta = \rho(Y^T)(-\alpha_{\rho(x_l)})$ whence $\rho(X)\alpha = \rho(Y)\beta = -\alpha_{\rho(x_l)} < 0$. Therefore $\alpha \in \mathcal{P}(c)$ and Theorem 2.12(a) says that $W_\alpha \preceq X$.

If $x_1 \in \operatorname{Supp} W_\alpha$, then $x_1 \preceq W_\alpha$ by Lemma 1.4(c) because x_1 is a sink in (Γ, c). Then $W_\alpha = U x_1$, for some $U \in \mathfrak{M}(\rho(x_1)c\rho(x_1))$, whence $\rho(U)\beta = \rho(U)[\rho(x_1)\alpha] = \rho(W_\alpha)\alpha < 0$. By Theorem 2.12(a), $Y \preceq U$ whence $X = Yx_1 \preceq Ux_1 = W_\alpha$. Thus $X = W_\alpha$ as claimed.

If $x_1 \notin \operatorname{Supp} W_\alpha$, then $x_1 \notin \operatorname{Supp} \alpha$ by Remark 2.7(a), and no edge of Γ joins x_1 with a vertex of $\operatorname{Supp} W_\alpha$ because x_1 is a sink, while $\operatorname{Supp} W_\alpha$ is a filter of (Γ_0, c) by Proposition 1.8(a). Hence $W_\alpha x_1 = x_1 W_\alpha \in \mathfrak{M}(c)$ so that $W_\alpha \in \mathfrak{M}(\rho(x_1)c\rho(x_1))$, and $W_\alpha x_1 \preceq X = Yx_1$ by Proposition 1.6. By Corollary 1.7(b), $W_\alpha \preceq Y$ and we have

$$\rho(W_\alpha)\beta = \rho(W_\alpha)[\rho(x_1)\alpha] = \rho(x_1)[\rho(W_\alpha)\alpha] < 0$$

whence $Y \preceq W_\alpha$ so that $Y = W_\alpha$. Thus $x_1 \notin \operatorname{Supp} Y = \operatorname{Supp} W_\alpha$, and no edge of Γ joins x_1 with a vertex of $\operatorname{Supp} Y$ as we noted above. Since $\operatorname{Supp} X = \operatorname{Supp} Y \cup \{x_1\}$, the full subgraph of Γ determined by $\operatorname{Supp} X$ is disconnected, which contradicts the assumption that $X \in \mathfrak{P}(c)$. □

DEFINITION 4.6. An element $w \in \mathcal{W}$ has a *c-admissible* (resp. *c-principal*) *reduced expression* if $w = \rho(X)$, where the word $\rho(X)$ is reduced and $X \in \mathfrak{M}(c)$ (resp. $X \in \mathfrak{P}(c)$). For any $\mathfrak{A} \subset \mathfrak{M}$, denote by $\operatorname{Red} \mathfrak{A}$ the set of elements $Z \in \mathfrak{A}$ for which the word $\rho(Z)$ is reduced.

We now characterize the words in $\operatorname{Red} \mathfrak{P}(c)$. Recall that a map $f : P \to Q$ of posets is *order-preserving* if $x \leq y$ implies $f(x) \leq f(y)$, and f is *order-reflecting* if

$f(x) \leq f(y)$ implies $x \leq y$. A map is *order-embedding* if it is both order-preserving and order-reflecting.

THEOREM 4.7. (a) *For any $X \in \mathfrak{P}(c)$, the word $\rho(X)$ is reduced if and only if $X = W_\alpha$ for some $\alpha \in \mathcal{P}(c)$. Hence* $\operatorname{Red}\mathfrak{P}(c)$ *is the image of the map $\nu : \mathcal{P}(c) \to \mathfrak{P}(c)$, and the map $\nu : \mathcal{P}(c) \to \operatorname{Red}\mathfrak{P}(c)$ is bijective.*

(b) *The map $\rho : \operatorname{Red}\mathfrak{P}(c) \to \mathcal{W}$ is order-embedding, hence, injective.*

(c) *The maps of (a) and (b) are bijections between $\mathcal{P}(c)$,* $\operatorname{Red}\mathfrak{P}(c)$, *and the set of elements of $\mathcal{W}$ having a c-principal reduced expression. If $\mathcal{W}$ is finite, $\mathcal{P}(c) = \Phi^+$.*

PROOF. (a) The sufficiency is Theorem 4.4(a). Let $X = x_l \dots x_1$ belong to $\mathfrak{P}(c)$ and suppose that $\rho(X)$ is reduced. We prove the necessity by induction on $l = \ell(X)$.

If $l = 1$ then $X = x_1 = W_{\alpha_{\rho(x_1)}}$ where $\alpha_{\rho(x_1)} \in \mathcal{P}(c)$ because x_1 is a sink. Let $l > 1$ and suppose that, for all Coxeter elements $d \in \mathcal{W}$, the statement holds for all words in $\mathfrak{P}(d)$ of length $< l$. By [**KP07**, Proposition 4.6], see also [**KT08**, Proposition 2.5], $Y = x_l \dots x_2$ belongs to $\mathfrak{P}\left(\rho(x_1)c\rho(x_1)\right)$. Since $\rho(X)$ is reduced, $\rho(Y)$ is reduced by Lemma 4.3(a). By the inductive hypothesis, $Y = W_\beta$ for some $\beta \in \mathcal{P}\left(\rho(x_1)c\rho(x_1)\right)$, so Theorem 2.12(c) says that $\beta = \rho\left(Y^T\right)\left(-\alpha_{\rho(x_l)}\right)$. In view of [**Hum90**, Theorem 5.4],

$$\rho(x_1)\beta = \rho(x_1)\rho\left(Y^T\right)\left(-\alpha_{\rho(x_l)}\right) = \rho\left(X^T\right)\left(-\alpha_{\rho(x_l)}\right) = \rho(x_1 \dots x_{l-1})\alpha_{\rho(x_l)} > 0$$

because the word $\rho\left(X^T\right)$ is reduced. Since $\beta \neq \alpha_{\rho(x_1)}$ by [**Hum90**, Proposition 5.6(a)], Lemma 4.5 says that $X = W_\alpha$ for some $\alpha \in \mathcal{P}(c)$.

We have proved that the map $\nu : \mathcal{P}(c) \to \operatorname{Red}\mathfrak{P}(c)$ is surjective. It is injective by Theorem 2.12(d).

(b) This is an immediate consequence of (a) and Theorem 4.4(b).

(c) This follows from (a) and (b). If $\mathcal{W}$ is finite, Theorem 3.1 says that $\mathcal{P}(c) = \Phi^+$. □

We finish with a characterization of the words in $\operatorname{Red}\mathfrak{M}(c)$.

THEOREM 4.8. (a) *For any $X \in \mathfrak{M}(c)$, the word $\rho(X)$ is reduced if and only if $X = W_\Psi$ for some $\Psi \in \mathfrak{i}(c)$. Hence* $\operatorname{Red}\mathfrak{M}(c)$ *is the image of the map $\xi : \mathfrak{i}(c) \to \mathfrak{M}(c)$, and the map $\xi : \mathfrak{i}(c) \to \operatorname{Red}\mathfrak{M}(c)$ is bijective.*

(b) *The map $\rho : \operatorname{Red}\mathfrak{M}(c) \to \mathcal{W}$ is order-embedding, hence, injective.*

(c) *The maps of (a) and (b) are bijections between $\mathfrak{i}(c)$,* $\operatorname{Red}\mathfrak{M}(c)$, *and the set of elements of $\mathcal{W}$ having a c-admissible reduced expression.*

PROOF. (a) The sufficiency is Theorem 4.4(a). By Proposition 1.21, $X = X_1 \vee \cdots \vee X_m$ where $\{X_1, \dots, X_m\}$ is an independent subset of $\mathfrak{P}(c)$. Suppose $\rho(X)$ is reduced. By Lemma 4.3(a), the word $\rho(X_j)$ is reduced for all j, so Theorem 4.7(a) says that $X_j = W_{\alpha_j}$ for some $\alpha_j \in \mathcal{P}(c)$. By Definition 2.15, $\Psi = \{\alpha_1, \dots, \alpha_m\}$ is an independent subset of $\mathcal{P}(c)$, so Proposition 2.16 says that $X = W_\Psi$ where Ψ is uniquely determined.

(b) This is an immediate consequence of (a) and Theorem 4.4(d).

(c) This follows from (a) and (b). □

REMARK 4.9. For any $\Theta \in \mathfrak{f}(c)$, Theorem 2.12(a) gives the word $W_\Theta \in \mathfrak{M}(c)$. By Proposition 1.21, there exists an independent subset $\{X_1, \dots, X_m\}$ of $\mathfrak{P}(c)$ for which $W_\Theta = X_1 \vee \cdots \vee X_m$ and the X_i are unique up to permutation. Since the word $\rho(W_\Theta)$ is reduced by Theorem 4.4(a), the word $\rho(X_i)$ is reduced for all i by Lemma 4.3(a), and Theorem 4.7(a) says that $X_i = W_{\alpha_i}$ for a unique $\alpha_i \in \mathcal{P}(c)$. By Definition 2.15, $\Psi = \{\alpha_1, \ldots, \alpha_m\}$ is an independent subset of $\mathcal{P}(c)$, and Proposition 2.16(a) says that $W_\Theta = W_\Psi$, while Proposition 2.16(b) says that Ψ is uniquely determined. Thus, for any $\Theta \in \mathfrak{f}(c)$ there exists a unique $\Psi \in \mathfrak{i}(c)$ satisfying $W_\Theta = W_\Psi$. One may view Ψ as an "approximation" of Θ.

References

[AIRT12] Claire Amiot, Osamu Iyama, Idun Reiten, and Gordana Todorov, *Preprojective algebras and c-sortable words*, Proc. Lond. Math. Soc. (3) **104** (2012), no. 3, 513–539, DOI 10.1112/plms/pdr020. MR2900235

[ARS97] Maurice Auslander, Idun Reiten, and Sverre O. Smalø, *Representation theory of Artin algebras*, Cambridge Studies in Advanced Mathematics, vol. 36, Cambridge University Press, Cambridge, 1997. Corrected reprint of the 1995 original. MR1476671

[BB05] Anders Björner and Francesco Brenti, *Combinatorics of Coxeter groups*, Graduate Texts in Mathematics, vol. 231, Springer, New York, 2005. MR2133266

[BGP73] I. N. Bernšteĭn, I. M. Gel′fand, and V. A. Ponomarev, *Coxeter functors, and Gabriel's theorem* (Russian), Uspehi Mat. Nauk **28** (1973), no. 2(170), 19–33. MR0393065

[Bri98] Brigitte Brink, *The set of dominance-minimal roots*, J. Algebra **206** (1998), no. 2, 371–412, DOI 10.1006/jabr.1997.7418. MR1637139

[CF69] P. Cartier and D. Foata, *Problèmes combinatoires de commutation et réarrangements* (French), Lecture Notes in Mathematics, No. 85, Springer-Verlag, Berlin-New York, 1969. MR0239978

[How82] Robert B. Howlett, *Coxeter groups and M-matrices*, Bull. London Math. Soc. **14** (1982), no. 2, 137–141, DOI 10.1112/blms/14.2.137. MR647197

[Hum90] James E. Humphreys, *Reflection groups and Coxeter groups*, Cambridge Studies in Advanced Mathematics, vol. 29, Cambridge University Press, Cambridge, 1990. MR1066460

[IS10] Kiyoshi Igusa and Ralf Schiffler, *Exceptional sequences and clusters*, J. Algebra **323** (2010), no. 8, 2183–2202, DOI 10.1016/j.jalgebra.2010.02.003. MR2596373

[IR11] Osamu Iyama and Idun Reiten, *2-Auslander algebras associated with reduced words in Coxeter groups*, Int. Math. Res. Not. IMRN **8** (2011), 1782–1803, DOI 10.1093/imrn/rnq101. MR2806521

[Kac80] V. G. Kac, *Infinite root systems, representations of graphs and invariant theory*, Invent. Math. **56** (1980), no. 1, 57–92, DOI 10.1007/BF01403155. MR557581

[KP07] Mark Kleiner and Allen Pelley, *Admissible sequences, preprojective representations of quivers, and reduced words in the Weyl group of a Kac-Moody algebra*, Int. Math. Res. Not. IMRN **4** (2007), Art. ID rnm013, 28, DOI 10.1093/imrn/rnm013. MR2338197

[KT05] Mark Kleiner and Helene R. Tyler, *Admissible sequences and the preprojective component of a quiver*, Adv. Math. **192** (2005), no. 2, 376–402, DOI 10.1016/j.aim.2004.04.006. MR2128704

[KT08] Mark Kleiner and Helene R. Tyler, *Sequences of reflection functors and the preprojective component of a valued quiver*, J. Pure Appl. Algebra **212** (2008), no. 4, 718–726, DOI 10.1016/j.jpaa.2007.06.014. MR2363487

[ORT15] Steffen Oppermann, Idun Reiten, and Hugh Thomas, *Quotient closed subcategories of quiver representations*, Compos. Math. **151** (2015), no. 3, 568–602, DOI 10.1112/S0010437X14007623. MR3320572

[Spe09] David E. Speyer, *Powers of Coxeter elements in infinite groups are reduced*, Proc. Amer. Math. Soc. **137** (2009), no. 4, 1295–1302, DOI 10.1090/S0002-9939-08-09638-X. MR2465651

Department of Mathematics, Syracuse University, 130 Sims Drive, Syracuse, New York 13244

Email address: mkleiner@syr.edu

Contemporary Mathematics
Volume **769**, 2021
https://doi.org/10.1090/conm/769/15415

Approximable triangulated categories

Amnon Neeman

ABSTRACT. In this survey we present the relatively new concept of *approximable triangulated categories*. We will show that the definition is natural, that it leads to powerful new results, and that it throws new light on old, familiar objects.

CONTENTS

1. Introduction

In this survey there is one major, recent technical tool—we present the concept of *approximable triangulated categories*. And then we will sketch some results, from the last few years, showing that this tool is useful.

2020 *Mathematics Subject Classification.* Primary 18E30, Secondary 18G55.

Key words and phrases. Derived categories, t–structures, homotopy limits.

The research was partly supported by the Australian Research Council.

The definition of approximable triangulated categories relies on the following building blocks: compact generators in triangulated categories and t-structures. We have an extensive background section introducing these—we recommend that beginners skip the remainder of the introduction and move on to Section 2. The introduction of an article is normally the author's attempt to persuade the expert to read on—hence it tends to assume some familiarity with the existing theory, the expert will not want to be bored with stuff she already knows, she will want to see if this article contains anything new and interesting. The introduction to an article is often more intimidating than the body of the manuscript.

In addition to Section 2 the beginners might wish to look at the three appendices, which were written to answer questions from students to whom the material was new. Especially relevant is Appendix C, where we draw up a dictionary between our approach and the more standard one in the literature—in the interest of efficiency we depart from the usual way to introduce derived categories. An interested beginner, who wants to explore this further by looking elsewhere, is encouraged to consult this dictionary.

Back to the experts: we plan to discuss approximability in triangulated categories and its applications, and we begin with a heuristic explanation of what approximability is all about.

Any t-structure on the triangulated category can be used to define a "metric": two objects are close to each other if they agree up to a small "difference". More precisely: the objects $x, y \in \mathcal{T}$ are close to each other if there exists in $\mathcal{T}$ a triangle $x \longrightarrow y \longrightarrow z \longrightarrow$ with $z \in \mathcal{T}^{\leq -n}$ for some large n.[1] We declare that, the larger the integer n, the closer the points x and y. In the world of metric spaces we are accustomed to the notion of equivalent metrics, and this naturally leads to the concept of equivalent t-structures.

We are also accustomed to expressing points in a metric space as limits of Cauchy sequences of simpler, more accessible points. For example the Taylor series approximates a function by polynomials, and the Fourier series approximates a function by finite sums of exponentials. There is a triangulated category version, we will explain it more fully in the body of the article. For the introduction the discussion below will give the gist of the construction, albeit a little vaguely and with details missing.

DISCUSSION 1.1. *If we plan to approximate objects of the triangulated category $\mathcal{T}$, by Cauchy sequences of simpler objects, then we first need to measure what we mean by "simplicity"—returning to the analogy of the previous paragraphs, we need to declare what will be the triangulated category replacement for the polynomials forming the partial sums in a Taylor series. In doing this we will slightly modify an idea due to Bondal and Van den Bergh* [**8**]. *We will start with a compact generator G for the triangulated category and, for each integer $n > 0$, we will define two classes of objects $\langle G\rangle_n^{[-n,n]} \subset \overline{\langle G\rangle}_n^{[-n,n]}$. These will be the objects obtainable from*

[1]The category $\mathcal{T}$ isn't quite a metric space, the obvious "metric" isn't symmetric. There is a rich literature on asymmetric metrics following Lawvere [**26, 27**], but the emphasis there goes in entirely different directions—it's the triangle inequality in Lawvere's asymmetric metrics that is emphasized, see for example Leinster [**28**]. Our metrics might be asymmetric, but they are decidedly non-archimedean—hence the triangle inequality is dumb, every triangle is isosceles.

G in n allowable steps—the difference between $\langle G\rangle_n^{[-n,n]}$ and $\overline{\langle G\rangle}_n^{[-n,n]}$ is that for the smaller $\langle G\rangle_n^{[-n,n]}$ there are fewer operations allowed.

Returning to the analogy with Taylor series: so far we have explained what will be our replacement for the polynomials of degree $\leq n$. We have already indicated the "metric" we plan to work with, it is the one determined by whatever t–structure we end up choosing. So it becomes interesting to know which objects in the triangulated category have Taylor series "converging" to them. And now we come to the (somewhat imprecise) definition: the triangulated category $\mathcal{T}$ is declared to be approximable *if it has coproducts, and there exist in $\mathcal{T}$*

(i) *a compact generator G*
(ii) *a t–structure $(\mathcal{T}^{\leq 0}, \mathcal{T}^{\geq 0})$*

and these t–structure and generator can be chosen to satisfy

(iii) *For some $n > 0$ we have $G[n] \in \mathcal{T}^{\leq 0}$ and* $\operatorname{Hom}\left(G[-n]\ ,\ \mathcal{T}^{\leq 0}\right) = 0$.
(iv) *In the metric induced by the t–structure $(\mathcal{T}^{\leq 0}, \mathcal{T}^{\geq 0})$ of (ii), every object in $\mathcal{T}^{\leq 0}$ can be expressed as the limit of a sequence whose terms belong to $\cup_n \overline{\langle G\rangle}_n^{[-n,n]}$.*

Remark 1.2. It is a formal consequence of the definition that an approximable triangulated category is complete with respect to the metric of Discussion 1.1(iv)—any Cauchy sequence converges. Also: by definition, if $\mathcal{T}$ is approximable then $\mathcal{T}^{\leq 0}$ is contained in the closure of $\cup_n \overline{\langle G\rangle}_n^{[-n,n]}$ with respect to the metric. It turns out that the closure is not much larger: it is nothing other than $\mathcal{T}^- = \cup_n \mathcal{T}^{\leq n}$.

One could also wonder what the closure of $\cup_n \langle G\rangle_n^{[-n,n]}$ might be—we will return to this later, it turns out to be a subcategory which, for many $\mathcal{T}$, has been studied extensively in the classical literature.

Now that we have a rough idea what approximability means, the first question we might ask ourselves is "Is the theory nonempty, are there any examples?"

Example 1.3. *It turns out there are plenty of examples. If R is any ring then $\mathbf{D}(R)$, the unbounded derived category of complexes of left R–modules, is an example. So is the homotopy category of spectra, and so is the category $\mathbf{D}(R\text{–Mod})$, provided R is a dga such that $H^i(R) = 0$ for all $i > 0$. Here $\mathbf{D}(R\text{–Mod})$ stands for the derived category whose objects are all left dg R–modules.*

All of these are easy examples. It is a nontrivial theorem that, when X is a quasicompact, separated scheme, the category $\mathbf{D_{qc}}(X)$ is approximable. Here $\mathbf{D_{qc}}(X)$ means the derived category, whose objects are cochain complexes of $\mathcal{O}_X$–modules with quasicoherent cohomology. It is also a nontrivial theorem that, under reasonable hypotheses, the recollement of two approximable categories is approximable.

Application 1.4. *We have introduced a new gadget—namely approximable triangulated categories—and mentioned that there are plenty of interesting examples out there. But the skeptical reader will want to know what use this new toy might have: are there applications? Do we learn anything new, about the familiar old categories of Example* 1.3*, because we now know them to be approximable?*

The answer is Yes. We list below five results we were recently able to prove, using the fact that $\mathbf{D_{qc}}(X)$ is approximable.

(i) *Suppose X is a quasicompact, separated scheme. Then the category $\mathbf{D}^{\mathrm{perf}}(X)$ is strongly generated, in the sense of Bondal and Van den*

Bergh [**8**], *if and only if X can be covered by open affine subsets $U_i = \mathrm{Spec}(R_i)$ with each R_i of finite global dimension.*

(ii) *Suppose X is a finite-dimensional, noetherian, separated scheme, and assume further that every closed, reduced, irreducible subscheme of X has a regular alteration. Then the category $\mathbf{D}^b_{\mathrm{coh}}(X)$ is strongly generated.*

(iii) *Suppose X is a scheme proper over a noetherian ring R, and let $\mathcal{Y} : \mathbf{D}^b_{\mathrm{coh}}(X) \longrightarrow \mathrm{Hom}_R\big[\mathbf{D}^{\mathrm{perf}}(X)^{\mathrm{op}}\ ,\ R\text{–}\mathrm{Mod}\big]$ be the Yoneda map. That is: $\mathcal{Y}$ is the map taking an object $B \in \mathbf{D}^b_{\mathrm{coh}}(X)$ to the functor $\mathcal{Y}(B) = \mathrm{Hom}(-, B)$, viewed as an R–linear homological functor $\mathbf{D}^{\mathrm{perf}}(X)^{\mathrm{op}} \longrightarrow R\text{–}\mathrm{Mod}$.*

Then $\mathcal{Y}$ is fully faithful, and the essential image of $\mathcal{Y}$ are the finite homological functors. A functor $H : \mathbf{D}^{\mathrm{perf}}(X)^{\mathrm{op}} \longrightarrow R\text{–}\mathrm{Mod}$ is finite *if, for any object $A \in \mathbf{D}^{\mathrm{perf}}(X)$, the R–modules $H\big(A[i]\big)$ are all finite and all but finitely many of them vanish.*

(iv) *Suppose X is a finite-dimensional scheme proper over a noetherian ring R, and assume further that every closed, reduced, irreducible subscheme of X has a regular alteration. Let $\widetilde{\mathcal{Y}} : \mathbf{D}^{\mathrm{perf}}(X)^{\mathrm{op}} \longrightarrow \mathrm{Hom}_R\big[\mathbf{D}^b_{\mathrm{coh}}(X)\ ,\ R\text{–}\mathrm{Mod}\big]$ be the Yoneda map. That is: $\widetilde{\mathcal{Y}}$ is the map taking an object $A \in \mathbf{D}^{\mathrm{perf}}(X)$ to the functor $\widetilde{\mathcal{Y}}(A) = \mathrm{Hom}(A, -)$, viewed as an R–linear homological functor $\mathbf{D}^b_{\mathrm{coh}}(X) \longrightarrow R\text{–}\mathrm{Mod}$.*

Then $\widetilde{\mathcal{Y}}$ is fully faithful, and the essential image of $\widetilde{\mathcal{Y}}$ are the finite homological functors.

(v) *Suppose X is a noetherian, separated scheme. Then the categories $\mathbf{D}^{\mathrm{perf}}(X)$ and $\mathbf{D}^b_{\mathrm{coh}}(X)$ determine each other. More explicitly: there is a recipe which takes a triangulated category $\mathcal{S}$ as input, and out of it cooks up another triangulated category denoted $\mathfrak{S}(\mathcal{S})$. If we apply this recipe to $\mathbf{D}^{\mathrm{perf}}(X)$ what comes out is $\mathbf{D}^b_{\mathrm{coh}}(X)$, and if we apply it to $\big[\mathbf{D}^b_{\mathrm{coh}}(X)\big]^{\mathrm{op}}$ the output is $\big[\mathbf{D}^{\mathrm{perf}}(X)\big]^{\mathrm{op}}$.*

In the body of the article we will say more about the theorems—for example we will remind the reader what it means for a triangulated category $\mathcal{T}$ to be "strongly generated". For now we note only that (i), (ii), (iii), (iv) and (v) above represent sharp improvements over the existing literature. More precisely

(vi) *There were versions of (i) and (ii) known before approximability, but they all assumed equal characteristic—the reader can find a sample of the known results in Bondal and Van den Bergh* [**8**, Theorem 3.1.4], *Iyengar and Takahashi* [**21**, Corollary 7.2], *Orlov* [**37**, Theorem 3.27] *and Rouquier* [**39**, Theorem 7.38].

(vii) *The only known versions of (iii) and (iv), prior to approximability, assumed that R is a field. See Bondal and Van den Bergh* [**8**, Theorem A.1] *for (iii), and Rouquier* [**39**, Corollary 7.51(ii)] *for (iv).*

(viii) *The only known versions of (v) prior to approximability assumed either that X is affine, see Rickard* [**38**, Theorem 6.4], *or that X is projective over a field, see Rouquier* [**39**, Remark 7.50].

The definition of approximability is the assumption that there exist a t–structure $(\mathcal{T}^{\leq 0}, \mathcal{T}^{\geq 0})$ and a compact generator $G \in \mathcal{T}$ with some properties. It becomes natural to wonder how free we are in the choice of t–structure and compact generator. This leads to a string of surprising results.

FACTS 1.5. *Let $\mathcal{T}$ be a triangulated category with coproducts, and assume it has a compact generator G. Then the following can be proved.*

(i) *There exists a preferred equivalence class of t–structures in $\mathcal{T}$. Here two t–structures are declared equivalent if they induce equivalent metrics.*

(ii) *Let us choose in $\mathcal{T}$ a compact generator G and a t–structure $(\mathcal{T}^{\leq 0}, \mathcal{T}^{\geq 0})$, and assume they satisfy the conditions in Discussion 1.1 (iii) and (iv)—that is: the pair G and $(\mathcal{T}^{\leq 0}, \mathcal{T}^{\geq 0})$ pass the test for checking the approximability of $\mathcal{T}$.*

Then it's automatic that $(\mathcal{T}^{\leq 0}, \mathcal{T}^{\geq 0})$ belongs to the preferred equivalence class of t–structures. Hence the metric defined by any t–structure $(\mathcal{T}^{\leq 0}, \mathcal{T}^{\geq 0})$, which can be used to test for approximability, is unique up to equivalence.

(iii) *Suppose $\mathcal{T}$ is approximable. Then any t–structure $(\mathcal{T}^{\leq 0}, \mathcal{T}^{\geq 0})$ in the preferred equivalence class and any compact generator G satisfy the conditions in Discussion 1.1 (iii) and (iv).*

Thus approximability is robust; it doesn't really matter which t–structure and compact generator one chooses, as long as the t–structure belongs to the preferred equivalence class. Furthermore the categories

$$\mathcal{T}^- = \cup_n \mathcal{T}^{\leq n}, \qquad \mathcal{T}^+ = \cup_n \mathcal{T}^{\geq -n}, \qquad \mathcal{T}^b = \mathcal{T}^- \cap \mathcal{T}^+$$

turn out to be intrinsic. They depend only on $\mathcal{T}$, not on the particular representative $(\mathcal{T}^{\leq 0}, \mathcal{T}^{\geq 0})$ in the preferred equivalence class.

Now that we know the metric is intrinsic (up to equivalence), it makes sense to return to the question raised in Remark 1.2. What is the closure of $\cup_n \langle G\rangle_n^{[-n,n]}$? In view of the above we should not be surprised to learn

(iv) *Define the category $\mathcal{T}_c^-$ to be the closure in $\mathcal{T}$ of $\cup_n \langle G\rangle_n^{[-n,n]}$. This category is intrinsic, it does not depend on the choice of compact generator G. And it follows that the category $\mathcal{T}_c^b = \mathcal{T}_c^- \cap \mathcal{T}^b$ must also be intrinsic.*

REMARK 1.6. It becomes interesting to figure out what these intrinsic subcategories are in the special cases of Example 1.3. Let us confine ourselves to just one case: assume $\mathcal{T} = \mathbf{D_{qc}}(X)$ with X a separated, noetherian scheme. In this special case one can prove:

(i) The standard t–structure is in the preferred equivalence class. Hence the categories $\mathcal{T}^-$, $\mathcal{T}^+$ and $\mathcal{T}^b$ have their usual meanings: that is $\mathcal{T}^- = \mathbf{D}^-_{\mathbf{qc}}(X)$, $\mathcal{T}^+ = \mathbf{D}^+_{\mathbf{qc}}(X)$ and $\mathcal{T}^b = \mathbf{D}^b_{\mathbf{qc}}(X)$.

(ii) The category $\mathcal{T}_c^-$ turns out to be $\mathbf{D}^-_{\mathrm{coh}}(X)$, hence the category $\mathcal{T}_c^b = \mathcal{T}_c^- \cap \mathcal{T}^b$ is nothing other than $\mathbf{D}^b_{\mathrm{coh}}(X)$.

It turns out that Applications 1.4 (iii), (iv) and (v) generalize greatly. The glorious, abstract versions for (iii) and (iv) go as follows. Let R be a noetherian ring, and let $\mathcal{T}$ be an R–linear, approximable triangulated category. Suppose there exists in $\mathcal{T}$ a compact generator G, such that $\mathrm{Hom}\big(G, G[n]\big)$ is a finite R–module for all $n \in \mathbb{Z}$. Consider the two functors

$$\mathcal{Y} : \mathcal{T}_c^- \longrightarrow \mathrm{Hom}_R\big([\mathcal{T}^c]^{\mathrm{op}}, R\text{–Mod}\big), \qquad \widetilde{\mathcal{Y}} : \big[\mathcal{T}_c^-\big]^{\mathrm{op}} \longrightarrow \mathrm{Hom}_R\big(\mathcal{T}_c^b, R\text{–Mod}\big)$$

defined by the formulas $\mathcal{Y}(B) = \mathrm{Hom}(-, B)$ and $\widetilde{\mathcal{Y}}(A) = \mathrm{Hom}(A, -)$. Note that, in these formulas, we permit all $A, B \in \mathcal{T}_c^-$. But the $(-)$ in the formula $\mathcal{Y}(B) =$

$\mathrm{Hom}(-,B)$ is assumed to belong to $\mathcal{T}^c$, whereas the $(-)$ in the formula $\widetilde{\mathcal{Y}}(A) = \mathrm{Hom}(A,-)$ must lie in $\mathcal{T}^b_c$. Now consider the following composites

$$\mathcal{T}^b_c \xhookrightarrow{\quad i \quad} \mathcal{T}^-_c \xrightarrow{\quad \mathcal{Y} \quad} \mathrm{Hom}_R\left([\mathcal{T}^c]^{\mathrm{op}}, R\text{–Mod}\right)$$

$$\left[\mathcal{T}^c\right]^{\mathrm{op}} \xhookrightarrow{\quad \widetilde{\imath} \quad} \left[\mathcal{T}^-_c\right]^{\mathrm{op}} \xrightarrow{\quad \widetilde{\mathcal{Y}} \quad} \mathrm{Hom}_R\left(\mathcal{T}^b_c, R\text{–Mod}\right)$$

where i and $\widetilde{\imath}$ are the obvious inclusions. We assert:

(iii) The functor $\mathcal{Y}$ is full, and the essential image consists of the locally finite homological functors. A functor $H : \left[\mathcal{T}^c\right]^{\mathrm{op}} \longrightarrow R$–Mod is *locally finite* if $H\big(A[n]\big)$ is a finite R–module for every $n \in \mathbb{Z}$, and vanishes if $n \ll 0$.

The composite $\mathcal{Y} \circ i$ is fully faithful, and the essential image consists of the finite homological functors.

(iv) Assume there exists an integer $N > 0$ and an object $G' \in \mathcal{T}^b_c$ with $\mathcal{T} = \overline{\langle G' \rangle}_N^{(-\infty,\infty)}$—this condition will be explained in the body of the paper, under the hypotheses placed on X in Application 1.4(iv) the condition is satisfied by $\mathcal{T} = \mathbf{D_{qc}}(X)$, this may be found in [**33**, Theorem 2.3].

Then the functor $\widetilde{\mathcal{Y}}$ is full, and the essential image consists of the locally finite homological functors. The composite $\widetilde{\mathcal{Y}} \circ \widetilde{\imath}$ is fully faithful, and the essential image consists of the finite homological functors.

As we have said, Application 1.4(v) also has a vast generalization, which goes as follows:

(v) With the notation as in Application 1.4(v) one has, for any approximable $\mathcal{T}$, a triangulated equivalence $\mathfrak{S}(\mathcal{T}^c) \cong \mathcal{T}^b_c$. If the triangulated category $\mathcal{T}$ is not only approximable but also *noetherian,* then one also has a triangulated equivalence $\mathfrak{S}\big(\left[\mathcal{T}^b_c\right]^{\mathrm{op}}\big) \cong \left[\mathcal{T}^c\right]^{\mathrm{op}}$.

The notion of noetherian triangulated categories in Remark 1.6(v) is new, and was inspired by the result. Noetherianness is a condition that seems natural, and guarantees that there will be plenty of objects in $\mathcal{T}^b_c$. Without some noetherian hypothesis, the only obvious object in $\mathcal{T}^b_c$ is zero.

2. Background

It's time to speak to the non-experts—the readers familiar with triangulated categories, compact generators and t–structures are advised to skip ahead to Section 3. In this section we will present a quick reminder of the three concepts in the sentence above. We plan to usually proceed from the concrete to the abstract: for most of the section we study first an example, actually four examples—all four examples will be derived categories $\mathbf{D}^{\mathfrak{C}'}_{\mathfrak{C}}(\mathcal{A})$, we list the four in Examaple 2.3—and only then do we move on to the general definitions. We should therefore begin by recalling what are the categories $\mathbf{D}^{\mathfrak{C}'}_{\mathfrak{C}}(\mathcal{A})$, first in generality that covers the four examples and more, and then narrowing down to the specific ones that will interest us.

EXAMPLE 2.1. *Let $\mathcal{A}$ be an abelian category. The derived category $\mathbf{D}^{\mathfrak{C}'}_{\mathfrak{C}}(\mathcal{A})$ is as follows:*

(i) *The objects are the cochain complexes in $\mathcal{A}$, that is diagrams in $\mathcal{A}$ of the form*

$$\cdots \longrightarrow A^{-2} \longrightarrow A^{-1} \longrightarrow A^{0} \longrightarrow A^{1} \longrightarrow A^{2} \longrightarrow \cdots$$

where the composites $A^i \longrightarrow A^{i+1} \longrightarrow A^{i+2}$ all vanish. The subscript $\mathfrak{C}$ and superscript $\mathfrak{C}'$ stand for conditions. We may choose not to allow all cochain complexes, when the mood strikes us we can capriciously impose any conditions on the cochain complexes that our heart desires—subject to the mild hypotheses that guarantee that the few operations we're about to perform take complexes satisfying the restrictions to complexes satisfying the restrictions.

(ii) *Cochain maps are morphisms in $\mathbf{D}^{\mathfrak{C}'}_{\mathfrak{C}}(\mathcal{A})$, that is any commutative diagram*

$$\begin{array}{ccccccccccc}
\cdots \longrightarrow & A^{-2} & \longrightarrow & A^{-1} & \longrightarrow & A^{0} & \longrightarrow & A^{1} & \longrightarrow & A^{2} & \longrightarrow \cdots \\
 & \downarrow & & \downarrow & & \downarrow & & \downarrow & & \downarrow & \\
\cdots \longrightarrow & B^{-2} & \longrightarrow & B^{-1} & \longrightarrow & B^{0} & \longrightarrow & B^{1} & \longrightarrow & B^{2} & \longrightarrow \cdots
\end{array}$$

is a morphism from the top to the bottom row—as long as the rows are cochain complexes satisfying the restrictions, that is objects in $\mathbf{D}^{\mathfrak{C}'}_{\mathfrak{C}}(\mathcal{A})$.

But then we formally invert the cohomology isomorphisms. In the literature the cohomology isomorphisms often go by the name "quasi-isomorphisms".

EXPLANATION 2.2. *Given a category $\mathcal{C}$ and a collection S of morphisms in $\mathcal{C}$, an old theorem of Gabriel and Zisman* [**15**] *tells us that there exists a functor $F : \mathcal{C} \longrightarrow S^{-1}\mathcal{C}$ so that*

(i) *If $s \in S \subset Mor(\mathcal{C})$ then $F(s)$ is invertible.*

(ii) *Any functor $F'' : \mathcal{C} \longrightarrow \mathcal{B}$, with $F''(S)$ contained in the isomorphisms of $\mathcal{B}$, factors uniquely as $\mathcal{C} \xrightarrow{F} S^{-1}\mathcal{C} \xrightarrow{F'} \mathcal{B}$.*

What we mean when we say that in $\mathbf{D}^{\mathfrak{C}'}_{\mathfrak{C}}(\mathcal{A})$ we "formally invert" the cohomology isomorphisms is: let $\mathcal{C}$ be the category with the same objects as $\mathbf{D}^{\mathfrak{C}'}_{\mathfrak{C}}(\mathcal{A})$ but where the morphisms are the cochain maps, and let S be the collection of cochain maps inducing cohomology isomorphisms. Then $\mathbf{D}^{\mathfrak{C}'}_{\mathfrak{C}}(\mathcal{A})$ is defined to be $S^{-1}\mathcal{C}$.

In principle categories of the form $S^{-1}\mathcal{C}$ can be dreadful—the morphisms are equivalence classes of composable strings, where each string is a sequence whose pieces are either morphisms in $\mathcal{C}$ or inverses of elements of S. The Hom-sets needn't be small, and in general it can be a nightmare to decide when two such strings are equivalent, meaning define the same morphism in $S^{-1}\mathcal{C}$. For categories like $\mathbf{D}^{\mathfrak{C}'}_{\mathfrak{C}}(\mathcal{A})$ the calculus of fractions happens not to be too bad, there is a literature dealing with it. The interested reader is referred to Hartshorne [**18**] *or Verdier* [**42**] *for the original presentations, or Gelfand and Manin* [**16**], *Kashiwara and Schapira* [**22**] *or Weibel* [**43**] *for more modern treatments.*

In this survey we skip the discussion of the calculus of fractions. This means that the reader will be asked to believe several computations along the way—when these occur there will be a footnote to the effect.

EXAMPLE 2.3. *In this survey, the key examples to keep in mind are:*

(i) *If R is a ring, $\mathbf{D}(R)$ will be our shorthand for $\mathbf{D}(R\text{–Mod})$; the abelian category $\mathcal{A}$ is the category of all left R–modules, and since there are no superscripts or subscripts decorating the $\mathbf{D}$ we impose no conditions. All cochain complexes of left R-modules are objects of $\mathbf{D}(R)$.*

Now let X be a scheme. The abelian category, in all three examples below, is the category of sheaves of $\mathcal{O}_X$–modules. It's customary to abbreviate what should be written $\mathbf{D}(\mathcal{O}_X\text{–Mod})$ to just $\mathbf{D}(X)$, and we will follow this custom.

But all three categories we will look at are decorated, there are restrictions. We list them below.

(ii) *The objects in $\mathbf{D_{qc}}(X)$ are cochain complexes of $\mathcal{O}_X$–modules, and the only condition we impose is that the cohomology sheaves must be quasicoherent.*

(iii) *The objects of $\mathbf{D}^{\text{perf}}(X)$ are the perfect complexes. A cochain complex of $\mathcal{O}_X$–modules is* perfect *if it is locally isomorphic to a bounded complex of vector bundles. More precisely: an object $P \in \mathbf{D_{qc}}(X)$ belongs to $\mathbf{D}^{\text{perf}}(X)$ if there exists an open cover of X of the form $X = \cup_i U_i$ such that, if $u_i : U_i \longrightarrow X$ is the inclusion, then the obvious functor $u_i^* : \mathbf{D_{qc}}(X) \longrightarrow \mathbf{D_{qc}}(U_i)$ takes $P \in \mathbf{D_{qc}}(X)$ to an object $u_i^*(P) \in \mathbf{D_{qc}}(U_i)$ which is isomorphic in $\mathbf{D_{qc}}(U_i)$ to a bounded complex of vector bundles.*

(iv) *Assume X is noetherian. The objects of $\mathbf{D}^b_{\text{coh}}(X)$ are the complexes of $\mathcal{O}_X$–modules with coherent cohomology—as indicated by the subscript—and this cohomology vanishes in all but finitely many degrees, the superscipt b stands for "bounded".*

REMARK 2.4. If we're going to be working with categories like $\mathbf{D}^{\mathfrak{C}'}_{\mathfrak{C}}(\mathcal{A})$, it is natural to wonder what useful structure they might have. The next definition spells out the answer.

The idea is simple enough: we started with the category $\mathcal{C}$ whose objects are the same as those of $\mathbf{D}^{\mathfrak{C}'}_{\mathfrak{C}}(\mathcal{A})$, but the morphisms were honest cochain maps. We then performed the construction of Explanation 2.2, formally inverting the class S of cohomology isomorphisms, to form $\mathbf{D}^{\mathfrak{C}'}_{\mathfrak{C}}(\mathcal{A}) = S^{-1}\mathcal{C}$, The information retained isn't much more than the cohomology of the complex. Ordinary homological algebra teaches us that there are really only two things you can do with cohomology:

(i) Shift the degrees.

(ii) Form the the long exact sequence in cohomology that comes from a short exact sequence of cochain complexes.

The structure of a *triangulated category,* formalized in Definition 2.5 (i) and (ii) below, encapsulates this: Definition 2.5(i) gives the shifting of degrees, while Definition 2.5(ii) is the abstract version of the long exact sequence in cohomology coming from a short exact sequence of cochain complexes. See Example 2.7 for more detail: we spell out the recipe that endows $\mathbf{D}^{\mathfrak{C}'}_{\mathfrak{C}}(\mathcal{A})$ with the structure of a triangulated category, and do so by steering as close as possible to the simple, motivating idea.

DEFINITION 2.5. *To give the additive category $\mathcal{T}$ the structure of a* triangulated category *we must:*

(i) *Specify an invertible additive endofunctor $\mathcal{T} \longrightarrow \mathcal{T}$. In this article we will denote it $[1]$ and have it act on the right: thus it takes the object X and the morphism f in $\mathcal{T}$ to $X[1]$ and $f[1]$, respectively.*

Before we continue the definition we set up

NOTATION 2.6. For the purpose of the current definition (Definition 2.5) we adopt the following convention. With $[1] : \mathcal{T} \longrightarrow \mathcal{T}$ the endofunctor of Definition 2.5(i), a *candidate triangle* is any three composable morphisms $X \xrightarrow{f} Y \xrightarrow{g}$

$Z \xrightarrow{h} X[1]$ in the category $\mathcal{T}$. The candidate triangles form a category, a *morphism* of candidate triangles is defined to be a commutative diagram in $\mathcal{T}$

$$\begin{array}{ccccccc} X & \xrightarrow{f} & Y & \xrightarrow{g} & Z & \xrightarrow{h} & X[1] \\ \downarrow{\scriptstyle u} & & \downarrow{\scriptstyle v} & & \downarrow{\scriptstyle w} & & \downarrow{\scriptstyle u[1]} \\ X' & \xrightarrow{f'} & Y' & \xrightarrow{g'} & Z' & \xrightarrow{h'} & X'[1] \end{array}$$

which we view as a morphism from the top to the bottom row. The composition of morphisms of candidate triangles is the obvious.

Continuation of Definition 2.5, now that the notation has been explained. In addition to the invertible endomorphism $[1] : \mathcal{T} \longrightarrow \mathcal{T}$ of (i) we must specify

(ii) *A full subcategory of the category of candidate triangles, whose objects will be called* exact triangles. *[In some parts of the literature they go by the name* distinguished triangles.*]*

For $\mathcal{T}$ to qualify as a triangulated category the data of (i) and (ii) above must satisfy the following axioms:

[TR1]: *Any candidate triangle isomorphic to an exact triangle is an exact triangle. For any object $X \in \mathcal{T}$ the diagram $0 \longrightarrow X \xrightarrow{\text{id}} X \longrightarrow 0$ is an exact triangle. Any morphism $f : X \longrightarrow Y$ may be completed to an exact triangle $X \xrightarrow{f} Y \xrightarrow{g} Z \xrightarrow{h} X[1]$.*

[TR2]: *Any rotation of an exact triangle is exact. That is: $X \xrightarrow{f} Y \xrightarrow{g} Z \xrightarrow{h} X[1]$ is an exact triangle if and only if $Y \xrightarrow{-g} Z \xrightarrow{-h} X[1] \xrightarrow{-f[1]} Y[1]$ is.*

[TR3+4]: *Given a commutative diagram, where the rows are exact triangles,*

$$\begin{array}{ccccccc} X & \xrightarrow{f} & Y & \xrightarrow{g} & Z & \xrightarrow{h} & X[1] \\ \downarrow{\scriptstyle u} & & \downarrow{\scriptstyle v} & & & & \\ X' & \xrightarrow{f'} & Y' & \xrightarrow{g'} & Z' & \xrightarrow{h'} & X'[1] \end{array}$$

we may complete it to a morphism of exact triangles, that is a commutative diagram

$$\begin{array}{ccccccc} X & \xrightarrow{f} & Y & \xrightarrow{g} & Z & \xrightarrow{h} & X[1] \\ \downarrow{\scriptstyle u} & & \downarrow{\scriptstyle v} & & \downarrow{\scriptstyle w} & & \downarrow{\scriptstyle u[1]} \\ X' & \xrightarrow{f'} & Y' & \xrightarrow{g'} & Z' & \xrightarrow{h'} & X'[1] \end{array}$$

Moreover: we can do it in such a way that

$$Y \oplus X' \xrightarrow{\begin{pmatrix} -g & 0 \\ v & f' \end{pmatrix}} Z \oplus Y' \xrightarrow{\begin{pmatrix} -h & 0 \\ w & g' \end{pmatrix}} X[1] \oplus Z' \xrightarrow{\begin{pmatrix} -f[1] & 0 \\ u[1] & h' \end{pmatrix}} Y[1] \oplus X'[1]$$

is an exact triangle.

Example 2.7. *We have asserted that the category $\mathbf{D}^{\mathfrak{C}'}_{\mathfrak{C}}(\mathcal{A})$ is triangulated. It is only fair to tell the reader what is the endofunctor $[1]:\mathbf{D}^{\mathfrak{C}'}_{\mathfrak{C}}(\mathcal{A})\longrightarrow\mathbf{D}^{\mathfrak{C}'}_{\mathfrak{C}}(\mathcal{A})$ and what are the exact triangles. The endofunctor $[1]$, called the shift or suspension, is easy: it takes the cochain complex A^*, that is the diagram*

$$\cdots \longrightarrow A^{-2} \xrightarrow{\partial^{-2}} A^{-1} \xrightarrow{\partial^{-1}} A^{0} \xrightarrow{\partial^{0}} A^{1} \xrightarrow{\partial^{1}} A^{2} \longrightarrow \cdots$$

to the cochain complex $\big(A[1]\big)^*$ below:

$$\cdots \longrightarrow A^{-1} \xrightarrow{-\partial^{-1}} A^{0} \xrightarrow{-\partial^{0}} A^{1} \xrightarrow{-\partial^{1}} A^{2} \xrightarrow{-\partial^{2}} A^{3} \longrightarrow \cdots$$

In words: we shift the complex to the left by one, that is $\big(A[1]\big)^n=A^{n+1}$, and the maps all change signs. This deals with objects.

If $f^*:A^*\longrightarrow B^*$ is a cochain map

$$\begin{array}{ccccccccccccc}
\cdots & \longrightarrow & A^{-2} & \xrightarrow{\partial_A^{-2}} & A^{-1} & \xrightarrow{\partial_A^{-1}} & A^{0} & \xrightarrow{\partial_A^{0}} & A^{1} & \xrightarrow{\partial_A^{1}} & A^{2} & \longrightarrow & \cdots \\
 & & \downarrow{\scriptstyle f^{-2}} & & \downarrow{\scriptstyle f^{-1}} & & \downarrow{\scriptstyle f^{0}} & & \downarrow{\scriptstyle f^{1}} & & \downarrow{\scriptstyle f^{2}} & & \\
\cdots & \longrightarrow & B^{-2} & \xrightarrow[\partial_B^{-2}]{} & B^{-1} & \xrightarrow[\partial_B^{-1}]{} & B^{0} & \xrightarrow[\partial_B^{0}]{} & B^{1} & \xrightarrow[\partial_B^{1}]{} & B^{2} & \longrightarrow & \cdots
\end{array}$$

then $\big(f[1]\big)^*$ is the cochain map

$$\begin{array}{ccccccccccccc}
\cdots & \longrightarrow & A^{-1} & \xrightarrow{-\partial_A^{-1}} & A^{0} & \xrightarrow{-\partial_A^{0}} & A^{1} & \xrightarrow{-\partial_A^{1}} & A^{2} & \xrightarrow{-\partial_A^{2}} & A^{3} & \longrightarrow & \cdots \\
 & & \downarrow{\scriptstyle f^{-1}} & & \downarrow{\scriptstyle f^{0}} & & \downarrow{\scriptstyle f^{1}} & & \downarrow{\scriptstyle f^{2}} & & \downarrow{\scriptstyle f^{3}} & & \\
\cdots & \longrightarrow & B^{-1} & \xrightarrow[-\partial_B^{-1}]{} & B^{0} & \xrightarrow[-\partial_B^{0}]{} & B^{1} & \xrightarrow[-\partial_B^{1}]{} & B^{2} & \xrightarrow[-\partial_B^{2}]{} & B^{3} & \longrightarrow & \cdots
\end{array}$$

This defines what the functor $[1]$ does to cochain maps, and we extend to arbitrary morphisms in $\mathbf{D}^{\mathfrak{C}'}_{\mathfrak{C}}(\mathcal{A})$ by the universal property of the localization process.

To spell this out a bit, as in Explanation 2.2: let $\mathcal{C}$ be the category with the same objects as $\mathbf{D}^{\mathfrak{C}'}_{\mathfrak{C}}(\mathcal{A})$ but where the morphisms are the cochain maps. We have defined a functor $[1]:\mathcal{C}\longrightarrow\mathcal{C}$, which takes the class $S\subset Mor(\mathcal{C})$ of cohomology isomorphisms to itself. The composite $\mathcal{C}\xrightarrow{[1]}\mathcal{C}\xrightarrow{F}S^{-1}\mathcal{C}$ is a functor from $\mathcal{C}$ to the category $S^{-1}\mathcal{C}=\mathbf{D}^{\mathfrak{C}'}_{\mathfrak{C}}(\mathcal{A})$, which takes the morphisms in S to isomorphisms. By the universal property it factors uniquely through F, that is there exists a commutative square

$$\begin{array}{ccc}
\mathcal{C} & \xrightarrow{[1]} & \mathcal{C} \\
{\scriptstyle F}\downarrow & & \downarrow{\scriptstyle F} \\
S^{-1}\mathcal{C} & \overset{\exists !}{\dashrightarrow} & S^{-1}\mathcal{C}
\end{array}$$

We declare $[1]:\mathbf{D}^{\mathfrak{C}'}_{\mathfrak{C}}(\mathcal{A})\longrightarrow\mathbf{D}^{\mathfrak{C}'}_{\mathfrak{C}}(\mathcal{A})$ to be the unique map $S^{-1}\mathcal{C}\longrightarrow S^{-1}\mathcal{C}$ making the square commute.

It remains to describe the exact triangles—Remark 2.4 provided the intuition, it told us that the exact triangles should be the formalization of the long exact sequence in cohomology coming from a short exact sequence of cochain complexes. We propose to give the skeleton of the construction below, and the reader interested in more detail is referred to the appendices.

Suppose therefore that we are given a commutative diagram in $\mathcal{A}$

$$\begin{CD}
\cdots @>>> X^{-2} @>>> X^{-1} @>>> X^{0} @>>> X^{1} @>>> X^{2} @>>> \cdots \\
@. @VVV @VVV @VVV @VVV @VVV \\
\cdots @>>> Y^{-2} @>>> Y^{-1} @>>> Y^{0} @>>> Y^{1} @>>> Y^{2} @>>> \cdots \\
@. @VVV @VVV @VVV @VVV @VVV \\
\cdots @>>> Z^{-2} @>>> Z^{-1} @>>> Z^{0} @>>> Z^{1} @>>> Z^{2} @>>> \cdots
\end{CD}$$

where the rows are objects of $\mathbf{D}^{\mathfrak{C}'}_{\mathfrak{C}}(\mathcal{A})$, that is cochain complexes satisfying the hypotheses. So far we may view the above as morphisms $X^* \xrightarrow{f^*} Y^* \xrightarrow{g^*} Z^*$ in the category $\mathbf{D}^{\mathfrak{C}'}_{\mathfrak{C}}(\mathcal{A})$. Assume further that, for each $i \in \mathbb{Z}$, the sequence $X^i \xrightarrow{f^i} Y^i \xrightarrow{g^i} Z^i$ is split exact—it's easier to deal with degreewise split short exact sequences, in Appendix B *the reader will see that, up to isomorphism in $\mathbf{D}(R)$, this suffices. We next want to mimic the process that produces the differential of the long exact sequence in cohomology. Choose, for each $i \in \mathbb{Z}$, a splitting $\theta^i : Z^i \longrightarrow Y^i$ of the map $g^i : Y^i \longrightarrow Z^i$. Now for each i we have the diagram*

$$\begin{CD}
Z^i @>{\theta^i}>> Y^i @>{g^i}>> Z^i \\
@V{\partial^i_Z}VV @VV{\partial^i_Y}V @VV{\partial^i_Z}V \\
Z^{i+1} @>{\theta^{i+1}}>> Y^{i+1} @>{g^{i+1}}>> Z^{i+1}
\end{CD}$$

If we delete the middle column the resulting square commutes—the composites of the horizontal maps are identities. If we delete the left column the square is commutative because it is part of the diagram defining the cochain map g^*. It follows that, in the diagram below,

$$\begin{CD}
Z^i @>{\theta^i}>> Y^i \\
@V{\partial^i_Z}VV @VV{\partial^i_Y}V \\
Z^{i+1} @>{\theta^{i+1}}>> Y^{i+1} @>{g^{i+1}}>> Z^{i+1}
\end{CD}$$

the two composites from top left to bottom right are equal. Thus the difference $\theta^{i+1}\partial^i_Z - \partial^i_Y\theta^i$ is annihilated by the map $g^{i+1} : Y^{i+1} \longrightarrow Z^{i+1}$, hence $\theta^{i+1}\partial^i_Z - \partial^i_Y\theta^i$ must factor uniquely through the kernel of g^{i+1}, it can be written uniquely as the composite $Z^i \xrightarrow{h^i} X^{i+1} \xrightarrow{f^{i+1}} Y^{i+1}$. In Appendix A, *the reader can see that the following is a cochain map*

$$\begin{CD}
\cdots @>>> Z^{-2} @>{\partial^{-2}_Z}>> Z^{-1} @>{\partial^{-1}_Z}>> Z^{0} @>{\partial^{-0}_Z}>> Z^{1} @>{\partial^{1}_Z}>> Z^{2} @>>> \cdots \\
@. @VV{h^{-2}}V @VV{h^{-1}}V @VV{h^{0}}V @VV{h^{1}}V @VV{h^{2}}V \\
\cdots @>>> X^{-1} @>>{-\partial^{-1}_X}> X^{0} @>>{-\partial^{0}_X}> X^{1} @>>{-\partial^{1}_X}> X^{2} @>>{-\partial^{2}_X}> X^{3} @>>> \cdots
\end{CD}$$

Thus we have constructed in the category $\mathbf{D}^{\mathfrak{C}'}_{\mathfrak{C}}(\mathcal{A})$ a diagram $X^* \xrightarrow{f^*} Y^* \xrightarrow{g^*} Z^* \xrightarrow{h^*} X^*[1]$. We declare

(i) *The exact triangles in* $\mathbf{D}_{\mathfrak{C}}^{\mathfrak{C}'}(\mathcal{A})$ *are all the isomorphs, in* $\mathbf{D}_{\mathfrak{C}}^{\mathfrak{C}'}(\mathcal{A})$*, of diagrams that come from our construction.*

It needs to be checked that [TR1], [TR2] and [TR3+4] of Definition 2.5 *are satisfied, the reader can amuse herself with this.*

For future reference we recall:

NOTATION 2.8. If $\mathcal{T}$ is a triangulated category and $n \in \mathbb{Z}$ is an integer, then $[n]$ will be our shorthand for the endofunctor $[1]^n : \mathcal{T} \longrightarrow \mathcal{T}$. Also: we will often lazily abbreviate "exact triangle" to just "triangle".

DEFINITION 2.9. *A full subcategory* $\mathcal{S} \subset \mathcal{T}$ *is called* triangulated *if* $0 \in \mathcal{S}$, *if* $\mathcal{S}[1] = \mathcal{S}$, *and if, whenever* $X, Y \in \mathcal{S}$ *and there exists in* $\mathcal{T}$ *a triangle* $X \longrightarrow Y \longrightarrow Z \longrightarrow X[1]$, *we must also have* $Z \in \mathcal{S}$. *The subcategory* $\mathcal{S}$ *is* thick *if it is triangulated, as well as closed in* $\mathcal{T}$ *under direct summands.*

Now that we have recalled the notion of triangulated categories, as well as thick and triangulated subcategories, it is time to remember the other two building blocks of the theory we plan to introduce: compact generators and t-structures. We begin with

DEFINITION 2.10. *Let* $\mathcal{T}$ *be a triangulated category with coproducts. An object* $C \in \mathcal{T}$ *is* compact *if the functor* $\mathrm{Hom}(C, -)$ *respects coproducts. A set of compact objects* $\{G_i,\ i \in I\}$ *is said to* generate *the category* $\mathcal{T}$ *if the following equivalent conditions hold*

(i) *If* $X \in \mathcal{T}$ *is an object, and if* $\mathrm{Hom}\big(G_i, X[n]\big) \cong 0$ *for all* $i \in I$ *and all* $n \in \mathbb{Z}$, *then* $X \cong 0$.

(ii) *If a triangulated subcategory* $\mathcal{S} \subset \mathcal{T}$ *is closed under coproducts and contains the objects* $\{G_i,\ i \in I\}$, *then* $\mathcal{S} = \mathcal{T}$.

If the category $\mathcal{T}$ *contains a set of compact generators it is called* compactly generated.

REMARK 2.11. The equivalence of (i) and (ii) in Definition 2.10 is not meant to be obvious, but it is a standard result. We will mostly be interested in the situation where the category $\mathcal{T}$ is compactly generated and, moreover, the set of compact generators may be chosen to consist of a single element. That is: for some compact object $G \in \mathcal{T}$ the set $\{G\}$ generates, as in Definition 2.10 (i) or (ii).

EXAMPLE 2.12. *The categories* $\mathbf{D}(R)$ *and* $\mathbf{D}_{\mathbf{qc}}(X)$ *of Example* 2.3 *(i) and (ii) both have coproducts*[2]*: the coproduct of a family of cochain complexes*

$$\cdots \longrightarrow A_\lambda^{-2} \longrightarrow A_\lambda^{-1} \longrightarrow A_\lambda^{0} \longrightarrow A_\lambda^{1} \longrightarrow A_\lambda^{2} \longrightarrow \cdots$$

turns out to be nothing other than

$$\cdots \longrightarrow \coprod_{\lambda\in\Lambda} A_\lambda^{-2} \longrightarrow \coprod_{\lambda\in\Lambda} A_\lambda^{-1} \longrightarrow \coprod_{\lambda\in\Lambda} A_\lambda^{0} \longrightarrow \coprod_{\lambda\in\Lambda} A_\lambda^{1} \longrightarrow \coprod_{\lambda\in\Lambda} A_\lambda^{2} \longrightarrow \cdots$$

It's clear that the formula above does not work for the categories $\mathbf{D}^{\mathrm{perf}}(X)$ *and* $\mathbf{D}^b_{\mathrm{coh}}(X)$ *of Example* 2.3 *(iii) and (iv), if we take a giant direct sum of complexes*

[2]In this example it helps to know the calulus of fractions of $\mathbf{D}_{\mathfrak{C}}^{\mathfrak{C}'}(\mathcal{A}) = S^{-1}\mathcal{C}$ mentioned in Example 2.2. After all we are making assertions about morphisms in $\mathbf{D}_{\mathfrak{C}}^{\mathfrak{C}'}(\mathcal{A})$: to say that an object is a coproduct is a universal property for certain morphisms. Moreover we also make an assertion about $\mathrm{Hom}_{\mathbf{D}(R)}(R, -)$.

satisfying the restrictions the resulting complex will fail to satisfy the restrictions. And it's not just that the formulas don't work, the categories $\mathbf{D}^{\mathrm{perf}}(X)$ *and* $\mathbf{D}^b_{\mathrm{coh}}(X)$ *don't have coproducts.*

Now for compact generators. If $R \in \mathbf{D}(R)$ *stands for the cochain complex*

$$\cdots \longrightarrow 0 \longrightarrow 0 \longrightarrow R \longrightarrow 0 \longrightarrow 0 \longrightarrow \cdots$$

that is the complex whose only nonzero entry is the module R *in degree 0, then it can be shown that there is an isomorphism of functors* $\mathrm{Hom}_{\mathbf{D}(R)}(R, -) \cong H^0(-)$. *The functor* $H^0(-)$ *obviously respects coproducts, hence so does* $\mathrm{Hom}_{\mathbf{D}(R)}(R, -)$*; that is the object* $R \in \mathbf{D}(R)$ *is compact.*

Next observe that, if $X \in \mathbf{D}(R)$ *is an object such that* $H^n(X) \cong \mathrm{Hom}\big(R, X[n]\big) \cong 0$ *for all* $n \in \mathbb{Z}$*, then* X *is acyclic; its cohomology all vanishes. The cochain map* $0 \longrightarrow X$ *is an isomorphism in cohomology, hence an isomorphism in* $\mathbf{D}(R)$*. That is:* $X \cong 0$*. Thus the compact object* $R \in \mathbf{D}(R)$ *satsifies Definition* 2.10*(i), it is a compact generator.*

The category $\mathbf{D}(R)$ *is compactly generated, and more precisely we have learned that the object* $R \in \mathbf{D}(R)$ *is a single compact generator.*

Not so easy is the fact that, if X *is a quasicompact, quasiseparated scheme, then the category* $\mathbf{D_{qc}}(X)$ *also has a single compact generator. This is a theorem, proved in Bondal and Van den Bergh* [**8**, Theorem 3.1.1(ii)].

NOTATION 2.13. Let $\mathcal{T}$ be a triangulated category with coproducts. It is standard to denote by $\mathcal{T}^c$ the full subcategory, whose objects are the compact objects in $\mathcal{T}$. It isn't difficult to show that $\mathcal{T}^c$ is always a thick subcategory of $\mathcal{T}$, as in Definition 2.9. In the case where $\mathcal{T} = \mathbf{D_{qc}}(X)$ the category $\mathcal{T}^c$ turns out to be the $\mathbf{D}^{\mathrm{perf}}(X)$ of Example 2.3(iii), the reader can find this fact in Bondal and Van den Bergh [**8**, Theorem 3.1.1(i)].

We also need to recall t-structures, and we plan to begin with the concrete. But first a reminder.

REMINDER 2.14. *Let* $\mathcal{T}$ *be an triangulated category, and* $\mathcal{A}$ *an abelian category. A functor* $H : \mathcal{T} \longrightarrow \mathcal{A}$ *is called* homological *if, for every triangle* $A \longrightarrow B \longrightarrow C \longrightarrow A[1]$ *in* $\mathcal{T}$*, the sequence* $H(A) \longrightarrow H(B) \longrightarrow H(C)$ *is exact in* $\mathcal{A}$*.*

From the axiom [TR2] of Definition 2.5*—the axiom telling us that any rotation of a triangle is a triangle—it follows that the functor* H *must take a triangle in* $\mathcal{T}$ *to a long exact sequence.*

EXAMPLE 2.15. *It follows from the axioms of triangulated categories that all representable functors are homological. That is: if* $\mathcal{T}$ *is a triangulated category and* $A \in \mathcal{T}$ *is an object, then* $\mathrm{Hom}(A, -)$ *and* $\mathrm{Hom}(-, A)$ *are, respectively, homological functors* $\mathcal{T} \longrightarrow \mathcal{A}b$ *and* $\mathcal{T}^{\mathrm{op}} \longrightarrow \mathcal{A}b$*, where* $\mathcal{A}b$ *is the category of abelian groups.*

The functor $H : \mathbf{D}(R) \longrightarrow R$–Mod*, taking a complex to its zeroth cohomology, is homological. In Example* 2.12 *we were told that* $H(-) \cong \mathrm{Hom}(R, -)$*, that is the functor* H *is a special case of the previous paragraph, it is a representable functor.*

On the categories $\mathbf{D_{qc}}(X)$*,* $\mathbf{D}^{\mathrm{perf}}(X)$ *and* $\mathbf{D}^b_{\mathrm{coh}}(X)$ *the homological functor we will usually consider is traditionally denoted* $\mathcal{H}$*, and takes its values in the abelian category* $\mathcal{O}_X$–Mod *of sheaves of* $\mathcal{O}_X$*–modules. Again: the functor* $\mathcal{H}$ *just takes a complex of sheaves to the zeroth cohomology sheaf.*[3]

[3]For the non-algebraic-geometers: the letter H is taken, it usually means another homological functor.

The fact that H and $\mathcal{H}$ are homological is by the construction of triangles, see Example 2.7—it comes down to the statement that the long exact sequence coming from a short exact sequence of cochain complexes is exact.

And finally we turn to t-structures, introducing them by example.

EXAMPLE 2.16. *In the category $\mathcal{T} = \mathbf{D}(R)$ we define two full subcategories by the formula*

$$\begin{array}{lcl}\text{(i) } \mathcal{T}^{\leq 0} & = & \{A \in \mathbf{D}(R) \mid H\big(A[i]\big) = 0 \text{ for all } i > 0\} \\ \text{(ii) } \mathcal{T}^{\geq 0} & = & \{A \in \mathbf{D}(R) \mid H\big(A[i]\big) = 0 \text{ for all } i < 0\}\end{array}$$

While in the case where $\mathcal{T}$ is either of the categories $\mathbf{D_{qc}}(X)$ or $\mathbf{D}^b_{\mathrm{coh}}(X)$, the formula is

$$\begin{array}{lcl}\text{(iii) } \mathcal{T}^{\leq 0} & = & \{A \in \mathcal{T} \mid \mathcal{H}\big(A[i]\big) = 0 \text{ for all } i > 0\} \\ \text{(iv) } \mathcal{T}^{\geq 0} & = & \{A \in \mathcal{T} \mid \mathcal{H}\big(A[i]\big) = 0 \text{ for all } i < 0\}\end{array}$$

These pairs of subcategories, in each of $\mathbf{D}(R)$, $\mathbf{D_{qc}}(X)$ and $\mathbf{D}^b_{\mathrm{coh}}(X)$, define a t-structure. For each of the three categories the particular t-structure above is traditionally called the standard t-structure. *The category $\mathbf{D}^{\mathrm{perf}}(X)$ does not usually have a nontrivial t-structure.*

Let us next give the formal definition:

DEFINITION 2.17. *A* t-structure *on a triangulated category $\mathcal{T}$ is a pair of full subcategories $\big(\mathcal{T}^{\leq 0}, \mathcal{T}^{\geq 0}\big)$ satisfying*

(i) $\mathcal{T}^{\leq 0}[1] \subset \mathcal{T}^{\leq 0}$ *and* $\mathcal{T}^{\geq 0} \subset \mathcal{T}^{\geq 0}[1]$

(ii) $\mathrm{Hom}\left(\mathcal{T}^{\leq 0}[1]\ ,\ \mathcal{T}^{\geq 0}\right) = 0$

(iii) *Every object $B \in \mathcal{T}$ admits a triangle $A \longrightarrow B \longrightarrow C \longrightarrow$ with $A \in \mathcal{T}^{\leq 0}[1]$ and $C \in \mathcal{T}^{\geq 0}$.*

REMARK 2.18. It can be checked that the pairs of subcategories of Example 2.16 satisfy parts (i), (ii) and (iii) of Definition 2.17, they do provide t-structures on each of $\mathbf{D}(R)$, $\mathbf{D_{qc}}(X)$ and $\mathbf{D}^b_{\mathrm{coh}}(X)$.

We have now introduced all the players: triangulated categories, compact generators and t-structures. We end the section recalling certain standard shorthand conventions.

NOTATION 2.19. Let $\mathcal{T}$ be a triangulated category with a t-structure $\big(\mathcal{T}^{\leq 0}, \mathcal{T}^{\geq 0}\big)$. Then

(i) For any integer $n \in \mathbb{Z}$ we set

$$\mathcal{T}^{\leq n} = \mathcal{T}^{\leq 0}[-n] \qquad \textit{and} \qquad \mathcal{T}^{\geq n} = \mathcal{T}^{\geq 0}[-n]$$

(ii) Furthermore, we adopt the conventions

$$\mathcal{T}^- = \bigcup_{n\in\mathbb{N}} \mathcal{T}^{\leq n}, \qquad \mathcal{T}^+ = \bigcup_{n\in\mathbb{N}} \mathcal{T}^{\geq -n}, \qquad \mathcal{T}^b = \mathcal{T}^- \cap \mathcal{T}^+.$$

3. Approximability—the intuition, which comes from $\mathbf{D}(R)$

In the last section we recalled, for the benefit of the non-expert, some standard facts about triangulated categories, compact generators and t-structures—as well as the special cases that play a big role in this article, namely $\mathbf{D}(R)$, $\mathbf{D_{qc}}(X)$, $\mathbf{D}^{\mathrm{perf}}(X)$ and $\mathbf{D}^b_{\mathrm{coh}}(X)$. It's time to move on to the subject matter of this article:

approximability. As we've tried to do throughout, we will proceed from the concrete to the abstract. Let us therefore first study what it all means for the category $\mathbf{D}(R)$, when R is a ring.

The category $\mathbf{D}(R)$ has a standard t–structure, see Example 2.16, Definition 2.17 and Remark 2.18. Suppose we are given an object $F^* \in \mathbf{D}(R)^{\leq 0}$, meaning a cochain complex

$$\cdots \longrightarrow F^{-2} \longrightarrow F^{-1} \longrightarrow F^{0} \longrightarrow F^{1} \longrightarrow F^{2} \longrightarrow \cdots$$

such that $H^i(F^*) = 0$ for all $i > 0$. Then F^* has a projective resolution. We can produce a cochain map

$$\begin{array}{ccccccccccc}
\cdots \longrightarrow & P^{-2} & \longrightarrow & P^{-1} & \longrightarrow & P^{0} & \longrightarrow & 0 & \longrightarrow & 0 & \longrightarrow \cdots \\
& \downarrow & & \downarrow & & \downarrow & & \downarrow & & \downarrow & \\
\cdots \longrightarrow & F^{-2} & \longrightarrow & F^{-1} & \longrightarrow & F^{0} & \longrightarrow & F^{1} & \longrightarrow & F^{2} & \longrightarrow \cdots
\end{array}$$

inducing an isomorphism in cohomology, and so that each P^i is a projective R–module. This gives us, in the category $\mathbf{D}(R)$, an isomorphism $P^* \longrightarrow F^*$. Now consider

$$\begin{array}{ccccccccccc}
\cdots \longrightarrow & 0 & \longrightarrow & P^{-n} & \longrightarrow \cdots \longrightarrow & P^{-1} & \longrightarrow & P^{0} & \longrightarrow & 0 & \longrightarrow \cdots \\
& \downarrow & & \downarrow & & \downarrow & & \downarrow & & \downarrow & \\
\cdots \longrightarrow & P^{-n-1} & \longrightarrow & P^{-n} & \longrightarrow \cdots \longrightarrow & P^{-1} & \longrightarrow & P^{0} & \longrightarrow & 0 & \longrightarrow \cdots \\
& \downarrow & & \downarrow & & \downarrow & & \downarrow & & \downarrow & \\
\cdots \longrightarrow & P^{-n-1} & \longrightarrow & 0 & \longrightarrow \cdots \longrightarrow & 0 & \longrightarrow & 0 & \longrightarrow & 0 & \longrightarrow \cdots
\end{array}$$

This yields a pair of cochain maps $E_n^* \xrightarrow{f_n^*} P^* \xrightarrow{g_n^*} D_n^*$ so that, in each degree i, the maps $E_n^i \xrightarrow{f_n^i} P^i \xrightarrow{g_n^i} D_n^i$ deliver a split exact sequence of R–modules. Example 2.7 constructs for us a cochain map $h_n^* : D_n^* \longrightarrow E_n^*[1]$ so that the diagram $E_n^* \xrightarrow{f_n^*} P^* \xrightarrow{g_n^*} D_n^* \xrightarrow{h_n^*} E_n^*[1]$ is an exact triangle. The isomorphism $P^* \longrightarrow F^*$ in the category $\mathbf{D}(R)$, coupled with the fact that any isomorph of a triangle is a triangle, produces in $\mathbf{D}(R)$ a triangle which we will write $E_n^* \xrightarrow{f_n} F^* \xrightarrow{g_n} D_n^* \xrightarrow{h_n^*} E_n^*[1]$.

SUMMARY 3.1. Given an object $F^* \in \mathbf{D}(R)^{\leq 0}$ and an integer $n \geq 0$ we have constructed, in $\mathbf{D}(R)$, a triangle $E_n^* \xrightarrow{f_n} F^* \xrightarrow{g_n} D_n^* \xrightarrow{h_n^*} E_n^*[1]$. This triangle is such that $D_n^* \in \mathbf{D}(R)^{\leq -n-1}$, while E_n^* is not too complicated.

In the Introduction we mentioned that we will view the objects D_n^* as "small" with respect to the metric induced by the t–structure. Up to an arbitrarily small "correction term" D_n^*, we have a way of approximating the object F^* by the object E_n^* which we view as simpler. In order to formalize the idea we need to make precise what we mean by saying that E_n^* is "not too complicated". We will do this in the next section.

4. Measuring the complexity of an object

As we have said in the Introduction, measuring how complicated an object is will involve a small tweak of an idea from Bondal and Van den Bergh [**8**]. We remind the reader.

REMINDER 4.1. *Let $\mathcal{T}$ be a triangulated category, possibly with coproducts, and let $\mathcal{A}, \mathcal{B} \subset \mathcal{T}$ be full subcategories. We define the full subcategories*

(i) $\mathcal{A} * \mathcal{B} \quad = \quad \left\{ x \in \mathcal{T} \;\middle|\; \begin{array}{c} \text{there exists in } \mathcal{T} \text{ a triangle } a \longrightarrow x \longrightarrow b \\ \text{with } a \in \mathcal{A} \text{ and } b \in \mathcal{B} \end{array} \right\}$

(ii) *add($\mathcal{A}$): this consists of all finite coproducts of objects of $\mathcal{A}$.*

(iii) *Assume $\mathcal{T}$ has coproducts. Define Add($\mathcal{A}$) to consist of all coproducts of objects of $\mathcal{A}$.*

(iv) *smd($\mathcal{A}$): the category of all direct summands of objects of $\mathcal{A}$.*

REMARK 4.2. Reminder 4.1(i) is as in [**6**, 1.3.9], while Reminder 4.1(iv) is identical with [**8**, beginning of 2.2]. Reminder 4.1 (ii) and (iii) follow the usual conventions in representation theory; in [**8**, beginning of 2.2] the authors adopt the (unconventional) notation that add($\mathcal{A}$) and Add($\mathcal{A}$) are also closed under the suspension—thus add($\mathcal{A}$) as defined in [**8**] is what we would denote $\text{add}\big(\bigcup_{n=-\infty}^{\infty} \mathcal{A}[n]\big)$. The definitions that follow are therefore slightly different from [**8**], and it is this small tweak that makes all the difference—with the tweaked definitions, approximations turn out to exist in great generality.

And now we come to the key definition: we're about to measure how much effort goes into constructing an object X out of some given full subcategory $\mathcal{A} \subset \mathcal{T}$. In practice our usual choice for $\mathcal{A}$ will be a $\mathcal{A} = \{G\}$, the subcategory with just a single object G, which we will often assume to be a compact generator.

DEFINITION 4.3. *Let $\mathcal{T}$ be a triangulated category, possibly with coproducts, let $\mathcal{A} \subset \mathcal{T}$ be a full subcategory and let $m \leq n$ be integers, possibly infinite. We define the full subcategories*

(i) $\mathcal{A}[m, n] = \cup_{i=m}^{n} \mathcal{A}[-i]$.

(ii) $\langle \mathcal{A} \rangle_1^{[m,n]} = \text{smd}\Big[\text{add}\big(\mathcal{A}[m, n]\big)\Big]$.

(iii) $\overline{\langle \mathcal{A} \rangle}_1^{[m,n]} = \text{smd}\Big[\text{Add}\big(\mathcal{A}[m, n]\big)\Big]$. [This definition assumes $\mathcal{T}$ has coproducts].

Now let $\ell > 0$ be an integer, and assume the categories $\langle \mathcal{A} \rangle_k^{[m,n]}$ and $\overline{\langle \mathcal{A} \rangle}_k^{[m,n]}$ have been defined for k in the range $1 \leq k \leq \ell$. We proceed inductively to set

(iv) $\langle \mathcal{A} \rangle_{\ell+1}^{[m,n]} = \text{smd}\Big[\langle \mathcal{A} \rangle_1^{[m,n]} * \langle \mathcal{A} \rangle_\ell^{[m,n]}\Big]$.

(v) $\overline{\langle \mathcal{A} \rangle}_{\ell+1}^{[m,n]} = \text{smd}\Big[\overline{\langle \mathcal{A} \rangle}_1^{[m,n]} * \overline{\langle \mathcal{A} \rangle}_\ell^{[m,n]}\Big]$. [This definition assumes $\mathcal{T}$ has coproducts].

EXAMPLE 4.4. *Let us go back to our favorite example $\mathbf{D}(R)$. Suppose $\mathcal{A} = \{R\}$ is the category with a single object R, and we will now proceed to say something about the subcategories $\langle R \rangle_\ell^{[-n,0]} \subset \mathbf{D}(R)$. Let us start with*

(i) $\langle R \rangle_1^{[-n,0]}$*: this turns out to be the category of all isomorphs in $\mathbf{D}(R)$ of the cochain complexes*

$$\cdots \longrightarrow 0 \longrightarrow P^{-n} \xrightarrow{\ 0\ } \cdots \xrightarrow{\ 0\ } P^{-1} \xrightarrow{\ 0\ } P^0 \longrightarrow 0 \longrightarrow \cdots$$

with P^i finitely generated and projective.

This much is basically true by construction. We start with the object R, and in Definition 4.3(i) we form the category $R[-n,0] = \{R[i],\ 0 \leq i \leq n\}$ with finitely many objects. And then Definition 4.3(ii) allows us to first form finite coproducts of objects in $R[-n,0]$, meaning cochain complexes

$$\cdots \longrightarrow 0 \longrightarrow P^{-n} \xrightarrow{\ 0\ } \cdots \xrightarrow{\ 0\ } P^{-1} \xrightarrow{\ 0\ } P^0 \longrightarrow 0 \longrightarrow \cdots$$

with each P^i a finitely generated free module, and then we are permitted direct summands in $\mathbf{D}(R)$ of the above. It may be shown that these are all isomorphic to complexes as above, but where we allow the P^i to be finitely generated and projective.

This was the easy part. Now the categories $\langle R\rangle_\ell^{[-n,0]}$ grow as ℓ grows, but it's a little unclear how fast. They all contain $\langle R\rangle_1^{[-n,0]}$, and are all contained in the subcategory $\mathcal{S} \subset \mathbf{D}(R)$ of objects isomorphic in $\mathbf{D}(R)$ to cochain complexes

$$\cdots \longrightarrow 0 \longrightarrow P^{-n} \longrightarrow \cdots \longrightarrow P^{-1} \longrightarrow P^0 \longrightarrow 0 \longrightarrow \cdots$$

with P^i finitely generated and projective. Unlike the category $\langle R\rangle_1^{[-n,0]}$ of (i) above, for objects in $\mathcal{S}$ the maps $P^i \longrightarrow P^{i+1}$ are unconstrained—beyond (of course) the standing assumption that all composites $P^i \longrightarrow P^{i+1} \longrightarrow P^{i+2}$ must vanish, the objects of $\mathcal{S} \subset \mathbf{D}(R)$ must be cochain complexes.

What turns out to be true is

(ii) *$\langle R\rangle_{n+1}^{[-n,0]} = \mathcal{S}$; hence $\langle R\rangle_\ell^{[-n,0]} = \mathcal{S}$ for all $\ell \geq n+1$.*

We leave to the reader the proofs of the assertions made in this Example[4].

EXAMPLE 4.5. *Let us stay with our favorite example $\mathbf{D}(R)$, and let us continue to put $\mathcal{A} = \{R\}$, that is $\mathcal{A}$ is the full subcategory of $\mathbf{D}(R)$ with the single object R. We now want to work out what are the categories $\overline{\langle R\rangle}_\ell^{[-n,0]}$. The discussion turns out to be much the same as in Example 4.4, and the summary of the results is*

(i) *The category $\overline{\langle R\rangle}_1^{[-n,0]}$ consists of all isomorphs in $\mathbf{D}(R)$ of cochain complexes*

$$\cdots \longrightarrow 0 \longrightarrow P^{-n} \xrightarrow{\ 0\ } \cdots \xrightarrow{\ 0\ } P^{-1} \xrightarrow{\ 0\ } P^0 \longrightarrow 0 \longrightarrow \cdots$$

with P^i projective.

(ii) *The category $\overline{\langle R\rangle}_{n+1}^{[-n,0]}$ consists of all isomorphs in $\mathbf{D}(R)$ of cochain complexes*

$$\cdots \longrightarrow 0 \longrightarrow P^{-n} \longrightarrow \cdots \longrightarrow P^{-1} \longrightarrow P^0 \longrightarrow 0 \longrightarrow \cdots$$

with P^i projective. Moreover: if $\ell \geq n+1$ then $\overline{\langle R\rangle}_{n+1}^{[-n,0]} = \overline{\langle R\rangle}_\ell^{[-n,0]}$.

Thus the objects in both $\overline{\langle R\rangle}_{n+1}^{[-n,0]}$ and $\langle R\rangle_{n+1}^{[-n,0]}$ are isomorphic in $\mathbf{D}(R)$ to complexes of projectives vanishing outside the interval $[-n,0]$, and the difference is that in $\overline{\langle R\rangle}_{n+1}^{[-n,0]}$ the projective modules are not constrained to be finitely generated.

---

[4]The proofs are easy for the reader familiar with the calculus of fractions of Explanation 2.2. Other readers are asked to accept the assertions on faith.

CONCLUSION 4.6. In the new notation we have introduced, Summary 3.1 and Example 4.5 combine to say: for any object $F \in \mathbf{D}(R)^{\leq 0}$ and any integer $n \geq 0$ there exists a triangle

$$E_n \xrightarrow{\ f\ } F \xrightarrow{\ g\ } D_n \xrightarrow{\ h\ } E_n[1]$$

with $D_n \in \mathbf{D}(R)^{\leq -n-1}$ and $E_n \in \overline{\langle R\rangle}_{n+1}^{[-n,0]}$.

REMARK 4.7. Let $\mathbf{D}(R\text{–proj})^{\leq 0} \subset \mathbf{D}(R)$ be the full subcategory, whose objects are the isomorphs in $\mathbf{D}(R)$ of cochain complexes

$$\cdots \longrightarrow P^{-n-1} \longrightarrow P^{-n} \longrightarrow \cdots \longrightarrow P^{-1} \longrightarrow P^0 \longrightarrow 0 \longrightarrow \cdots$$

with P^i finitely generated and projective. Summary 3.1 and Example 4.4 combine to say: for any object $F \in \mathbf{D}(R\text{–proj})^{\leq 0}$ and any integer $n \geq 0$ there exists a triangle

$$E_n \xrightarrow{\ f\ } F \xrightarrow{\ g\ } D_n \xrightarrow{\ h\ } E_n[1]$$

with $D_n \in \mathbf{D}(R)^{\leq -n-1}$ and $E_n \in \langle R\rangle_{n+1}^{[-n,0]}$.

We will return to the category $\mathbf{D}(R\text{–proj})^{\leq 0}$ and to its relative

$$\mathbf{D}^-(R\text{–proj}) \quad = \quad \bigcup_{n\in\mathbb{N}} \mathbf{D}(R\text{–proj})^{\leq 0}[-n]$$

much later in the article.

5. The formal definition of approximability

Now that we are thoroughly prepared, approximability becomes easy to formulate precisely:

DEFINITION 5.1. *Let $\mathcal{T}$ be a triangulated category with coproducts. It is* approximable *if there exists a compact generator $G \in \mathcal{T}$, a t–structure $(\mathcal{T}^{\leq 0}, \mathcal{T}^{\geq 0})$, and an integer $A > 0$ so that*

(i) $G[A] \in \mathcal{T}^{\leq 0}$ *and* $\mathrm{Hom}\big(G[-A]\,,\,\mathcal{T}^{\leq 0}\big) = 0$.

(ii) *For every object $F \in \mathcal{T}^{\leq 0}$ there exists a triangle $E \longrightarrow F \longrightarrow D \longrightarrow E[1]$, with $D \in \mathcal{T}^{\leq -1}$ and $E \in \overline{\langle G\rangle}_A^{[-A,A]}$.*

EXAMPLE 5.2. *Let R be a ring. In Example 2.12 we learned that the object $R \in \mathbf{D}(R)$ is a compact generator, we will take this to be our G of Definition 5.1. For the t–structure we choose the standard one, see Example 2.16. And for our integer we set $A = 1$.*

It's clear that $R[1] \in \mathbf{D}(R)^{\leq 0}$ and that $\mathrm{Hom}\big(R[-1], \mathbf{D}(R)^{\leq 0}\big) = 0$. This establishes Definition 5.1(i). Finally suppose we are given an object $F \in \mathbf{D}(R)^{\leq 0}$. By Conclusion 4.6, with $n = 0$, there must exist a triangle $E \longrightarrow F \longrightarrow D \longrightarrow E[1]$ with $D \in \mathbf{D}(R)^{\leq -1}$ and $E \in \overline{\langle R\rangle}_1^{[0,0]} \subset \overline{\langle R\rangle}_1^{[-1,1]}$. This proves that Definition 5.1(ii) holds.

Thus the category $\mathbf{D}(R)$ is approximable.

REMARK 5.3. The reader might be disappointed: until now we have been stressing that approximability will allow us to obtain arbitrarily good estimates of the objects in any approximable category $\mathcal{T}$, and in Example 5.2 we see that the definition only involves a zero-order approximation.

Don't let this disturb you, it's easy to iterate and estimate the given object F to arbitrarily high order. This will manifest itself in our theorems.

REMARK 5.4. In Example 2.3 we told the reader that, in this survey, the key examples of triangulated categories will be $\mathbf{D}(R)$, $\mathbf{D_{qc}}(X)$, $\mathbf{D}^{\mathrm{perf}}(X)$ and $\mathbf{D}^b_{\mathrm{coh}}(X)$. The definition of approximability is tailored so that the category $\mathbf{D}(R)$ is obviously approximable, as we have seen in Example 5.2. What about the other three?

The categories $\mathbf{D}^{\mathrm{perf}}(X)$ and $\mathbf{D}^b_{\mathrm{coh}}(X)$ cannot possibly be approximable, in Example 2.12 we learned that they don't even have coproducts.

It is a non-obvious theorem that, as long as the scheme X is quasicompact and separated, the category $\mathbf{D_{qc}}(X)$ is approximable. And it should come as a surprise—after all approximability was modeled on the idea of taking a projective resolution of a bounded-above cochain complex and then truncating, and this is a construction that can only work in the presence of enough projectives. There aren't enough projectives in either the category of sheaves of $\mathcal{O}_X$–modules, or in its subcategory of quasicoherent sheaves.

In Example 2.12 we mentioned that even the existence of a single compact generator in $\mathbf{D_{qc}}(X)$ isn't obvious, it's a theorem of Bondal and Van den Bergh. The existence proof isn't particularly constructive—it doesn't give us much of a handle on this compact generator. And the definition of approximability is the assertion that the compact generator may be chosen to satisfy several useful properties; it decidedly isn't clear how to prove any of them.

Given that it is going to entail real effort to prove that $\mathbf{D_{qc}}(X)$ is approximable, it shouldn't come as a surprise that there are far-reaching consequences.

And now it's time for

6. The main theorems

REMARK 6.1. The theorems break up into three groups, namely

(i) Theorems that produce more examples of approximable categories. So far we have discussed in some detail the example $\mathbf{D}(R)$, and then made some passing comments about $\mathbf{D_{qc}}(X)$. See Remark 5.4.
(ii) Formal consequences of approximability—that is structure that comes for free, which every approximable category has.
(iii) Applications to concrete examples, which teach us new and interesting facts about old and familiar categories.

In this section we will list the results of type (i) and (iii) by group, doing little more than giving formal statements. In the remainder of the article we will first expand on the results in group (iii), saying something about what was known before and about the proofs, both of the new and the old versions—presumably the reader is most likely to be persuaded by the theory if she can see applications that matter.

And then, towards the end of the article, we will give results in group (ii). We hope that by then, with the reader's interest piqued by the group (iii) applications, she will have the patience to also read the structural theorems.

FACTS 6.2. **(The main theorems—sources of more examples).** *The following statements are true:*

(i) *If* $\mathcal{T}$ *has a compact generator* G, *so that* $\mathrm{Hom}\big(G, G[n]\big) = 0$ *for all* $n \geq 1$, *then* $\mathcal{T}$ *is approximable.*

Special cases of (i) include:

The category $\mathcal{T} = \mathbf{D}(R)$ *and the compact generator* $G = R$*, in other words we recover Example* 5.2 *as a special case of (i). More generally: if* R *is a dga, and* $H^n(R) = 0$ *for all* $n > 0$*, then the category* $\mathcal{T} = \mathbf{D}(R\text{–Mod})$ *with* $G = R$ *is an example. Further examples come from topology, for instance we can let* $\mathcal{T}$ *be the homotopy category of spectra and let* $G = \mathbb{S}^0$ *be the zero-sphere.*

The proof of (i) is basically trivial, there is a brief discussion in [**34**, Remark 3.3].

(ii) *Let* X *be a quasicompact, separated scheme. Then the category* $\mathbf{D_{qc}}(X)$ *is approximable.*

In Remark 5.4 *we noted that this isn't an easy fact, it is after all counterintuitive—it says that, in the category* $\mathbf{D_{qc}}(X)$*, one can pretend to have enough projectives—at least for some purposes. The proof isn't trivial.*

If X *is a separated scheme, of finite type over a noetherian ring, then the reader can find a proof* [**33**, Theorem 5.8]. *It constitutes the main technical lemma of the paper, the rest amounts to applications. The generalization to quasicompact, separated schemes is by a trick which may be found in* [**34**, Example 3.6].

(iii) *Suppose we are given a recollement of triangulated categories*

$$\mathcal{R} \rightleftarrows \mathcal{S} \rightleftarrows \mathcal{T}$$

with $\mathcal{R}$ *and* $\mathcal{T}$ *approximable. Assume further that the category* $\mathcal{S}$ *is compactly generated, and any compact object* $H \in \mathcal{S}$ *has the property that* $\operatorname{Hom}\big(H, H[i]\big) = 0$ *for* $i \gg 0$*. Then the category* $\mathcal{S}$ *is also approximable.*

Once again this isn't obvious, it requires proof. The reader can find it in [**9**, Theorem 4.1]*—it is the main theorem of the article.*

So far the majority of the interesting applications has been to algebraic geometry—it's the example in Fact 6.2(ii) that has proved useful. But the subject is in its infancy, it is to be hoped that there will be applications to come, in other contexts.

FACTS 6.3. **(The main theorems—applications).** *Assertions (i) and (ii) below are* [**33**, Theorem 0.5 and Theorem 0.15]*, respectively. Assertion (iii) follows from* [**34**, Corollary 0.5]*, while assertion (iv) follows from* [**32**, Theorem 0.2]*. Assertion (v) is a consequence of* [**31**, Proposition 0.15]*, together with the elaboration and discussion in the couple of paragraphs immediately following the statement of the Proposition. Anyway: all of (v) may be found in* [**31**].

In Explanation 6.4 *the reader is reminded what the various technical terms in the statements below mean.*

(i) *Let* X *be a quasicompact, separated scheme. The category* $\mathbf{D}^{\mathrm{perf}}(X)$ *is strongly generated if and only if* X *has an open cover by affine schemes* $\operatorname{Spec}(R_i)$*, with each* R_i *of finite global dimension.*

(ii) *Let* X *be a separated scheme, and assume it is noetherian, finite-dimensional, and that every closed, reduced, irreducible subscheme of* X *has a regular alteration. Then the category* $\mathbf{D}^b_{\mathrm{coh}}(X)$ *is strongly generated.*

(iii) *Let* X *be a scheme proper over a noetherian ring* R*. Let* $\mathcal{Y}$ *be the Yoneda map*

$$\mathbf{D}^b_{\mathrm{coh}}(X) \xrightarrow{\quad \mathcal{Y} \quad} \operatorname{Hom}_R\Big(\big[\mathbf{D}^{\mathrm{perf}}(X)\big]^{\mathrm{op}}, R\text{–Mod}\Big)$$

That is: the map $\mathcal{Y}$ *sends the object* $B \in \mathbf{D}^b_{\mathrm{coh}}(X)$ *to the functor* $\mathcal{Y}(B) = \mathrm{Hom}(-, B)$*, viewed as an* R*–linear homological functor* $\big[\mathbf{D}^{\mathrm{perf}}(X)\big]^{\mathrm{op}} \longrightarrow R\text{–Mod}$.

Then $\mathcal{Y}$ *is fully faithful, and the essential image is the set of finite* R*–linear homological functors* $H : \mathbf{D}^{\mathrm{perf}}(X)^{\mathrm{op}} \longrightarrow R\text{–mod}$*. An* R*–linear homological functor is* finite *if, for all objects* $C \in \mathbf{D}^{\mathrm{perf}}(X)$*, the* R*–module* $\oplus_n H\big(C[n]\big)$ *is finite.*

(iv) *Suppose* X *is finite dimensional scheme proper over a noetherian ring* R*, and assume further that every closed, reduced, irreducible subscheme of* X *has a regular alteration. Let* $\widetilde{\mathcal{Y}}$ *be the Yoneda map*

$$\big[\mathbf{D}^{\mathrm{perf}}(X)\big]^{\mathrm{op}} \xrightarrow{\quad\widetilde{\mathcal{Y}}\quad} \mathrm{Hom}_R\Big(\mathbf{D}^b_{\mathrm{coh}}X, R\text{–Mod}\Big)$$

That is: the map $\widetilde{\mathcal{Y}}$ *takes an object* $A \in \mathbf{D}^{\mathrm{perf}}(X)$ *to the functor* $\widetilde{\mathcal{Y}}(A) = \mathrm{Hom}(A, -)$*, viewed as an* R*–linear homological functor* $\mathbf{D}^b_{\mathrm{coh}}(X) \longrightarrow ModR$.

Then $\widetilde{\mathcal{Y}}$ *is fully faithful, and the essential image of* $\widetilde{\mathcal{Y}}$ *are the finite homological functors.*

(v) *Let* X *be a noetherian, separated scheme. There is a recipe which takes the triangulated category* $\mathbf{D}^{\mathrm{perf}}(X)$ *as input, and out of it constructs the triangulated category* $\mathbf{D}^b_{\mathrm{coh}}(X)$*. And there is a recipe going back: from the triangulated category* $\big[\mathbf{D}^b_{\mathrm{coh}}(X)\big]^{\mathrm{op}}$ *as input the machine spews out* $\big[\mathbf{D}^{\mathrm{perf}}(X)\big]^{\mathrm{op}}$.

EXPLANATION 6.4. *We remind the reader what the terms used in the theorems mean. Let* $\mathcal{S}$ *be a triangulated category, and let* $G \in \mathcal{S}$ *be an object. Then*

(i) G *is a* classical generator *if* $\mathcal{S} = \cup_n \langle G\rangle_n^{[-n,n]}$.

(ii) G *is a* strong generator *if there exists an integer* $\ell > 0$ *with* $\mathcal{S} = \cup_n \langle G\rangle_\ell^{[-n,n]}$*. The category* $\mathcal{S}$ *is called* strongly generated *if it has a strong generator.*

(iii) *Suppose* X *is a noetherian scheme, finite-dimensional, reduced and irreducible. A* regular alteration *of* X *is a generically finite, proper, surjective morphism* $\widetilde{X} \longrightarrow X$ *with* $\widetilde{X}$ *regular.*

The non-expert deserves some explanation of (iii): we all know what a resolution of singularities is, but the known existence theorems are too restrictive (for our purposes). Of course resolutions of singularities are conjectured *to exist quite generally, unfortunately what has been proved so far is limited to equal characteristic zero, or to schemes of very low dimension. Regular alterations are less restrictive, and the known existence theorems are much more general—see de Jong* [**13, 14**].

As it turns out, in the proofs of Facts 6.3 *(ii) and (iv) regular alterations suffice. The non-expert should therefore view the condition imposed on the noetherian scheme* X*, in Facts* 6.3 *(ii) and (iv), as a mild technical hypothesis.*

REMARK 6.5. The reader should note that Fact 6.2 asserts that the category $\mathbf{D_{qc}}(X)$ is approximable, and now we're telling the reader that the consequences—Facts 6.3 (i), (ii), (iii), (iv) and (v)—are all assertions about the categories $\mathbf{D}^{\mathrm{perf}}(X)$ and $\mathbf{D}^b_{\mathrm{coh}}(X)$. A technical, formal statement, about the huge category $\mathbf{D_{qc}}(X)$, turns out to have a string of powerful consequences about the much smaller categories $\mathbf{D}^{\mathrm{perf}}(X)$ and $\mathbf{D}^b_{\mathrm{coh}}(X)$, that many people have been studying for decades.

7. More about the strong generation of $\mathbf{D}^{\mathrm{perf}}(X)$ and $\mathbf{D}^b_{\mathrm{coh}}(X)$

As promised, we will now say a little more about Facts 6.3 (i) and (ii). In this section we will survey what was known before, the basic idea of the old proofs, and how the proof based on approximability departs from the older methods.

The non-algebraic-geometers are advised to skip the discussions of the proofs. The brief summary is that the proofs based on approximability are short, simple, sweet and work in great generality—the hard work goes into proving that the category $\mathbf{D_{qc}}(X)$ is approximable. After that it's all downhill.

Let us begin with Facts 6.3(i), we recall the statement for the reader's convenience:

THEOREM 7.1. *Assume X is quasicompact, separated scheme. Then $\mathbf{D}^{\mathrm{perf}}(X)$ is strongly generated if and only if X may be covered by open affine subsets* $\mathrm{Spec}(R_i)$, *with each R_i of finite global dimension.*

REMARK 7.2. If X is noetherian and separated, this simplifies to saying that $\mathbf{D}^{\mathrm{perf}}(X)$ is strongly generated if and only if X is regular and finite dimensional.

HISTORICAL SURVEY 7.3. *When $X = \mathrm{Spec}(R)$ is affine Theorem* 7.1 *is old: it was first proved by Kelly* [**25**], *see also Street* [**40**]. *The result was rediscovered by Christensen* [**10**, Corollary 8.4] *and later Rouquier* [**39**, Proposition 7.25].

Bondal and Van den Bergh [**8**, Theorem 3.1.4] *proved the first global version: if X is a separated scheme, smooth over a field k, then the category $\mathbf{D}^{\mathrm{perf}}(X)$ is strongly generated. The case where X is assumed of finite type over a field and regular [regularity is weaker than smoothness] follows from either Rouquier* [**39**, Theorem 7.38] *or Orlov* [**37**, Theorem 3.27].

This summarizes the results known before approximability. Note that, with the exception of Kelly's, the old results all assumed equal characteristic and that X is noetherian. By contrast Theorem 7.1 *works fine in the mixed characteristic, non-noetherian situation.*

DISCUSSION OF THE PROOFS, OLD AND NEW 7.4. *By combining Kelly's old theorem* [**25**] *with the main theorem of Thomason and Trobaugh* [**41**], *one easily deduces one of the implications in Theorem* 7.1*: if X is quasicompact and separated, and* $\mathrm{Spec}(R)$ *embeds in X as an open, affine subset, then R must be of finite global dimension. The reader can find the argument spelled out in more detail in (for example)* [**33**, Remark 0.11].

Now for the tricky direction of Theorem 7.1, *the direction saying that, if X is quasicompact and separated, and admits a cover by open affines* $\mathrm{Spec}(R_i)$ *with each R_i of finite global dimension, then it follows that $\mathbf{D}^{\mathrm{perf}}(X)$ is strongly generated. As we have already said: the case where X is affine is contained in Kelly's old theorem.*

We remind the reader: Bondal and Van den Bergh [**8**, Theorem 3.1.4] *proved the first global version. They proved that, if X is a separated scheme, smooth over a field k, then the category $\mathbf{D}^{\mathrm{perf}}(X)$ is strongly generated. Their proof relies on the fact that, if $\delta : X \longrightarrow X \times_k X$ is the diagonal embedding, then the functor $\mathbf{R}\delta_*$ respects perfect complexes. It is a characterization of smoothness for $\mathbf{R}\delta_*$ to respect perfect complexes—hence the argument isn't one that readily lends itself to generalizations.*

Nevertheless there were improvements. The case where X is assumed of finite type over a field and regular follows from either Rouquier [**39**, Theorem 7.38] *or*

Orlov [**37**, Theorem 3.27]. *Both proofs still use a diagonal argument—Rouquier's approach refines Bondal and Van den Bergh's by stratifying X, while the refinement in Orlov's article is not quite so easy to sum up briefly. It was Orlov's clever approach to the problem that inspired the idea of approximability.*

It remains to give the reader some idea how approximability helps in the proof of Theorem 7.1. And the main point is that approximability allows us to reduce the general case to the case of an affine scheme, where we can use the old theorem of Kelly's. For the reader's convenience and because the proof is so easy, we prove below the variant of Kelly's theorem we will actually use.

THEOREM 7.5. *Suppose R is an associative ring, and $\mathbf{D}(R)$ its derived category. Let $n \geq 0$ be an integer, and let $F \in \mathbf{D}(R)$ be an object such that the projective dimension of $H^i(F)$ is $\leq n$ for all $i \in \mathbb{Z}$. Then $F \in \overline{\langle R \rangle}_{n+1}^{(-\infty,\infty)}$.*

Before proving the theorem we remind the reader [who is familiar with the calculus of fractions in derived categories]: any morphism $P \longrightarrow H^i(E)$ in $\mathbf{D}(R)$, for any projective R–module P and any $E \in \mathbf{D}(R)$, lifts uniquely to a cochain map

$$\begin{array}{ccccccccccc}
\cdots & \longrightarrow & 0 & \longrightarrow & 0 & \longrightarrow & P & \longrightarrow & 0 & \longrightarrow & 0 & \longrightarrow & \cdots \\
 & & \downarrow & & \downarrow & & \downarrow & & \downarrow & & \downarrow & & \\
\cdots & \longrightarrow & E^{i-2} & \longrightarrow & E^{i-1} & \longrightarrow & E^{i} & \longrightarrow & E^{i+1} & \longrightarrow & E^{i+2} & \longrightarrow & \cdots
\end{array}$$

PROOF. We prove the theorem by induction on n. Suppose first that $n = 0$; hence $H^i(F)$ is projective for every $i \in \mathbb{Z}$. The identity map $H^i(F) \longrightarrow H^i(F)$ lifts to a cochain map

$$\begin{array}{ccccccccccc}
\cdots & \longrightarrow & 0 & \longrightarrow & 0 & \longrightarrow & H^i(F) & \longrightarrow & 0 & \longrightarrow & 0 & \longrightarrow & \cdots \\
 & & \downarrow & & \downarrow & & \downarrow & & \downarrow & & \downarrow & & \\
\cdots & \longrightarrow & F^{i-2} & \longrightarrow & F^{i-1} & \longrightarrow & F^{i} & \longrightarrow & F^{i+1} & \longrightarrow & F^{i+2} & \longrightarrow & \cdots
\end{array}$$

and when we combine, for every $i \in \mathbb{Z}$, we obtain a cochain map

$$\begin{array}{ccccccccccc}
\cdots & \longrightarrow & H^{-2}(F) & \xrightarrow{0} & H^{-1}(F) & \xrightarrow{0} & H^{0}(F) & \xrightarrow{0} & H^{1}(F) & \xrightarrow{0} & H^{2}(F) & \longrightarrow & \cdots \\
 & & \downarrow & & \downarrow & & \downarrow & & \downarrow & & \downarrow & & \\
\cdots & \longrightarrow & F^{-2} & \longrightarrow & F^{-1} & \longrightarrow & F^{0} & \longrightarrow & F^{1} & \longrightarrow & F^{2} & \longrightarrow & \cdots
\end{array}$$

This is an isomorphism in cohomology, hence an isomorphism in $\mathbf{D}(R)$.

Now suppose $n \geq 0$, and we know the result for every ℓ with $0 \leq \ell \leq n$. We wish to show it for $n+1$. Suppose therefore that we are given an object $F \in \mathbf{D}(R)$ with $H^i(F)$ of projective dimension $\leq n+1$ for every i. Choose for every i a projective module P^i and a surjection $P^i \longrightarrow H^i(F)$. Now form the corresponding cochain map

$$\begin{array}{ccccccccccc}
\cdots & \longrightarrow & 0 & \longrightarrow & 0 & \longrightarrow & P^i & \longrightarrow & 0 & \longrightarrow & 0 & \longrightarrow & \cdots \\
 & & \downarrow & & \downarrow & & \downarrow & & \downarrow & & \downarrow & & \\
\cdots & \longrightarrow & F^{i-2} & \longrightarrow & F^{i-1} & \longrightarrow & F^{i} & \longrightarrow & F^{i+1} & \longrightarrow & F^{i+2} & \longrightarrow & \cdots
\end{array}$$

and combine over i to form

$$\begin{array}{ccccccccccccc}
\cdots & \longrightarrow & P^{-2} & \xrightarrow{0} & P^{-1} & \xrightarrow{0} & P^{0} & \xrightarrow{0} & P^{1} & \xrightarrow{0} & P^{2} & \longrightarrow & \cdots \\
 & & \downarrow & & \downarrow & & \downarrow & & \downarrow & & \downarrow & & \\
\cdots & \longrightarrow & F^{-2} & \longrightarrow & F^{-1} & \longrightarrow & F^{0} & \longrightarrow & F^{1} & \longrightarrow & F^{2} & \longrightarrow & \cdots
\end{array}$$

giving a map $P \longrightarrow F$, which we complete to a triangle $P \longrightarrow F \longrightarrow Q$. Clearly $P \in \overline{\langle R\rangle}_1^{(-\infty,\infty)}$, and the long exact sequence in cohomology gives that $H^i(Q)$ is of projective dimension $\leq n$. Hence F belongs to $\overline{\langle R\rangle}_1^{(-\infty,\infty)} * \overline{\langle R\rangle}_{n+1}^{(-\infty,\infty)} \subset \overline{\langle R\rangle}_{n+2}^{(-\infty,\infty)}$. □

At this point the non-algebraic-geometer (who hasn't yet done so) is advised to skip ahead to Theorem 7.12. What will come between now and then is largely aimed to show that the approximability of $\mathbf{D_{qc}}(X)$ makes the reduction to Kelly's old theorem straightforward and easy—hopefully the sketch we give will make this transparent to the experts, but for non-algebraic-geometers it might be mystifying. Anyway: the reduction depends on the following little lemma—the reader should note the way approximability enters the proof of the lemma, this is the only point where approximability will be used.

LEMMA 7.6. *Let X be a quasicompact, separated scheme, let $G \in \mathbf{D_{qc}}(X)$ be a compact generator, and let $u : U \longrightarrow X$ be an open immersion with U quasicompact. Then the object $\mathbf{R}u_*\mathcal{O}_U \in \mathbf{D_{qc}}(X)$ belongs to $\overline{\langle G\rangle}_n^{[-n,n]}$ for some integer $n > 0$.*

PROOF. It is relatively easy to show that, for some sufficiently large integer $\ell > 0$, we have $\mathrm{Hom}\big(\mathbf{R}u_*\mathcal{O}_U\,,\ \mathbf{D_{qc}}(X)^{\leq -\ell}\big) = 0$. By the approximability[5] of $\mathbf{D_{qc}}(X)$ we may choose an integer n and a triangle $E \longrightarrow \mathbf{R}u_*\mathcal{O}_U \longrightarrow D$ with $D \in \mathbf{D_{qc}}(X)^{\leq -\ell}$ and $E \in \overline{\langle G\rangle}_n^{[-n,n]}$.

But the map $\mathbf{R}u_*\mathcal{O}_U \longrightarrow D$ must vanish by the choice of ℓ, making $\mathbf{R}u_*\mathcal{O}_U$ a direct summand of the object $E \in \overline{\langle G\rangle}_n^{[-n,n]}$. □

SKETCH 7.7. *We should indicate how Theorem* 7.1 *follows from the combination of Lemma* 7.6 *and Kelly's old theorem. Let X and the open affine cover by $U_i = \mathrm{Spec}(R_i)$ be as in the hypotheses of Theorem* 7.1. *Because X is quasicompact we may, possibly after passing to a subcover, assume that our cover is finite; write the cover as $\{U_i,\ 1 \leq i \leq r\}$.*

Now choose a compact generator $G \in \mathbf{D_{qc}}(X)$. The Lemma allows us to choose, for each open subset U_i, an integer n_i so that $\mathbf{R}u_{i*}\mathcal{O}_{U_i} \in \overline{\langle G\rangle}_{n_i}^{[-n_i,n_i]}$. Let n be the maximum of the finitely many n_i; then $\mathbf{R}u_{i*}\mathcal{O}_{U_i} \in \overline{\langle G\rangle}_n^{[-n,n]} \subset \overline{\langle G\rangle}_n^{(-\infty,\infty)}$ for every i in the finite set.

Next, for each i we know that $U_i = \mathrm{Spec}(R_i)$ with R_i is of finite global dimension, and Theorem 7.5 *tells us that we may choose an integer $\ell > 0$ so that, for*

[5]This isn't immediate from the definition of approximability, but follows from the structural theorems. We are using the fact that $\mathbf{R}u_*\mathcal{O}_U \in \mathbf{D_{qc}}(X)^{\leq m}$ for some $m > 0$, coupled with the fact that one can approximate objects in $\mathbf{D_{qc}}(X)^{\leq 0}$ to arbitrary order, not just to order zero as given in the definition. See Sketch 8.19(i) for more detail.

every one of the finitely many i with $1 \le i \le r$, we have $\mathbf{D_{qc}}(U_i) = \overline{\langle \mathcal{O}_{U_i} \rangle}_\ell^{(-\infty,\infty)}$. It follows that

$$\mathbf{R}u_{i*}\mathbf{D_{qc}}(U_i) = \mathbf{R}u_{i*}\left[\overline{\langle \mathcal{O}_{U_i} \rangle}_\ell^{(-\infty,\infty)}\right] \subset \overline{\langle \mathbf{R}u_{i*}\mathcal{O}_{U_i} \rangle}_\ell^{(-\infty,\infty)} \subset \overline{\langle G \rangle}_{\ell n}^{(-\infty,\infty)}$$

Let $\mathcal{V} = \text{add}\big[\cup_{i=1}^r \mathbf{R}u_{i*}\mathbf{D_{qc}}(U_i)\big]$, with the notation as in Reminder 4.1(ii). By the displayed inclusion above $\big[\cup_{i=1}^r \mathbf{R}u_{i*}\mathbf{D_{qc}}(U_i)\big] \subset \overline{\langle G \rangle}_{\ell n}^{(-\infty,\infty)}$, and as $\overline{\langle G \rangle}_{\ell n}^{(-\infty,\infty)}$ is closed under (finite) coproducts it follows that $\mathcal{V} \subset \overline{\langle G \rangle}_{\ell n}^{(-\infty,\infty)}$.

It's an exercise to show that

$$\mathbf{D_{qc}}(X) \quad = \quad \underbrace{\mathcal{V} * \mathcal{V} * \cdots * \mathcal{V}}_{r \text{ copies}}$$

with the notation as in Reminder 4.1(i). Hence $\mathbf{D_{qc}}(X) = \overline{\langle G \rangle}_{\ell n r}^{(-\infty,\infty)}$. We have proved a statement about $\mathbf{D_{qc}}(X)$, and in Notation 2.13 we learned that $\mathbf{D}^{\text{perf}}(X)$ is equal to the subcategory of compact objects in $\mathbf{D_{qc}}(X)$. Standard compactness arguments tell us that from the equality $\mathbf{D_{qc}}(X) = \overline{\langle G \rangle}_{\ell n r}^{(-\infty,\infty)}$ we can formally deduce the equality $\mathbf{D}^{\text{perf}}(X) = \cup_{m>0}\langle G \rangle_{\ell n r}^{[-m,m]}$. □

We want to highlight the power of approximability. Sketch 7.7 was meant to show the expert that Theorem 7.1 is easy to deduce by combining Kelly's old theorem with Lemma 7.6, and the proof of Lemma 7.6 displays how the lemma follows immediately from the fact that $\mathbf{D_{qc}}(X)$ is approximable.

While we're into exhibiting the power of approximability, let us mention another corollary of Lemma 7.6—and therefore another easy consequence of approximability.

THEOREM 7.8. *Suppose $f : X \longrightarrow Y$ is a separated morphism of quasicompact, quasiseparated schemes. If $\mathbf{R}f_* : \mathbf{D_{qc}}(X) \longrightarrow \mathbf{D_{qc}}(Y)$ takes perfect complexes to complexes of bounded-below Tor-amplitude then f must be of finite Tor-dimension.*

REMINDER 7.9. *We owe the reader a glossary of the technical terms in the statement of Theorem 7.8.*

(i) *Given a morphism of schemes $f : X \longrightarrow Y$, for any $x \in X$ there is an induced ring homomorphism $\mathcal{O}_{Y,f(x)} \longrightarrow \mathcal{O}_{X,x}$ of the stalks. The map f is* of finite Tor-dimension *at x if $\mathcal{O}_{X,x}$ has a finite flat resolution over $\mathcal{O}_{Y,f(x)}$.*
(ii) *The map f is* of finite Tor-dimension *if it is of finite Tor-dimension at every $x \in X$.*
(iii) *The complex $C \in \mathbf{D_{qc}}(Y)$ is* of bounded-below Tor-amplitude *if, for every open immersion $u : U \longrightarrow Y$ with $U =$ Spec(R) affine, the complex $u^*C \in \mathbf{D_{qc}}(U) \cong \mathbf{D}(R)$ is isomorphic to a bounded-below K-flat complex.*

HISTORICAL SURVEY 7.10. *We should tell the reader what was known in the direction of Theorem 7.8. If the schemes X and Y are noetherian and $f : X \longrightarrow Y$ is of finite type, then the converse of Theorem 7.8 is known and old—the reader may find it in Illusie* [**19**, Corollaire 4.8.1]. *The direction proved in Theorem 7.8 was open for a long time, the first progress was in* [**29**]. *But the statement in* [**29**] *is much narrower than Theorem 7.8, it is confined to the situation where f is proper.*

SKETCH 7.11. *We should give the reader some idea why Theorem 7.8 follows easily from Lemma 7.6—this discussion is for algebraic geometers, the non-specialists are advised to skip ahead to Theorem 7.12.*

It's obviously local in Y to determine if f is of finite Tor-dimension. Using the main theorem of Thomason and Trobaugh [**41**], *it's also local in Y to determine whether $\mathbf{R}f_*$ takes perfect complexes to complexes of bounded-below Tor-amplitude. Hence we may assume Y is affine, therefore separated. As f is separated we deduce that X must be separated.*

We are given that $\mathbf{R}f_*$ takes perfect complexes to complexes of bounded-below Tor-amplitude, and wish to show that f is of finite Tor-dimension. Being of finite Tor-dimension is local in X; it suffices to show that, for each of open immersion $u : U \longrightarrow X$ with U affine, the composite $U \xrightarrow{u} X \xrightarrow{f} Y$ is of finite Tor-dimension. By Lemma 7.6 there exists a perfect complex $G \in \mathbf{D_{qc}}(X)$ and an integer $n > 0$ with $\mathbf{R}u_*\mathcal{O}_U \in \overline{\langle G\rangle}_n^{[-n,n]}$. Therefore

$$\begin{array}{ccccc} (fu)_*\mathcal{O}_U & \cong & \mathbf{R}(fu)_*\mathcal{O}_U & \cong & \mathbf{R}f_*\mathbf{R}u_*\mathcal{O}_U \\ & \in & \mathbf{R}f_*\left[\overline{\langle G\rangle}_n^{[-n,n]}\right] & \subset & \overline{\langle \mathbf{R}f_*G\rangle}_n^{[-n,n]} \end{array}$$

But $\mathbf{R}f_*G$ is of bounded below Tor-amplitude by hypothesis, and in forming $\overline{\langle \mathbf{R}f_*G\rangle}_n^{[-n,n]}$ we only allow $\mathbf{R}f_*G[i]$ with $-n \leq i \leq n$, coproducts, extensions and direct summands. Hence the objects of $\overline{\langle \mathbf{R}f_*G\rangle}_n^{[-n,n]}$ have Tor-amplitude uniformly bounded below.

It's time to turn our attention to Fact 6.3(ii), we remind the reader of the statement:

THEOREM 7.12. *Let X be a separated, noetherian, finite-dimensional scheme, and assume that every closed, reduced, irreducible subscheme of X has a regular alteration. Then the category $\mathbf{D}^b_{\mathrm{coh}}(X)$ is strongly generated.*

HISTORICAL SURVEY 7.13. *We should tell the reader what was known in the direction of Theorem 7.12. We have already alluded to the fact that, when X is regular and finite-dimensional, the inclusion $\mathbf{D}^{\mathrm{perf}}(X) \longrightarrow \mathbf{D}^b_{\mathrm{coh}}(X)$ is an equivalence and Theorem 7.1 tells us that the equivalent categories $\mathbf{D}^{\mathrm{perf}}(X) \cong \mathbf{D}^b_{\mathrm{coh}}(X)$ are strongly generated. Using a stratification, of a possibly singular X, Rouquier* [**39**, Theorem 7.38] *built and substantially extended on the argument in Bondal and Van den Bergh* [**8**, Theorem 3.1.4] *to show that $\mathbf{D}^b_{\mathrm{coh}}(X)$ is strongly generated whenever X is a separated scheme of finite type over a perfect field k. The preprint by Keller and Van den Bergh* [**24**, Proposition 5.1.2] *generalized to separated schemes of finite type over arbitrary fields, but this Proposition disappeared in the passage to the published version* [**23**]. *The reader might also wish to look at Lunts* [**30**, Theorem 6.3] *for a different approach to the proof, but still using stratifications. If we specialize the result of Rouquier, extended by Keller and Van den Bergh, to the case where $X = \mathrm{Spec}(R)$ is an affine scheme, we learn that $\mathbf{D}^b(R\text{–mod})$ is strongly generated whenever R is of finite type over a field k.*

Note that, while Theorem 7.1 is easy and classical in the case where X is affine, Theorem 7.12 is neither easy nor classical for affine X. *In recent years there has been interest among commutative algebraists in understanding this better: the reader is referred to Aihara and Takahashi* [**2**], *Bahlekeh, Hakimian, Salarian* *and Takahashi* [**5**] *and Iyengar and Takahashi* [**21**] *for a sample of the literature. There is also a connection with the concept of the radius of the (abelian) category of modules over R; see Dao and Takahashi* [**11, 12**] *and Iyengar and Takahashi* [**21**]. *The union of the known results seems to be that $\mathbf{D}^b(R\text{–mod})$ is strongly generated if R is an equicharacteristic excellent local ring, or essentially of finite type over a field—see* [**21**, Corollary 7.2]. *In* [**21**, Remark 7.3] *it is observed that there are examples of commutative, noetherian rings for which $\mathbf{D}^b(R\text{–mod})$ is* not *strongly generated.*

The structure of the proof of Theorem 7.12 *(see Sketch* 7.14*) is that one passes to regular alterations of X and its closed subschemes. Assuming X affine is no help with the approximability proof of Theorem* 7.12*—when X is affine and singular we end up proving a result in commutative algebra, but the technique of the proof passes through non-affine schemes.*

Unlike all the pre-approximability results, except Kelly's, Theorems 7.1 *and* 7.12 *do not assume equal characteristic.*

SKETCH 7.14. *We should tell the reader a little about the proof. But first we should make it clear that Theorem* 7.12 *will not be proved using approximability directly, instead we will prove it as a corollary of Theorem* 7.1*, which followed from approximability. Precisely: Theorem* 7.1 *and Theorem* 7.12 *are identical when X is a finite-dimensional, regular, noetherian, separated scheme. And the idea is to reduce to this case.*

Resolutions of singularities might look tempting, but in mixed characteristic they are known to exist only in low dimension. So the key is that we can get by with regular alterations—the hypotheses of the theorem say that they exist for every closed subvariety of X, and it turns out that Theorem 7.12 *can be deduced from this using induction on the dimension of X and two old theorems of Thomason's.*

This survey has been stressing that the hard work goes into proving approximability, the consequences are all easy corollaries. Theorem 7.12 *must count as an exception, the argument is tricky. It might be relevant to note that in this field—noncommutative algebraic geometry—there are quite a number of theorems that are known in characteristic zero with proofs that rely on resolutions of singularities, and conjectured in positive characteristic. I wasn't the first to come up with the idea of trying to use de Jong's theorem, in other words trying to prove these conjectures using regular alterations. So far Theorem* 7.12 *is the only success story. It isn't regular alterations alone that do the trick, it's the combination of regular alterations and support theory—in this case support theory manifests itself as the two old theorems of Thomason's.*

PROBLEM 7.15. *There is a non-commutative version—Kelly's old theorem doesn't assume commutativity. This raises the obvious question: to what extent do the more recent theorems extend beyond commutative algebraic geometry?*

Perhaps we should explain, and for simplicity let us stick to the case where $X = \text{Spec}(R)$ is affine. As we have presented the theory, up to now, we have implicitly been assuming that the ring R is commutative. But what Kelly proved doesn't depend on commutativity—the reader can see this for herself, just look at the proof of Theorem 7.5. *Let R be any associative ring and let $\mathbf{D}^b(R\text{–proj})$ be the derived category of bounded complexes of finitely generated, projective R–modules. Kelly's 1965 theorem says that $D^b(R\text{–proj})$ has a strong generator if and only if R is of finite global dimension.*

All the later theorems listed above, including the recent ones whose proof relies on approximability, assume commutativity. In particular: assume R is a commutative, noetherian ring, of finite type over an excellent ring of dimension ≤ 2. Theorem 7.12*, in the special case where* $X = \mathrm{Spec}(R)$*, tells us that the category* $\mathbf{D}^b_{\mathrm{coh}}(X) \cong \mathbf{D}^b(R\text{–mod})$ *is strongly generated. The category* $\mathbf{D}^b(R\text{–mod})$ *has for objects the bounded complexes of finite R–modules.*

Is the commutativity hypothesis necessary in the above? Is there some large class of noncommutative, noetherian rings for which $\mathbf{D}^b(R\text{–mod})$ *is strongly generated? The proof in the commutative case, which goes by way of the regular alterations of de Jong, doesn't seem capable of a noncommutative extension.*

REMARK 7.16. Recall: a strong generator in $\mathcal{S}$ is an object $G \in \mathcal{S}$ such that, for some integer $\ell > 0$, we have $\mathcal{S} = \cup_n \langle G \rangle_\ell^{[-n,n]}$. One can ask for estimates on ℓ. This leads to the definitions

(i) Given objects $G, F \in \mathcal{S}$, the *G–level of F* is the smallest integer ℓ such that $F \in \cup_n \langle G \rangle_{\ell+1}^{[-n,n]}$. This notion is due to Avramov, Buchweitz and Iyengar [**4**].
(ii) Let G be an object of $\mathcal{S}$. The *generation time* of G is the smallest ℓ for which $\mathcal{S} = \cup_n \langle G \rangle_{\ell+1}^{[-n,n]}$. The set of all possible generation times, taken over all strong generators $G \in \mathcal{S}$, is known as the *Orlov spectrum* of $\mathcal{S}$. These notions first appeared in Orlov [**36**].
(iii) The *Rouquier dimension* of $\mathcal{S}$ is the smallest integer ℓ such that there exists a G with $\mathcal{S} = \cup_n \langle G \rangle_{\ell+1}^{[-n,n]}$. This integer first appeared in Rouquier [**39**]—Rouquier's name for this number was just plain "dimension".

There are several conjectures, and many papers estimating these numbers—almost all in the equal charateristic case, after all until recently there was no existence theorem of strong generators in mixed characteristic. One can ask if the theorems surveyed in this article give good bounds in mixed characteristic—and the short answer is No. In more detail:

(iv) If we assume that X is regular and quasiprojective, then the proof of Application 1(i) is effective. It gives an explicit upper bound on the Rouquier dimension of $\mathbf{D}^{\mathrm{perf}}(X) = \mathbf{D}^b_{\mathrm{coh}}(X)$. But the bound is dreadful.
(v) If we drop the quasiprojectivity hypothesis, and/or if we allow singularities, then the proof becomes ineffective. It proves the existence of an integer $\ell > 0$ and a generator G with $\mathbf{D}^b_{\mathrm{coh}}(X) = \cup_n \langle G \rangle_\ell^{[-n,n]}$, but there is no estimate on ℓ.

ELABORATION 7.17. *The following is for the benefit of the readers who would like Remark* 7.16 *spelt out a little more.*

In [**33**, Section 4] *the reader can find the argument that proves the approximability of* $\mathbf{D}_{\mathbf{qc}}(X)$ *when X is quasiprojective—and in* [**33**, Proposition 4.4] *it's made clear that the estimates are explicit. And, assuming X is not only quasiprojective but also regular, Sketch* 7.7 *shows us how to pass from the estimates given by approximability to explicit estimates on the ℓ for which* $\mathbf{D}^{\mathrm{perf}}(X) = \cup_n \langle G \rangle_{\ell+1}^{[-n,n]}$. *Still assuming that X is quasiprojective and regular, a careful reading of the proof of* [**33**, Proposition 4.4] *will show us that these crude estimates can easily be improved. But our point for now is that the proof is effective, it gives bounds.*

The general proof of approximability, for quasicompact, separated X, is by reduction to the quasiprojective case. This reduction goes by

(i) *We first use noetherian approximation to reduce to the case where X is of finite type over $\mathbb{Z}$.*
(ii) *Next we use induction on the dimension coupled with Chow's Lemma to reduce to the quasiprojective case.*

The way to use noetherian approximation is straightforward enough—we don't go into detail, leaving this to the reader's imagination. In principle this part could be made effective, but not in a way that yields good bounds. Still: so far there are bounds of some sort.

But the true subtlety arises in (ii), with the way we use Chow's Lemma. Chow's Lemma produces for us a birational, projective morphism $\pi : \widetilde{X} \longrightarrow X$ with $\widetilde{X}$ quasiprojective. And two remarks are in order

(iii) *Chow's Lemma is where we have to assume X separated. This is the point of the proof where it doesn't suffice for X to be quasiseparated.*
(iv) *Since $\widetilde{X}$ is quasiprojective the category $\mathbf{D_{qc}}(\widetilde{X})$ is approximable. To deduce from this useful information about $\mathbf{D_{qc}}(X)$ we need to take a compact generator $\widetilde{G} \in \mathbf{D_{qc}}(\widetilde{X})$, and approximate $\mathbf{R}\pi_*\widetilde{G}$ using a compact generator $G \in \mathbf{D_{qc}}(X)$.*

There is a way to achieve (iv), it relies on [**29**, Theorem 4.1]. *But the proof of* [**29**, Theorem 4.1] *is homotopy-theoretic, it hinges on the fact that, for a compact object G, the functor* $\mathrm{Hom}(G,-)$ *commutes with homotopy colimits. Ignoring the technicality that we work with homotopy colimits and not ordinary colimits, the point is the following. Any map from a compact object G to an object $\operatorname{colim}\limits_{\longrightarrow} T_i$ factors through some T_i*, but we have no control over how large $i \in \mathbb{N}$ will have to be. *Thus any argument which has, embedded in it, the appeal to such homotopy colimit arguments, is inherently and hopelessly ineffective. This explains why we lose control when X isn't assumed quasiprojective.*

If we allow X to become singular, then Sketch 7.14 *hints how to reduce to the regular case using regular alterations. As was said in Sketch* 7.14*: the proof appeals to two old theorems of Thomason's, both of which depend on homotopy-colimit arguments. Hence this passage is also ineffective beyond salvation.*

8. More about finite R–linear functors $H : \left[\mathbf{D}^{\mathrm{perf}}(X)\right]^{\mathrm{op}} \longrightarrow R\text{–Mod}$ and $\widetilde{H} : \mathbf{D}^b_{\mathrm{coh}}(X) \longrightarrow R\text{–Mod}$

It's time to expand on Facts 6.3 (iii) and (iv). We begin by recalling the statements.

THEOREM 8.1. *Let X be a finite-dimensional scheme proper over a noetherian ring R. Let $\mathcal{Y}$ the Yoneda map*

$$\mathbf{D}^b_{\mathrm{coh}}(X) \xrightarrow{\quad \mathcal{Y} \quad} \mathrm{Hom}_R\Big(\mathbf{D}^{\mathrm{perf}}(X)^{\mathrm{op}}, R\text{–Mod}\Big)$$

taking $B \in \mathbf{D}^b_{\mathrm{coh}}(X)$ to the functor $\mathrm{Hom}(-,B)$, *and let $\widetilde{\mathcal{Y}}$ be the Yoneda map*

$$\left[\mathbf{D}^{\mathrm{perf}}(X)\right]^{\mathrm{op}} \xrightarrow{\quad \widetilde{\mathcal{Y}} \quad} \mathrm{Hom}_R\Big(\mathbf{D}^b_{\mathrm{coh}}X, R\text{–Mod}\Big)$$

taking $A \in \mathbf{D}^{\text{perf}}(X)$ *to the functor* $\text{Hom}(A,-)$. *Assuming every closed subvariety of* X *admits a regular alteration both functors are fully faithful, and in each case the essential image is the set of finite* R*–linear homological functors. Recall: an* R*–linear homological functor* $H : \mathcal{S} \longrightarrow R$–Mod *is* finite *if, for all objects* $C \in \mathcal{S}$, *the* R*–module* $\oplus_n H\big(C[n]\big)$ *is finite.*

For the functor $\mathcal{Y}$ *the assertion is true even without the hypotheses of finite-dimensionality and the existence of regular alterations.*

Historical Survey 8.2. *We remind the reader what was known before.*

(i) *If* X *is proper over* R, *if* $A \in \mathbf{D}^{\text{perf}}(X)$ *and if* $B \in \mathbf{D}^b_{\text{coh}}(X)$, *then*

$$\text{Hom}(A[i], B) \quad \cong \quad H^{-i}(A^\vee \otimes B)$$

is a finite R*–module for every* i *and vanishes outside a bounded range. This much was proved by Grothendieck* [**17**, Théorème 3.2.1].

Translating to the language of Theorem 8.1*: given objects* $A \in \mathbf{D}^{\text{perf}}(X)$ *and* $B \in \mathbf{D}^b_{\text{coh}}(X)$, *then* $\mathcal{Y}(B) = \text{Hom}(-, B)$ *is a finite homological functor on* $\big[\mathbf{D}^{\text{perf}}(X)\big]^{\text{op}}$, *while* $\widetilde{\mathcal{Y}}(A) = \text{Hom}(A,-)$ *is a finite homological functor on* $\mathbf{D}^b_{\text{coh}}(X)$. *This much has been known since 1961.*

(ii) As long as R is a field, *Bondal and Van den Bergh* [**8**, Theorem A.1] *proved that every finite homological functor on* $\big[\mathbf{D}^{\text{perf}}(X)\big]^{\text{op}}$ *is* $\text{Hom}(-, B)$ *for some* $B \in \mathbf{D}^b_{\text{coh}}(X)$. *In the language of Theorem* 8.1*: they proved that the essential image of* $\mathcal{Y}$ *consists of the finite homological functors.*

(iii) Still assuming R is a field, *the assertion of Theorem* 8.1 *about the functor* $\widetilde{\mathcal{Y}}$ *can be found in Rouquier* [**39**, Corollary 7.51(ii)]*—although the author of the present article doesn't follow the argument in* [**39**] *that briefly outlines how a proof might go, it's too skimpy.*

If R *is a field Theorem* 8.1 *improves on what was known about the functor* $\mathcal{Y}$ *by showing that it's fully faithful. And for* R *more general Theorem* 8.1 *is new, for both the functor* $\mathcal{Y}$ *and the functor* $\widetilde{\mathcal{Y}}$.

And now the time has come to tell the reader something about the proof of Theorem 8.1. It turns out that the theorem is an immediate corollary of a far more general fact, and the discussion of this result brings us naturally to the structure that all approximable categories share. Let us begin in even greater generality, not assuming all the hypotheses of approximability.

Definition 8.3. *Let* $\mathcal{T}$ *be a triangulated category, and let* $\big(\mathcal{T}_1^{\leq 0}, \mathcal{T}_1^{\geq 0}\big)$ *and* $\big(\mathcal{T}_2^{\leq 0}, \mathcal{T}_2^{\geq 0}\big)$ *be two t–structures on* $\mathcal{T}$. *We declare them* equivalent *if there exists an integer* $A > 0$ *with* $\mathcal{T}_1^{\leq -A} \subset \mathcal{T}_2^{\leq 0} \subset \mathcal{T}_1^{\leq A}$.

The definition agrees with the intuition of the Introduction: each t–structure defines a kind of (directed) metric, and we'd like declare t–structures equivalent whenever they induce equivalent metrics. And now we recall

Remark 8.4. Let $\mathcal{T}$ be a triangulated category with coproducts, and let $G \in \mathcal{T}$ be a compact generator. From Alonso, Jeremías and Souto [**3**, Theorem A.1] we learn that $\mathcal{T}$ has a unique t–structure $\big(\mathcal{T}_G^{\leq 0}, \mathcal{T}_G^{\geq 0}\big)$ *generated by* G.

It is not difficult to show that, if G and H are two compact generators for $\mathcal{T}$, then the t–structures $\big(\mathcal{T}_G^{\leq 0}, \mathcal{T}_G^{\geq 0}\big)$ and $\big(\mathcal{T}_H^{\leq 0}, \mathcal{T}_H^{\geq 0}\big)$ are equivalent. Thus *up to equivalence* there is a preferred t–structure on $\mathcal{T}$, namely $\big(\mathcal{T}_G^{\leq 0}, \mathcal{T}_G^{\geq 0}\big)$ where G is a compact generator. We say that a t-structure $\big(\mathcal{T}^{\leq 0},\mathcal{T}^{\geq 0}\big)$ is in the *preferred equivalence class* if it is equivalent to $\big(\mathcal{T}_G^{\leq 0},\mathcal{T}_G^{\geq 0}\big)$ for some compact generator G, hence for every compact generator.

DISCUSSION 8.5. *Given a t–structure* $\big(\mathcal{T}^{\leq 0},\mathcal{T}^{\geq 0}\big)$ *it is customary to define the categories*

$$\mathcal{T}^- = \cup_n \mathcal{T}^{\leq n}\,, \qquad \mathcal{T}^+ = \cup_n \mathcal{T}^{\geq -n}\,, \qquad \mathcal{T}^b = \mathcal{T}^- \cap \mathcal{T}^+$$

as in Notation 2.19. *It's obvious from Definition* 8.3 *that equivalent t–structures yield identical* $\mathcal{T}^-$, $\mathcal{T}^+$ *and* $\mathcal{T}^b$.

Now assume we are in the situation of Remark 8.4, *that is* $\mathcal{T}$ *has coproducts and there exists a single compact generator* G. *Then there is a preferred equivalence class of t–structures and, correspondingly, preferred* $\mathcal{T}^-$, $\mathcal{T}^+$ *and* $\mathcal{T}^b$. *These are intrinsic, they're independent of any choice. In the remainder of the article we only consider the "preferred"* $\mathcal{T}^-$, $\mathcal{T}^+$ *and* $\mathcal{T}^b$.

Slightly more sophisticated is the category $\mathcal{T}_c^-$ below.

DEFINITION 8.6. *Let* $\mathcal{T}$ *be a triangulated category with coproducts, and assume it has a compact generator* G. *Choose a t–structure* $\big(\mathcal{T}^{\leq 0},\mathcal{T}^{\geq 0}\big)$ *in the preferred equivalence class. The full subcategory* $\mathcal{T}_c^-$ *is defined by*

$$\mathcal{T}_c^- \quad = \quad \left\{ F\in\mathcal{T} \,\middle|\, \begin{array}{c} \textit{For all integers } n>0 \textit{ there exists a triangle} \\ E\longrightarrow F\longrightarrow D\longrightarrow E[1] \\ \textit{with } E \textit{ compact and } D\in\mathcal{T}^{\leq -n-1} \end{array} \right\}$$

We furthermore define $\mathcal{T}_c^b = \mathcal{T}^b \cap \mathcal{T}_c^-$.

REMARK 8.7. Intuitively the category $\mathcal{T}_c^-$ is the closure, with respect to the metric induced by the t-structure $\big(\mathcal{T}^{\leq 0},\mathcal{T}^{\geq 0}\big)$, of the subcategory $\mathcal{T}^c$ of all compact objects in $\mathcal{T}$. It's obvious that the category $\mathcal{T}_c^-$ is intrinsic, after all equivalent metrics will lead to the same closure. And as $\mathcal{T}_c^-$ and $\mathcal{T}^b$ are both intrinsic, so is their intersection $\mathcal{T}_c^b$.

We have defined all this intrinsic structure, assuming only that $\mathcal{T}$ is a triangulated category with coproducts and with a single compact generator. In this generality we know that the subcategories $\mathcal{T}^-$, $\mathcal{T}^+$ and $\mathcal{T}^b$ are thick. For the subcategories $\mathcal{T}_c^-$ and $\mathcal{T}_c^b$ one proves

PROPOSITION 8.8. *If* $\mathcal{T}$ *has a compact generator* G, *such that* $\operatorname{Hom}\big(G,G[n]\big)=0$ *for* $n\gg 0$, *then the subcategories* $\mathcal{T}_c^-$ *and* $\mathcal{T}_c^b$ *are thick.*

REMARK 8.9. If $\mathcal{T}$ is approximable then, by Definition 5.1(i), there is an integer $A>0$, a compact generator $G\in\mathcal{T}$ and a t-structure $\big(\mathcal{T}^{\leq 0},\mathcal{T}^{\geq 0}\big)$, so that $\operatorname{Hom}\big(G[-A],\mathcal{T}^{\leq 0}\big)=0$ and $G[A]\in\mathcal{T}^{\leq 0}$. Hence $G[n]\in\mathcal{T}^{\leq 0}$ for all $n\geq A$ and $\operatorname{Hom}\big(G,G[n]\big)=0$ for all $n\geq 2A$; Proposition 8.8 therefore tells us that the subcategories $\mathcal{T}_c^-$ and $\mathcal{T}_c^b$ are thick whenever $\mathcal{T}$ is approximable.

Of course it would be nice to be able to work out examples: what does all of this intrinsic structure come down to in special cases? This is where approximability helps. We first note

PROPOSITION 8.10. *Assume the category* $\mathcal{T}$ *is approximable; see Definition* 5.1. *We recall part of the definition: the category* $\mathcal{T}$ *is approximable if it has a compact* *generator G, a t–structure $\big(\mathcal{T}^{\leq 0},\mathcal{T}^{\geq 0}\big)$ and an integer $A>0$ satisfying some properties, see Definition* 5.1 *(i) and (ii) for the properties.*

Then any t–structure, which comes as part of a triad satisfying the properties of Definition 5.1 *(i) and (ii), must be in the preferred equivalence class. Furthermore: for any compact generator G' and any t–structure $\big(\mathcal{T}^{\leq 0},\mathcal{T}^{\geq 0}\big)$ in the preferred equivalence class, there is an integer $A'>0$ so that the properties of Definition* 5.1 *(i) and (ii) hold.*

In practice this means that, in proving that $\mathcal{T}$ is approximable, we must produce at least one useful t–structure that we know belongs to the preferred equivalence class. After all: this t–structure must be manageable enough to lend itself to a proof that the conditions in Definition 5.1 (i) and (ii) hold. Note that the proof of Alonso, Jeremías and Souto [**3**, Theorem A.1] yields a t–structure $\big(\mathcal{T}_G^{\leq 0},\mathcal{T}_G^{\geq 0}\big)$ in the preferred equivalence class, but the construction is a little opaque—it shows existence and uniqueness, but usually doesn't give us much of a handle on $\big(\mathcal{T}_G^{\leq 0},\mathcal{T}_G^{\geq 0}\big)$. So while we know that t–structures in the preferred equivalence class exist, this needn't be especially useful in working with them.

Let X be a quasicompact, separated scheme. We have told the reader that [**33**, Theorem 5.8] combined with [**34**, Example 3.6] prove that $\mathcal{T}=\mathbf{D_{qc}}(X)$ is approximable; the t–structure used in the proof happens to be the standard t–structure of Example 2.16 (iii) and (iv). We remind the reader: the standard t–structure on $\mathbf{D_{qc}}(X)$ has

$$\begin{aligned}\mathbf{D_{qc}}(X)^{\leq 0} &= \{F\in\mathbf{D_{qc}}(X)\mid \mathcal{H}^i(F)=0 \text{ for all } i>0\}\\ \mathbf{D_{qc}}(X)^{\geq 0} &= \{F\in\mathbf{D_{qc}}(X)\mid \mathcal{H}^i(F)=0 \text{ for all } i<0\}\end{aligned}$$

where $\mathcal{H}^i$ is the functor taking a cochain complex to its ith cohomology sheaf. Proposition 8.10 now informs us that the standard t–structure must belong to the preferred equivalence class. Hence the categories $\mathcal{T}^-$, $\mathcal{T}^+$ and $\mathcal{T}^b$ are the usual: we have $\mathcal{T}^-=\mathbf{D}^-_{\mathbf{qc}}(X)$, $\mathcal{T}^+=\mathbf{D}^+_{\mathbf{qc}}(X)$ and $\mathcal{T}^b=\mathbf{D}^b_{\mathbf{qc}}(X)$. The subcategories $\mathbf{D}^-_{\mathbf{qc}}(X)$, $\mathbf{D}^+_{\mathbf{qc}}(X)$ and $\mathbf{D}^b_{\mathbf{qc}}(X)$ of $\mathbf{D_{qc}}(X)$ are traditionally defined to be

$$\begin{aligned}\mathbf{D}^-_{\mathbf{qc}}(X) &= \{F\in\mathbf{D_{qc}}(X)\mid \mathcal{H}^i(F)=0 \text{ for all } i\gg 0\}\\ \mathbf{D}^+_{\mathbf{qc}}(X) &= \{F\in\mathbf{D_{qc}}(X)\mid \mathcal{H}^i(F)=0 \text{ for all } i\ll 0\}\\ \mathbf{D}^b_{\mathbf{qc}}(X) &= \mathbf{D}^-_{\mathbf{qc}}(X)\cap\mathbf{D}^+_{\mathbf{qc}}(X)\end{aligned}$$

Next we ask ourselves: what about $\mathcal{T}^-_c$ and $\mathcal{T}^b_c$? We begin with the affine case.

EXERCISE 8.11. *Let R be a ring. Prove that, in the category $\mathcal{T}=\mathbf{D}(R)$, the subcategory $\mathcal{T}^-_c$ agrees with the $\mathbf{D}^-(R\text{–proj})$ of Remark* 4.7.

OBSERVATION 8.12. Now assume R is a commutative ring, and let $X=\operatorname{Spec}(R)$. Then [**7**, Theorem 5.1] tells us that the natural functor $\mathbf{D}(R)\longrightarrow\mathbf{D_{qc}}(X)$ is an equivalence of categories. Putting $\mathcal{T}=\mathbf{D_{qc}}(X)\cong\mathbf{D}(R)$, we learn from Exercise 8.11 what the category $\mathcal{T}^-_c$ is.

Now let X be any quasicompact, separated scheme. If $u:U\longrightarrow X$ is an open immersion, then the functor $u^*:\mathbf{D_{qc}}(X)\longrightarrow\mathbf{D_{qc}}(U)$ respects the standard t–structure and sends compact objects in $\mathbf{D_{qc}}(X)$ to compact objects in $\mathbf{D_{qc}}(U)$. Hence $u^*\mathbf{D_{qc}}(X)^-_c\subset\mathbf{D_{qc}}(U)^-_c$. Thus every object in $\mathbf{D_{qc}}(X)^-_c$ must be "locally in $\mathbf{D}^-(R\text{–proj})$", meaning that for every open immersion $u:\operatorname{Spec}(R)\longrightarrow X$ we must have that $u^*\mathbf{D_{qc}}(X)^-_c\subset\mathbf{D}^-(R\text{–proj})$. The objects "locally in $\mathbf{D}^-(R\text{–proj})$"

were first studied by Illusie [**19**, **20**] in SGA6. They have a name, they are the pseudocoherent complexes.

The next result is not so obvious. In [**29**, Theorem 4.1] the reader can find a proof that

PROPOSITION 8.13. *Let X be a quasicompact, separated scheme. Then every pseudocoherent complex belongs to $\mathbf{D_{qc}}(X)_c^-$. Coupled with Observation 8.12 this teaches us that the objects of $\mathbf{D_{qc}}(X)_c^-$ are precisely the pseudocoherent complexes.*

REMARK 8.14. From now on we will assume the scheme X noetherian and separated. In this case pseudocoherence simplifies: we have $\mathbf{D_{qc}}(X)_c^- = \mathbf{D}^-_{\text{coh}}(X)$. The objects $F \in \mathbf{D}^-_{\text{coh}}(X)$ are the complexes whose cohomology sheaves $\mathcal{H}^n(F)$ are coherent for all n, and vanish if $n \gg 0$. And $\mathbf{D_{qc}}(X)_c^b$ is also explicit: it is our old friend, the category traditionally denoted $\mathbf{D}^b_{\text{coh}}(X)$—we first met $\mathbf{D}^b_{\text{coh}}(X)$ in Example 2.3(iv), and it figures prominently in the statement of Theorem 8.1.

REMARK 8.15. In Remark 8.14 we observed that the category $\mathbf{D}^b_{\text{coh}}(X)$ has an intrinsic description as a subcategory of $\mathcal{T} = \mathbf{D_{qc}}(X)$, it is $\mathcal{T}^b_c$. The category $\mathbf{D}^{\text{perf}}(X)$ also has an intrinsic description, it's the subcategory $\mathcal{T}^c$ of all compact objects in $\mathcal{T} = \mathbf{D_{qc}}(X)$, see Notation 2.13. With $\mathcal{T} = \mathbf{D_{qc}}(X)$, Theorem 8.1 is a statement about the categories $\mathcal{T}^c = \mathbf{D}^{\text{perf}}(X)$ and $\mathcal{T}^b_c = \mathbf{D}^b_{\text{coh}}(X)$—and it turns out to be a special case of the following, infinitely more general assertion.

THEOREM 8.16. *Let R be a noetherian ring, and let $\mathcal{T}$ be an R–linear, approximable triangulated category. Suppose there exists in $\mathcal{T}$ a compact generator G so that* $\text{Hom}\big(G, G[n]\big)$ *is a finite R–module for all $n \in \mathbb{Z}$. Consider the two functors*

$$\mathcal{Y} : \mathcal{T}^-_c \longrightarrow \text{Hom}_R\left([\mathcal{T}^c]^{\text{op}}, R\text{–Mod}\right), \qquad \widetilde{\mathcal{Y}} : \left[\mathcal{T}^-_c\right]^{\text{op}} \longrightarrow \text{Hom}_R\left(\mathcal{T}^b_c, R\text{–Mod}\right)$$

defined by the formulas $\mathcal{Y}(B) = \text{Hom}(-,B)$ *and* $\widetilde{\mathcal{Y}}(A) = \text{Hom}(A,-)$. *Note that, in these formulas, we permit all $A, B \in \mathcal{T}^-_c$. But the $(-)$ in the formula* $\mathcal{Y}(B) = \text{Hom}(-,B)$ *is assumed to belong to $\mathcal{T}^c$, whereas the $(-)$ in the formula* $\widetilde{\mathcal{Y}}(A) = \text{Hom}(A,-)$ *must lie in $\mathcal{T}^b_c$. Now consider the following composites*

$$\mathcal{T}^b_c \overset{i}{\hookrightarrow} \mathcal{T}^-_c \xrightarrow{\ \mathcal{Y}\ } \text{Hom}_R\left([\mathcal{T}^c]^{\text{op}}, R\text{–Mod}\right)$$

$$\left[\mathcal{T}^c\right]^{\text{op}} \overset{\widetilde{\imath}}{\hookrightarrow} \left[\mathcal{T}^-_c\right]^{\text{op}} \xrightarrow{\ \widetilde{\mathcal{Y}}\ } \text{Hom}_R\left(\mathcal{T}^b_c, R\text{–Mod}\right)$$

We assert:

(i) *The functor $\mathcal{Y}$ is full, and the essential image consists of the locally finite homological functors [see Explanation 8.17 for the definition of locally finite functors]. The composite $\mathcal{Y} \circ i$ is fully faithful, and the essential image consists of the finite homological functors.*

(ii) *Assume there exists an integer $N > 0$ and an object $G' \in \mathcal{T}^b_c$ with $\mathcal{T} = \overline{\langle G' \rangle}_N^{(-\infty,\infty)}$. Then the functor $\widetilde{\mathcal{Y}}$ is full, and the essential image consists of the locally finite homological functors. The composite $\widetilde{\mathcal{Y}} \circ \widetilde{\imath}$ is fully faithful, and the essential image consists of the finite homological functors.*

EXPLANATION 8.17. *In the statement of Theorem 8.16, the* locally finite *functors* $\left[\mathcal{T}^c\right]^{\text{op}} \longrightarrow R\text{–Mod}$ *are those functors H such that*

(i) $H\big(A[i]\big)$ *is a finite R–module for every $i \in \mathbb{Z}$ and every $A \in \mathcal{T}^c$.*

(ii) *For fixed $A \in \mathcal{T}^c$ we have* $H\big(A[i]\big) = 0$ *if $i \ll 0$.*

while the locally finite functors $H : \mathcal{T}_c^b \longrightarrow R\text{–Mod}$ *are those satisfying*

(iii) $H\big(A[i]\big)$ *is a finite* R*–module for every* $i \in \mathbb{Z}$ *and every* $A \in \mathcal{T}_c^b$.

(iv) *For fixed* $A \in \mathcal{T}_c^b$ *we have* $H\big(A[i]\big) = 0$ *if* $i \ll 0$.

The careful reader will observe that these definitions aren't *dual. The finiteness of* $H\big(A[i]\big)$ *for every* A *and every* i *is, of course, obviously self-dual. But the vanishing isn't. We might be tempted to unify the definitions into*

(v) *Let* $\mathcal{S}$ *be a triangulated category. A homological functor* $H : \mathcal{S} \longrightarrow R\text{–Mod}$ *is locally finite if* $H\big(A[i]\big)$ *is a finite* R*–module for every* $i \in \mathbb{Z}$ *and every* $A \in \mathcal{S}$*, and for fixed* A *it vanishes when* $i \ll 0$.

But this definition is wrong for $\big[\mathcal{T}^c\big]^{\mathrm{op}}$*, because its suspension functor is the negative of that of* $\mathcal{T}^c$.

The way to unify the two definitions is to think of them as continuity with respect to a metric. This might become a little clearer in the next section.

REMARK 8.18. From what we have said already it's clear that the statement about $\mathcal{Y}$ in Theorem 8.1 is a special case of Theorem 8.16(i). To deduce the assertion of Theorem 8.1 about the functor $\widetilde{\mathcal{Y}}$, from Theorem 8.16(ii), we need to know that the category $\mathbf{D}^b_{\mathrm{coh}}(X)$ contains an object G' with $\mathbf{D_{qc}}(X) = \overline{\langle G' \rangle}_N^{(-\infty,\infty)}$. This is proved in [**33**, Theorem 2.3].

SKETCH 8.19. *We should say something about the proof of Theorem* 8.16*— to keep the discussion focused let us restrict our attention to the proof of Theorem* 8.16*(i). For the purpose of this discussion let us fix a compact generator* $G \in \mathcal{T}$ *and a t–structure* $\big(\mathcal{T}^{\leq 0}, \mathcal{T}^{\geq 0}\big)$ *in the preferred equivalence class. Proposition* 8.10 *tells us that we may choose an integer* $A > 0$ *so that the properties of Definition* 5.1 *(i) and (ii) hold. An easy induction on the integer* m *leads to the following consequence of Definition* 5.1*(ii):*

(i) *For every integer* $m > 0$ *and every object* $F \in \mathcal{T}^{\leq 0}$ *there exists a triangle* $E_m \longrightarrow F \longrightarrow D_m \longrightarrow E[1]$ *with* $D_m \in \mathcal{T}^{\leq -m}$ *and* $E \in \overline{\langle G \rangle}_{mA}^{[1-m-A,A]}$.

This much is easy. Not quite so straightforward is the following:

(ii) *There exists an integer* B*, depending only on* A*, with the following property. For any integer* $m > 0$ *and any object* $F \in \mathcal{T}^{\leq 0} \cap \mathcal{T}_c^-$ *there exists a triangle* $E_m \longrightarrow F \longrightarrow D_m \longrightarrow E[1]$ *with* $D \in \mathcal{T}^{\leq -m}$ *and* $E \in \langle G \rangle_{mB}^{[1-m-B,B]}$.

(iii) *In fact more is true: the objects* E_m*, in either (i) or (ii) above, can be chosen to form a sequence* $E_1 \longrightarrow E_2 \longrightarrow E_3 \longrightarrow \cdots$ *mapping to* F*, and such that* F *is the homotopy colimit of the sequence. It is in this sense that the Introduction should be understood: we have expressed* F *as the homotopy colimit of the (directed) Cauchy sequence* $\{E_m\}$.

The reader might wish to go back to Examples 4.4 *and* 4.5*, in which we explicitly worked out what the abstract theory comes down to in the special case where* $\mathcal{T} = \mathbf{D}(R)$*, the t–structure is the standard one, and the compact generator* G *is the object* $R \in \mathbf{D}(R)$*. In the terminology of (i) and (ii) above, Examples* 4.4 *and* 4.5 *amount to showing that* $A = B = 1$ *works for the special case.*

Now back to the proof of Theorem 8.16*(i). The fact that* $\mathcal{Y}$ *is full on the category* $\mathcal{T}_c^-$ *and fully faithful on the category* $\mathcal{T}_c^b$ *turns out to be a straightforward consequence of (iii) above. It remains to show that the essential image of* $\mathcal{Y}$ *is as claimed. One* *containment is easy: the fact that the essential image is contained in the locally finite (respectively finite) functors follows directly from the the hypothesis that* $\mathcal{T}$ *has a compact generator* G *so that* $\mathrm{Hom}\big(G, G[n]\big)$ *is a finite* R*–module for all* $n \in \mathbb{Z}$*, coupled with the fact that approximability implies the vanishing of* $\mathrm{Hom}\big(G, G[n]\big)$ *for* $n \gg 0$*.*

It remains to prove the opposite containment. Fix a locally finite homological functor $H : (\mathcal{T}^c)^{\mathrm{op}} \longrightarrow R\text{–Mod}$*. We need to exhibit an object* $F \in \mathcal{T}_c^-$ *and an isomorphism* $H \cong \mathcal{Y}(F)$*. This actually suffices: it's easy to show that if* $\mathcal{Y}(F)$ *is finite then* $F \in \mathcal{T}_c^b$*.*

Modifying an old idea of Adams [**1**] *one can produce, for each integer* $m > 0$*, an object* $F_m \in \mathcal{T}_c^-$*, a morphism* $\mathcal{Y}(F_m) \longrightarrow H$*, and show that this morphism is surjective when restricted to* $\cup_n \langle G\rangle_m^{[-n,n]} \subset \mathcal{T}^c$*. Shifting* H *if necessary, we may furthermore assume that* $F_m \in \mathcal{T}_c^- \cap \mathcal{T}^{\leq 0}$ *for all* $m > 0$*. By (ii) above we may, for each integer* m*, construct a triangle* $E_m \longrightarrow F_m \longrightarrow D_m \longrightarrow E_m[1]$ *with* $E_m \in \langle G\rangle_{mB}^{[1-m-B,B]}$ *and* $D_m \in \mathcal{T}^{\leq -m}$*. One then needs to show the existence of an increasing sequence of integers* $\{m_1 < m_2 < m_3 < \cdots\}$ *such that there is a Cauchy sequence* $E_{m_1} \longrightarrow E_{m_2} \longrightarrow E_{m_3} \longrightarrow \cdots$ *converging in* $\mathcal{T}_c^-$ *to an object* E *with a surjection* $\mathcal{Y}(E) \longrightarrow H$*.*

This is the hard part. Once we have produced a surjection $\mathcal{Y}(E) \longrightarrow H$*, one performs some tricks to deduce an isomorphism* $\mathcal{Y}(F) \cong H$*.*

9. The categories $\mathbf{D}^{\mathrm{perf}}(X)$ and $\mathbf{D}^b_{\mathrm{coh}}(X)$ determine each other

It remains to discuss Fact 6.3(v). We remind the reader: this is the assertion that $\mathbf{D}^{\mathrm{perf}}(X)$ and $\mathbf{D}^b_{\mathrm{coh}}(X)$ determine each other, as triangulated categories.

Historical Survey 9.1. *Probably the first result in this direction deals with the case of affine schemes and may be found in Rickard* [**38**, Theorem 6.4]. *Rickard's result tells us that, if* R *and* S *are noetherian rings, then*

$$\mathbf{D}^b(R\text{–proj}) \cong \mathbf{D}^b(S\text{–proj}) \qquad \Longleftrightarrow \qquad \mathbf{D}^b(R\text{–mod}) \cong \mathbf{D}^b(S\text{–mod})$$

The way Rickard's theorem was proved was by analyzing the triangulated equivalences—in other words Rickard developed a Morita theory, and showed that the data that produces an equivalence on the left is the same as the data producing an equivalence on the right. Thus the proof does guarantee that a triangulated equivalence on the left will produce a triangulated equivalence on the right, and vice versa. But the new result is better in giving explicit recipes for passing back and forth between $\mathbf{D}^b(R\text{–proj})$ *and* $\mathbf{D}^b(R\text{–mod})$*; there is only one scheme and two derived categories, not two schemes and four derived categories. Moreover: the new result doesn't assume the schemes to be affine.*

A different result by Rouquier [**39**, Remark 7.50] *asserts that, if* X *and* Y *are projective over a field* k*, then*

$$\mathbf{D}^b_{\mathrm{coh}}(X) \cong \mathbf{D}^b_{\mathrm{coh}}(Y) \qquad \Longleftrightarrow \qquad \mathbf{D}^{\mathrm{perf}}(X) \cong \mathbf{D}^{\mathrm{perf}}(Y)$$

And we know both the result and the proof from the previous section: the category $\mathbf{D}^b_{\mathrm{coh}}(X)$ *can be described as the category of finite homological functors on* $\big[\mathbf{D}^{\mathrm{perf}}(X)\big]^{\mathrm{op}}$*, and the category* $\big[\mathbf{D}^{\mathrm{perf}}(X)\big]^{\mathrm{op}}$ *can be described as the finite homological functors on* $\mathbf{D}^b_{\mathrm{coh}}(X)$*. This time there is no Morita theory as background, we explicitly know how to construct the categories out of each other. The drawbacks of the result, compared to the new one, are*

(i) *In Rouquier's statement the schemes were assumed proper over a field, and even the improved representability theorems of the last section only work under the assumption of properness over some commutative, noetherian ring R.*
(ii) *The triangulated structure isn't obvious. The category of (finite) homological functors on an R–linear triangulated category doesn't usually carry a triangulated structure.*

Thus the old results only work if the schemes are either affine or proper—whereas the new result has no such restrictions—and even in the special circumstances where one of the old results holds, the conclusion is less powerful than what's in Fact 6.3(v).

DISCUSSION 9.2. *In the Introduction we already mentioned that a heuristic way to think of approximability is to consider the "metric" defined by a t–structure $(\mathcal{T}^{\leq 0}, \mathcal{T}^{\geq 0})$, and maybe study the limits of Cauchy sequences with respect to this metric. The reader is referred back to Discussion 1.1. Phrased in this language, the category $\mathcal{T}_c^-$ can be defined as the closure in $\mathcal{T}$ of the subcategory $\mathcal{T}^c$ with respect to the metric—see Fact 1.5(iv). This suggests the recipe for constructing $\mathcal{T}_c^b$ out of $\mathcal{T}^c$, it remains to flesh it out a little.*

DEFINITION 9.3. *Let $\mathcal{S}$ be a triangulated category. A* metric *on $\mathcal{S}$ is a sequence of additive subcategories $\{\mathcal{M}_i,\, i \in \mathbb{N}\}$, satisfying*

(i) $\mathcal{M}_{i+1} \subset \mathcal{M}_i$ *for every* $i \in \mathbb{N}$.
(ii) $\mathcal{M}_i * \mathcal{M}_i = \mathcal{M}_i$, *with the notation as in Reminder 4.1.*

A metric $\{\mathcal{M}_i\}$ is declared to be finer than the metric *$\{\mathcal{N}_i\}$ if, for every integer $i > 0$, there exists an integer $j > 0$ with $\mathcal{M}_j \subset \mathcal{N}_i$; we denote this partial order by $\{\mathcal{M}_i\} \preceq \{\mathcal{N}_i\}$. The metrics $\{\mathcal{M}_i\}$, $\{\mathcal{N}_i\}$ are* equivalent *if $\{\mathcal{M}_i\} \preceq \{\mathcal{N}_i\} \preceq \{\mathcal{M}_i\}$.*

EXAMPLE 9.4. *Suppose $\mathcal{T}$ is an approximable triangulated category, and let $(\mathcal{T}^{\leq 0}, \mathcal{T}^{\geq 0})$ be a t–structure in the preferred equivalence class. Out of this data we can construct two examples of $\mathcal{S}$'s with metrics:*

(i) *Let $\mathcal{S}$ be the subcategory $\mathcal{T}^c \subset \mathcal{T}$, and put $\mathcal{M}_i = \mathcal{T}^c \cap \mathcal{T}^{\leq -i}$.*
(ii) *Let $\mathcal{S}$ be the subcategory $\left[\mathcal{T}_c^b\right]^{\mathrm{op}}$, and put $\mathcal{M}_i^{\mathrm{op}} = \mathcal{T}_c^b \cap \mathcal{T}^{\leq -i}$.*

It's obvious that equivalent t–structures define equivalent metrics. Thus up to equivalence we have a canonical metric on $\mathcal{T}^c$ and a canonical metric on $\left[\mathcal{T}_c^b\right]^{\mathrm{op}}$. But the definition depends on the embedding into $\mathcal{T}$, which is the category with the t–structure.

DEFINITION 9.5. *Let $\mathcal{S}$ be a triangulated category with a metric $\{\mathcal{M}_i\}$. A* Cauchy sequence *in $\mathcal{S}$ is a sequence $E_1 \longrightarrow E_2 \longrightarrow E_3 \longrightarrow \cdots$ such that, for every pair of integers $i > 0$, $j \in \mathbb{Z}$, there exists an integer $M > 0$ such that, in any triangle $E_m \longrightarrow E_{m'} \longrightarrow D_{m,m'}$ with $M \leq m < m'$, the object $\Sigma^j D_{m,m'}$ lies in $\mathcal{M}_i$.*

It is clear that the Cauchy sequences depend only on the equivalence class of the metric.

CONSTRUCTION 9.6. *Now suppose $\mathcal{S}$ is an essentially small triangulated category with a metric. If we write* Mod-$\mathcal{S}$ *for the category of additive functors $\mathcal{S}^{\mathrm{op}} \longrightarrow \mathcal{A}b$, then the Yoneda functor is a fully faithful embedding $Y : \mathcal{S} \longrightarrow$* Mod-$\mathcal{S}$. *We remind the reader: the formula is $Y(A) = \mathrm{Hom}(-, A)$.*

In the abelian category Mod-$\mathcal{S}$ *we can form colimits. We define* $\mathfrak{L}(\mathcal{S}) \subset$ Mod-$\mathcal{S}$ *to be the full subcategory of all colimits in* Mod-$\mathcal{S}$ *of Cauchy sequences in* $\mathcal{S}$*, and define* $\mathfrak{S}(\mathcal{S}) \subset \mathfrak{L}(\mathcal{S})$ *to be the full subcategory*

$$\mathfrak{S}(\mathcal{S}) \quad = \quad \mathfrak{L}(\mathcal{S}) \bigcap \left[\bigcap_{j\in\mathbb{Z}} \bigcup_{i\in\mathbb{N}} \left[Y(\Sigma^j \mathcal{M}_i)\right]^\perp \right]$$

Recall: an object $M \in$ Mod-$\mathcal{S}$ *belongs to* $\left[Y(\Sigma^j\mathcal{M}_i)\right]^\perp$ *if, for every object* $m \in \Sigma^j\mathcal{M}_i$*, we have* $\mathrm{Hom}\big(Y(m), M\big) = 0$.

It's clear that the subcategories $\mathfrak{S}(\mathcal{S}) \subset \mathfrak{L}(\mathcal{S}) \subset$ Mod-$\mathcal{S}$ *depend only on the equivalence class of the metric.*

The first result, which may be found in [**31**, Theorem 2.11], asserts:

THEOREM 9.7. *The category* $\mathfrak{S}(\mathcal{S})$ *is a triangulated category, with the triangles being the colimits in* Mod-$\mathcal{S}$ *of Cauchy sequences of triangles in* $\mathcal{S}$*.*

Thus the triangulated structure on $\mathfrak{S}(\mathcal{S})$ also depends only on the equivalence class of the metric.

And the second result, which is in [**31**, Example 4.2 and Proposition 5.6], gives:

THEOREM 9.8. *With the metrics as in Example* 9.4 *(i) and (ii), we have triangulated equivalences*

(i) $\mathfrak{S}(\mathcal{T}^c) = \mathcal{T}^b_c$.

(ii) *If* $\mathcal{T}$ *is noetherian then* $\mathfrak{S}\big([\mathcal{T}^b_c]^{\mathrm{op}}\big) = [\mathcal{T}^c]^{\mathrm{op}}$.

NOTATION 9.9. We owe the reader an explanation of the hypothesis in Theorem 9.8(ii). Let $\mathcal{T}$ be an approximable triangulated category, and let $\big(\mathcal{T}^{\le 0}, \mathcal{T}^{\ge 0}\big)$ be a t–structure in the preferred equivalence class. The category $\mathcal{T}$ is *noetherian* if there exists an integer $N > 0$ so that, for every object $F \in \mathcal{T}^-_c$, there exists a triangle $F' \longrightarrow F \longrightarrow F''$ in $\mathcal{T}^-_c$ with $F' \in \mathcal{T}^-_c \cap \mathcal{T}^{\le N}$ and $F'' \in \mathcal{T}^-_c \cap \mathcal{T}^{\ge 0}$.

It's clear that replacing the t–structure $\big(\mathcal{T}^{\le 0}, \mathcal{T}^{\ge 0}\big)$ by an equivalent one will only have the effect of changing the integer N; the definition doesn't depend on the choice of t–structure in the preferred equivalence class.

REMARK 9.10. The way the metrics were presented, in Example 9.4, depends on the embedding of $\mathcal{T}^c$ and $\mathcal{T}^b_c$ in $\mathcal{T}$. After all they are defined in terms of the preferred equivalence class of t–structures, which makes sense only in $\mathcal{T}$.

But there are recipes that cook up equivalent metrics directly from $\mathcal{T}^c$ and $\mathcal{T}^b_c$. The reader is referred to [**31**, Remark 4.7 as well as Propositions 4.8 and 6.5].

REMARK 9.11. In this survey we have often said that, so far, the main applications of the theory have been to algebraic geometry—it's the fact that $\mathbf{D_{qc}}(X)$ is approximable which has had the far-reaching consequences to date. In this context Theorem 9.8 is the first exception, it has striking consequences for categories not coming from algebraic geometry. From the theorem we learn the following (among other things).

(i) Let R be any ring, possibly noncommutative. The recipe takes the triangulated category $\mathbf{D}^b(R\text{–proj})$ and out of it constructs the triangulated category $\mathbf{D}^-(R\text{–proj}) \cap \mathbf{D}^b(R\text{–Mod})$. The objects of this intersection are the bounded-above cochain complexes of finitely generated, projective R–modules, with bounded cohomology.

If R is a coherent ring this category is equivalent to $\mathbf{D}^b(R\text{–mod})$.

(ii) If R is a coherent (possibly noncommutative) ring, then the recipe takes $\big[\mathbf{D}^b(R\text{–mod})\big]^{\mathrm{op}}$ and out of it constructs $\big[\mathbf{D}^b(R\text{–proj})\big]^{\mathrm{op}}$.

(iii) Out of the homotopy category of finite spectra we construct the homotopy category of spectra with finitely many nonzero stable homotopy groups, all of them finitely generated.

(iv) Out of the homotopy category of spectra with finitely many nonzero stable homotopy groups, all of them finitely generated, we construct the homotopy category of finite spectra.

REMARK 9.12. When I posted the current article on the archive Steve Lack wrote to inform me that metrics in categories aren't new. From Richard Garner I later learned that completing with respect to such metrics has also been done before. Hence I wrote the survey [**35**], which offers a much-expanded treatment of the material discussed in Section 9 and includes some remarks about the category-theory literature that preceded.

Appendix A. Some dumb maps in $\mathbf{D}^{\mathfrak{C}'}_{\mathfrak{C}}(\mathcal{A})$, and the proof that the third map of the triangle is a cochain map

For any object A in the abelian category $\mathcal{A}$, we will write $\widetilde{A}$ for the cochain complex

$$\cdots \longrightarrow 0 \longrightarrow A \;=\!=\!=\; A \longrightarrow 0 \longrightarrow 0 \longrightarrow \cdots$$

with A in degrees -1 and 0. Since the cohomology of the complex $\widetilde{A}$ vanishes, in the derived category $\mathbf{D}^{\mathfrak{C}'}_{\mathfrak{C}}(\mathcal{A})$ the morphism $0 \longrightarrow \widetilde{A}$ is an isomorphism. We reiterate: $\widetilde{A}$ is nothing more than a complicated representative of the isomorphism class of the zero object in $\mathbf{D}^{\mathfrak{C}'}_{\mathfrak{C}}(\mathcal{A})$. With the conventions of Example 2.7 and Notation 2.8, for any integer $i \in \mathbb{Z}$ we may form the object $\widetilde{A}[i]$, which is the cochain complex

$$\cdots \longrightarrow 0 \longrightarrow A \xrightarrow{\;\varepsilon\;} A \longrightarrow 0 \longrightarrow 0 \longrightarrow \cdots$$

with A in degrees $-i-1$ and $-i$, and where $\varepsilon = (-1)^i$.

Let $X^* \in \mathbf{D}^{\mathfrak{C}'}_{\mathfrak{C}}(\mathcal{A})$ be any object, meaning X^* is a cochain complex

$$\cdots \longrightarrow X^{-2} \xrightarrow{\partial_X^{-2}} X^{-1} \xrightarrow{\partial_X^{-1}} X^0 \xrightarrow{\partial_X^{0}} X^1 \xrightarrow{\partial_X^{1}} X^2 \longrightarrow \cdots$$

The cochain maps $X^* \longrightarrow \widetilde{A}[-i]$ are in bijection with morphisms $\theta^i : X^i \longrightarrow A$ in $\mathcal{A}$, the bijection takes θ^i to the cochain map

$$\begin{array}{ccccccccccc}
\cdots \longrightarrow & X^{i-2} & \xrightarrow{\partial_X^{i-2}} & X^{i-1} & \xrightarrow{\partial_X^{i-1}} & X^{i} & \xrightarrow{\partial_X^{i}} & X^{i+1} & \xrightarrow{\partial_X^{i+1}} & X^{i+2} & \longrightarrow \cdots \\
 & \big\downarrow & & \big\downarrow{\scriptstyle \varepsilon\theta^i\partial_X^{i-1}} & & \big\downarrow{\scriptstyle \theta^i} & & \big\downarrow & & \big\downarrow & \\
\cdots \longrightarrow & 0 & \longrightarrow & A & \xrightarrow[\varepsilon]{} & A & \longrightarrow & 0 & \longrightarrow & 0 & \longrightarrow \cdots
\end{array}$$

Given two cochain complexes X^* and Y^*, as well as a morphism $\theta^i : X^i \longrightarrow Y^{i-1}$ in $\mathcal{A}$, we may form the composite

$$\begin{array}{ccccccccccccc}
\cdots & \longrightarrow & X^{i-2} & \xrightarrow{\partial_X^{i-2}} & X^{i-1} & \xrightarrow{\partial_X^{i-1}} & X^{i} & \xrightarrow{\partial_X^{i}} & X^{i+1} & \xrightarrow{\partial_X^{i+1}} & X^{i+2} & \longrightarrow & \cdots \\
 & & \downarrow & & \downarrow{\scriptstyle \varepsilon\partial_X^{i-1}} & & \downarrow{\scriptstyle \mathrm{id}} & & \downarrow & & \downarrow & & \\
\cdots & \longrightarrow & 0 & \longrightarrow & X^{i} & \xrightarrow{\varepsilon} & X^{i} & \longrightarrow & 0 & \longrightarrow & 0 & \longrightarrow & \cdots \\
 & & \downarrow & & \downarrow{\scriptstyle \varepsilon\theta^{i}} & & \downarrow{\scriptstyle \partial_Y^{i-1}\theta^i} & & \downarrow & & \downarrow & & \\
\cdots & \longrightarrow & Y^{i-2} & \xrightarrow{\partial_Y^{i-2}} & Y^{i-1} & \xrightarrow{\partial_Y^{i-1}} & Y^{i} & \xrightarrow{\partial_Y^{i}} & Y^{i+1} & \xrightarrow{\partial_Y^{i+1}} & Y^{i+2} & \longrightarrow & \cdots
\end{array}$$

which is manifestly a cochain map; we will denote it by $\widetilde{\theta^i}$. Of course in the category $\mathbf{D}_{\mathfrak{C}}^{\mathfrak{C}'}(\mathcal{A})$ the cochain map $\widetilde{\theta^i}$ must be equal to the zero map, all we have done is produced many representatives of the zero map. We may also sum over $i \in \mathbb{Z}$; given any collection of morphisms $\theta^i : X^i \longrightarrow Y^{i-1}$ in $\mathcal{A}$ we may form $\oplus_{i=-\infty}^{\infty}\widetilde{\theta^i}$, which is a cochain map $\widetilde{\theta^*} : X^* \longrightarrow Y^*$. Needless to say the map $\widetilde{\theta^*}$ also vanishes in $\mathbf{D}_{\mathfrak{C}}^{\mathfrak{C}'}(\mathcal{A})$.

Now go back to Example 2.7: in the example we construct a diagram

$$\begin{array}{ccccccccccccc}
\cdots & \longrightarrow & Z^{-2} & \xrightarrow{\partial_Z^{-2}} & Z^{-1} & \xrightarrow{\partial_Z^{-1}} & Z^{0} & \xrightarrow{\partial_Z^{-0}} & Z^{1} & \xrightarrow{\partial_Z^{1}} & Z^{2} & \longrightarrow & \cdots \\
 & & \downarrow{\scriptstyle h^{-2}} & & \downarrow{\scriptstyle h^{-1}} & & \downarrow{\scriptstyle h^{0}} & & \downarrow{\scriptstyle h^{1}} & & \downarrow{\scriptstyle h^{2}} & & \\
\cdots & \longrightarrow & X^{-1} & \xrightarrow[-\partial_X^{-1}]{} & X^{0} & \xrightarrow[-\partial_X^{0}]{} & X^{1} & \xrightarrow[-\partial_X^{1}]{} & X^{2} & \xrightarrow[-\partial_X^{2}]{} & X^{3} & \longrightarrow & \cdots \\
 & & \downarrow{\scriptstyle f^{-1}} & & \downarrow{\scriptstyle f^{0}} & & \downarrow{\scriptstyle f^{1}} & & \downarrow{\scriptstyle f^{2}} & & \downarrow{\scriptstyle f^{3}} & & \\
\cdots & \longrightarrow & Y^{-1} & \xrightarrow[-\partial_Y^{-1}]{} & Y^{0} & \xrightarrow[-\partial_Y^{0}]{} & Y^{1} & \xrightarrow[-\partial_Y^{1}]{} & Y^{2} & \xrightarrow[-\partial_Y^{2}]{} & Y^{3} & \longrightarrow & \cdots
\end{array}$$

The construction is such that, if we delete the middle row, then the composite is $\widetilde{\theta^*}$ for the explicit θ^* chosen in the Example. In particular: deleting the middle row gives a commutative diagram. Deleting the top row yields a commutative diagram, we are left with nothing other than the given cochain map $f^*[1]$. We conclude that, for each i, the composites from top left to bottom right in the diagram

$$\begin{array}{ccc}
Z^{i} & \xrightarrow{\partial_Z^{i}} & Z^{i+1} \\
\downarrow{\scriptstyle h^{i}} & & \downarrow{\scriptstyle h^{i+1}} \\
X^{i+1} & \xrightarrow[-\partial_X^{i+1}]{} & X^{i+2} \\
 & & \downarrow{\scriptstyle f^{i+2}} \\
 & & Y^{i+2}
\end{array}$$

must agree. Since the map f^{i+2} is a (split) monomorphism the square commutes, and we conclude that the diagram

$$\begin{array}{ccccccccccccc}
\cdots & \longrightarrow & Z^{-2} & \xrightarrow{\partial_Z^{-2}} & Z^{-1} & \xrightarrow{\partial_Z^{-1}} & Z^{0} & \xrightarrow{\partial_Z^{-0}} & Z^{1} & \xrightarrow{\partial_Z^{1}} & Z^{2} & \longrightarrow & \cdots \\
 & & \downarrow{\scriptstyle h^{-2}} & & \downarrow{\scriptstyle h^{-1}} & & \downarrow{\scriptstyle h^{0}} & & \downarrow{\scriptstyle h^{1}} & & \downarrow{\scriptstyle h^{2}} & & \\
\cdots & \longrightarrow & X^{-1} & \xrightarrow[-\partial_X^{-1}]{} & X^{0} & \xrightarrow[-\partial_X^{0}]{} & X^{1} & \xrightarrow[-\partial_X^{1}]{} & X^{2} & \xrightarrow[-\partial_X^{2}]{} & X^{3} & \longrightarrow & \cdots
\end{array}$$

commutes. Thus h^* is indeed a cochain map, as promised in Example 2.7.

Appendix B. The assumption that the short exact sequences of cochain complexes are degreewise split is harmless

Given a cochain complex X^*, an object $A \in \mathcal{A}$ and a morphism $\theta^i : X^i \longrightarrow A$, in Appendix A we constructed a corresponding cochain map $X^* \longrightarrow \widetilde{A}[-i]$. In the special case where $A = X^i$ and θ^i is the identity $\mathrm{id} : X^i \longrightarrow X^i$ the general recipe specializes to the cochain map ρ^i below

$$\begin{array}{ccccccccccccc}
\cdots & \longrightarrow & X^{i-2} & \xrightarrow{\partial_X^{i-2}} & X^{i-1} & \xrightarrow{\partial_X^{i-1}} & X^i & \xrightarrow{\partial_X^{i}} & X^{i+1} & \xrightarrow{\partial_X^{i+1}} & X^{i+2} & \longrightarrow & \cdots \\
 & & \big\downarrow & & \big\downarrow{\scriptstyle \varepsilon\partial_X^{i-1}} & & \Big\| & & \big\downarrow & & \big\downarrow & & \\
\cdots & \longrightarrow & 0 & \longrightarrow & X^i & \xrightarrow[\varepsilon]{} & X^i & \longrightarrow & 0 & \longrightarrow & 0 & \longrightarrow & \cdots
\end{array}$$

For this section the key point is that, in degree i, the cochain map $\rho^i : X^* \longrightarrow \widetilde{X}^i[-i]$ is a split monomorphism. Taking the direct sum over i produces

$$X^* \longrightarrow \bigoplus_{i\in\mathbb{Z}} \widetilde{X}^i[-i]$$

We denote this map $\rho^* : X^* \longrightarrow \widetilde{X}^*$ and observe that

(i) The object $\widetilde{X}^* = \oplus_{i\in\mathbb{Z}}\widetilde{X}^i[-i]$ vanishes in $\mathbf{D}_{\mathfrak{C}}^{\mathfrak{C}'}(\mathcal{A})$.

(ii) The cochain map ρ^* is a split monomorphism in each degree.

Now suppose we're given a short exact sequence of cochain complexes $X^* \xrightarrow{f^*} Y^* \xrightarrow{g^*} Z^*$. We may form the commutative diagram of cochain complexes, where the rows are exact

$$\begin{array}{ccccccccc}
0 & \longrightarrow & X^* & \xrightarrow{\begin{pmatrix} \rho^* \\ f^* \end{pmatrix}} & \widetilde{X}^* \oplus Y^* & \xrightarrow{\widetilde{g}^*} & \widetilde{Z}^* & \longrightarrow & 0 \\
 & & \Big\| & & \big\downarrow{\scriptstyle \pi} & & \big\downarrow{\scriptstyle \varphi^*} & & \\
0 & \longrightarrow & X^* & \xrightarrow{f^*} & Y^* & \xrightarrow{g^*} & Z^* & \longrightarrow & 0
\end{array}$$

We know that the vertical maps $\mathrm{id} : X \longrightarrow X$ and $\pi : \widetilde{X}^* \oplus Y^* \longrightarrow Y^*$ both induce isomorphisms in cohomology. The 5-lemma, applied to the long exact sequences in cohomology that come from the short exact sequences of cochain complexes in the rows, tells us that the vertical morphism φ^* also induces an isomorphism in cohomology. Hence the top row is isomorphic in $\mathbf{D}_{\mathfrak{C}}^{\mathfrak{C}'}(\mathcal{A})$ to the bottom row, and the top row is degreewise split. This is the sense in which we said, back in Example 2.7, that up to isomorphism in $\mathbf{D}_{\mathfrak{C}}^{\mathfrak{C}'}(\mathcal{A})$ we may assume our short exact sequence of cochain complexes is degreewise split.

Appendix C. Translating the approach to derived categories we presented here to the more standard one in the literature

Our presentation of derived categories has been minimalist—we have tried to be accurate without providing any information that wasn't absolutely indispensable. This means that the student who is seeing this for the first time, and would like to look for more detail elsewhere in the literature, might have a hard time reconciling what's here with other, more expansive accounts. This appendix was written to help.

DEFINITION C.1. *Suppose $\mathcal{A}$ is an abelian category. Following tradition we define the two categories $\mathbf{C}_{\mathfrak{C}}^{\mathfrak{C}'}(\mathcal{A})$ and $\mathbf{D}_{\mathfrak{C}}^{\mathfrak{C}'}(\mathcal{A})$ to have the same objects: cochain complexes in $\mathcal{A}$ subject to identical restrictions. The morphisms are*

(i) *In the category $\mathbf{C}_{\mathfrak{C}}^{\mathfrak{C}'}(\mathcal{A})$ a morphism is a cochain map.*
(ii) *We have already met $\mathbf{D}_{\mathfrak{C}}^{\mathfrak{C}'}(\mathcal{A})$: it is obtained from $\mathbf{C}_{\mathfrak{C}}^{\mathfrak{C}'}(\mathcal{A})$ by formally inverting the morphisms in $\mathbf{C}_{\mathfrak{C}}^{\mathfrak{C}'}(\mathcal{A})$ inducing isomorphisms in cohomology.*

REMARK C.2. We give the following list, translating the constructions of the previous two appendices to more standard language.

(i) Given a pair of objects X^* and Y^* in $\mathbf{C}_{\mathfrak{C}}^{\mathfrak{C}'}(\mathcal{A})$, that is a pair of cochain complexes X^* and Y^*, as well as a sequence of morphisms $\{\theta^i : X^i \longrightarrow Y^{i-1},\ i \in \mathbb{Z}\}$ in the category $\mathcal{A}$, Appendix A constructed for us (among other things) a morphism $\widetilde{\theta}^* : X^* \longrightarrow Y^*$ in the category $\mathbf{C}_{\mathfrak{C}}^{\mathfrak{C}'}(\mathcal{A})$. It is traditional to say that the cochain map $\widetilde{\theta}^*$ is *null homotopic,* and the sequence of maps $\{\theta^i : X^i \longrightarrow Y^{i-1},\ i \in \mathbb{Z}\}$ is called a *homotopy of $\widetilde{\theta}^*$ with the zero map.* More generally: two cochain maps $f^*, g^* : X^* \longrightarrow Y^*$ are declared to be *homotopic to each other* if there exists a homotopy of $f^* - g^*$ with the zero map. That is: if $f^* - g^* = \widetilde{\theta}^*$ with $\widetilde{\theta}^*$ as in Appendix A.
(ii) In Appendix B we constructed, for every cochain complex X^*, another cochain complex $\widetilde{X}^*$ and a map $\rho : X^* \longrightarrow \widetilde{X}^*$ in the category $\mathbf{C}_{\mathfrak{C}}^{\mathfrak{C}'}(\mathcal{A})$. Note that, by the construction of the map $\widetilde{\theta}^* : X^* \longrightarrow Y^*$ of (i) above (see Appendix A), the map $\widetilde{\theta}^*$ comes with a factorization $X^* \xrightarrow{\rho} \widetilde{X}^* \xrightarrow{\theta} Y^*$ in the category $\mathbf{C}_{\mathfrak{C}}^{\mathfrak{C}'}(\mathcal{A})$.
(iii) Given a cochain map $f^* : X^* \longrightarrow Y^*$, the object $\widetilde{X}^* \oplus Y^*$ is isomorphic in $\mathbf{C}_{\mathfrak{C}}^{\mathfrak{C}'}(\mathcal{A})$ to a complex traditionally called the *mapping cylinder* of f^*. This is obviously dumb terminology since $\widetilde{X}^* \oplus Y^*$ manifestly doesn't depend on f^*. But the isomorph in $\mathbf{C}_{\mathfrak{C}}^{\mathfrak{C}'}(\mathcal{A})$, that is traditionally presented as the mapping cylinder, seems to depend on f^*—needless to to say, the isomorphism with $\widetilde{X}^* \oplus Y^*$ involves f^*.
(iv) In Appendix B we considered the degreewise split short exact sequence of cochain complexes

$$0 \longrightarrow X^* \xrightarrow{\begin{pmatrix} \rho^* \\ f^* \end{pmatrix}} \widetilde{X}^* \oplus Y^* \xrightarrow{\quad \widetilde{g}^* \quad} \widetilde{Z}^* \longrightarrow 0$$

The cochain complex $\widetilde{Z}^*$ is isomorphic in $\mathbf{C}_{\mathfrak{C}}^{\mathfrak{C}'}(\mathcal{A})$ to a complex traditionally known as the *mapping cone* on f^*.
(v) When $f^* : X^* \longrightarrow Y^*$ is the identity $\mathrm{id} : X^* \longrightarrow X^*$, the mapping cone is isomorphic in $\mathbf{C}_{\mathfrak{C}}^{\mathfrak{C}'}(\mathcal{A})$ to $\widetilde{X}^*$. Thus the object we call $\widetilde{X}^*$ is traditionally called the *mapping cone on the identity.* Once again it is traditional to give an isomorphic complex, which is more complicated-looking than the $\widetilde{X}^*$ of Appendix B.

DEFINITION C.3. *Suppose $\mathcal{A}$ is an abelian category. Following tradition we define, in addition to the two categories of Definition C.1, yet another category $\mathbf{K}_{\mathfrak{C}}^{\mathfrak{C}'}(\mathcal{A})$. It has the same objects as $\mathbf{C}_{\mathfrak{C}}^{\mathfrak{C}'}(\mathcal{A})$ or $\mathbf{D}_{\mathfrak{C}}^{\mathfrak{C}'}(\mathcal{A})$. However:*

(i) *In the category $\mathbf{K}_{\mathfrak{C}}^{\mathfrak{C}'}(\mathcal{A})$ a morphism is a homotopy equivalence class of cochain maps. Thus two cochain maps are declared to be equal in $\mathbf{K}_{\mathfrak{C}}^{\mathfrak{C}'}(\mathcal{A})$ if they're homotopic.*

LEMMA C.4. *The natural functor $F : \mathbf{C}_{\mathfrak{C}}^{\mathfrak{C}'}(\mathcal{A}) \longrightarrow \mathbf{D}_{\mathfrak{C}}^{\mathfrak{C}'}(\mathcal{A})$, that is the universal functor out of $\mathbf{C}_{\mathfrak{C}}^{\mathfrak{C}'}(\mathcal{A})$ which sends every cohomology isomorphism to an isomorphism, factors uniquely through $\mathbf{K}_{\mathfrak{C}}^{\mathfrak{C}'}(\mathcal{A})$.*

PROOF. Let X^* be any object in $\mathbf{C}_{\mathfrak{C}}^{\mathfrak{C}'}(\mathcal{A})$, and consider the following morphisms in $\mathbf{C}_{\mathfrak{C}}^{\mathfrak{C}'}(\mathcal{A})$

$$X^* \underset{\begin{pmatrix} 0 \\ \mathrm{id} \end{pmatrix}}{\overset{\begin{pmatrix} \rho^* \\ \mathrm{id} \end{pmatrix}}{\rightrightarrows}} \widetilde{X}^* \oplus X^* \xrightarrow{\quad (0,\mathrm{id}) \quad} X^*$$

The two composites are equal, in each case the composite is the identity map $\mathrm{id} : X^* \longrightarrow X^*$. But the map $(0, \mathrm{id}) : \widetilde{X}^* \oplus X^* \longrightarrow X^*$ is an isomorphism in cohomology [because $\widetilde{X}^*$ is acyclic], therefore $F : \mathbf{C}_{\mathfrak{C}}^{\mathfrak{C}'}(\mathcal{A}) \longrightarrow \mathbf{D}_{\mathfrak{C}}^{\mathfrak{C}'}(\mathcal{A})$ takes $(0, \mathrm{id})$ to an isomorphism in $\mathbf{D}_{\mathfrak{C}}^{\mathfrak{C}'}(\mathcal{A})$. Hence the functor $F : \mathbf{C}_{\mathfrak{C}}^{\mathfrak{C}'}(\mathcal{A}) \longrightarrow \mathbf{D}_{\mathfrak{C}}^{\mathfrak{C}'}(\mathcal{A})$ must take the two cochain maps[6]

$$X^* \underset{\begin{pmatrix} 0 \\ \mathrm{id} \end{pmatrix}}{\overset{\begin{pmatrix} \rho^* \\ \mathrm{id} \end{pmatrix}}{\rightrightarrows}} \widetilde{X}^* \oplus X^*$$

to equal morphisms in $\mathbf{D}_{\mathfrak{C}}^{\mathfrak{C}'}(\mathcal{A})$. But then, for any homotopy $\{\theta^i : X^i \longrightarrow Y^{i-1}, i \in \mathbb{Z}\}$ as in Remark C.2(i), the two composites

$$X^* \underset{\begin{pmatrix} 0 \\ \mathrm{id} \end{pmatrix}}{\overset{\begin{pmatrix} \rho^* \\ \mathrm{id} \end{pmatrix}}{\rightrightarrows}} \widetilde{X}^* \oplus X^* \xrightarrow{\quad (\theta,g^*) \quad} Y^*$$

[6]Recall Remark C.2(iii): for any cochain map $f^* : X^* \longrightarrow X^*$ there is a cochain complex traditionally called the *mapping cylinder of* f^*, and all it is, as an object of the category $\mathbf{C}_{\mathfrak{C}}^{\mathfrak{C}'}(\mathcal{A})$, is a complicated isomorph of $\widetilde{X}^* \oplus X^*$. If $f^* : X^* \longrightarrow X^*$ is the identity map then the two morphisms $X^* \rightrightarrows \widetilde{X}^* \oplus X^*$, studied in the proof of Lemma C.4, traditionally go by the name "the inclusions of the front and back faces" of the mapping cylinder.

Note that, although the mapping cylinder of f^* is independent of f^* up to isomorphism in $\mathbf{C}_{\mathfrak{C}}^{\mathfrak{C}'}(\mathcal{A})$, the inclusion of the back face depends on f^*.

must map under $F : \mathbf{C}_{\mathfrak{C}}^{\mathfrak{C}'}(\mathcal{A}) \longrightarrow \mathbf{D}_{\mathfrak{C}}^{\mathfrak{C}'}(\mathcal{A})$ to equal morphisms in $\mathbf{D}_{\mathfrak{C}}^{\mathfrak{C}'}(\mathcal{A})$. If $\{\theta^i : X^i \longrightarrow Y^{i-1}, i \in \mathbb{Z}\}$ is a homotopy with $f^* - g^* = \widetilde{\theta}^* = \theta\rho^*$ we discover that, in the category $\mathbf{D}_{\mathfrak{C}}^{\mathfrak{C}'}(\mathcal{A})$, we must have $F(f^*) = F(g^*)$. □

REMARK C.5. It now follows easily that $\mathbf{D}_{\mathfrak{C}}^{\mathfrak{C}'}(\mathcal{A})$ can also be obtained from $\mathbf{K}_{\mathfrak{C}}^{\mathfrak{C}'}(\mathcal{A})$ by formally inverting the maps inducing isomorphisms in cohomology. This brings us to the traditional approach: one proves first that the category $\mathbf{K}_{\mathfrak{C}}^{\mathfrak{C}'}(\mathcal{A})$ is triangulated, then formally inverts. And there is a general theorem of Verdier, giving conditions under which the process of formally inverting morphisms takes one triangulated category to another.

REMARK C.6. In Example 2.7 we started in $\mathbf{C}_{\mathfrak{C}}^{\mathfrak{C}'}(\mathcal{A})$ with a degreewise split short exact sequence of cochain complexes $X^* \xrightarrow{f^*} Y^* \xrightarrow{g^*} Z^*$ and, after choosing for every $i \in \mathbb{Z}$ a degreewise splitting $\theta^i : Z^i \longrightarrow Y^i$, we extended in $\mathbf{C}_{\mathfrak{C}}^{\mathfrak{C}'}(\mathcal{A})$ to form the sequence of morphisms $X^* \xrightarrow{f^*} Y^* \xrightarrow{g^*} Z^* \xrightarrow{h^*} \Sigma X^*$. The triangles in $\mathbf{K}_{\mathfrak{C}}^{\mathfrak{C}'}(\mathcal{A})$ can be defined to be the isomorphs in $\mathbf{K}_{\mathfrak{C}}^{\mathfrak{C}'}(\mathcal{A})$ of the $X^* \xrightarrow{f^*} Y^* \xrightarrow{g^*} Z^* \xrightarrow{h^*} \Sigma X^*$ that come from the construction of Example 2.7.

We leave it to the reader to check (if she wishes) that the triangles we construct are isomorphic in $\mathbf{K}_{\mathfrak{C}}^{\mathfrak{C}'}(\mathcal{A})$ to the standard ones in the literature. The interested reader can amuse herself by furthermore checking that the axioms of triangulated categories are satisfied, using either the description of the triangles in the paragraph above or the more standard one in the literature.

Even if the reader is feeling lazy today, the honest truth is that the existing literature won't help—it tends to leave this verification to the reader, there would be no benefit in working with the standard description of triangles. Doing this will not reduce the amount of labor the indolent reader is asked to perform, the checking will still be her task, only the details that need to be proved will shift a little.

The good news is that, once the reader has verified that the category $\mathbf{K}_{\mathfrak{C}}^{\mathfrak{C}'}(\mathcal{A})$ is triangulated, the passage from $\mathbf{K}_{\mathfrak{C}}^{\mathfrak{C}'}(\mathcal{A})$ to $\mathbf{D}_{\mathfrak{C}}^{\mathfrak{C}'}(\mathcal{A})$ becomes well-documented. The theorem of Verdier, alluded to in Remark C.5, is explicit enough to provide helpful information about the calculus of fractions involved—and this may be found in any of the standard treatments in the literature.

Acknowledgments

The author would like to thank Jesse Burke, Anthony Kling, Steve Lack, Luke Mitchelson, Bregje Pauwels, Geordie Williamson and an anonymous referee for questions and comments that led to improvements on earlier versions of the manuscript. These comments were made both about earlier drafts, and during talks presenting parts or all of the material.

References

[1] J. F. Adams, *A variant of E. H. Brown's representability theorem*, Topology **10** (1971), 185–198, DOI 10.1016/0040-9383(71)90003-6. MR283788

[2] Takuma Aihara and Ryo Takahashi, *Generators and dimensions of derived categories of modules*, Comm. Algebra **43** (2015), no. 11, 5003–5029, DOI 10.1080/00927872.2014.957384. MR3422380

[3] Leovigildo Alonso Tarrío, Ana Jeremías López, and María José Souto Salorio, *Construction of t-structures and equivalences of derived categories*, Trans. Amer. Math. Soc. **355** (2003), no. 6, 2523–2543, DOI 10.1090/S0002-9947-03-03261-6. MR1974001

[4] Luchezar L. Avramov, Ragnar-Olaf Buchweitz, and Srikanth Iyengar, *Class and rank of differential modules*, Invent. Math. **169** (2007), no. 1, 1–35, DOI 10.1007/s00222-007-0041-6. MR2308849

[5] Abdolnaser Bahlekeh, Ehsan Hakimian, Shokrollah Salarian, and Ryo Takahashi, *Annihilation of cohomology, generation of modules and finiteness of derived dimension*, Q. J. Math. **67** (2016), no. 3, 387–404, DOI 10.1093/qmath/haw015. MR3556492

[6] Alexander A. Beĭlinson, Joseph Bernstein, and Pierre Deligne, *Analyse et topologie sur les éspaces singuliers*, Astérisque, vol. 100, Soc. Math. France, 1982 (French).

[7] Marcel Bökstedt and Amnon Neeman, *Homotopy limits in triangulated categories*, Compositio Math. **86** (1993), no. 2, 209–234. MR1214458

[8] A. Bondal and M. van den Bergh, *Generators and representability of functors in commutative and noncommutative geometry* (English, with English and Russian summaries), Mosc. Math. J. **3** (2003), no. 1, 1–36, 258, DOI 10.17323/1609-4514-2003-3-1-1-36. MR1996800

[9] Jesse Burke, Amnon Neeman, and Bregje Pauwels, *Gluing approximable triangulated categories*, `arXiv:1806.05342`.

[10] J. Daniel Christensen, *Ideals in triangulated categories: phantoms, ghosts and skeleta*, Adv. Math. **136** (1998), no. 2, 284–339, DOI 10.1006/aima.1998.1735. MR1626856

[11] Hailong Dao and Ryo Takahashi, *The radius of a subcategory of modules*, Algebra Number Theory **8** (2014), no. 1, 141–172, DOI 10.2140/ant.2014.8.141. MR3207581

[12] Hailong Dao and Ryo Takahashi, *The dimension of a subcategory of modules*, Forum Math. Sigma **3** (2015), Paper No. e19, 31, DOI 10.1017/fms.2015.19. MR3482266

[13] A. J. de Jong, *Smoothness, semi-stability and alterations*, Inst. Hautes Études Sci. Publ. Math. **83** (1996), 51–93. MR1423020

[14] A. Johan de Jong, *Families of curves and alterations*, Ann. Inst. Fourier (Grenoble) **47** (1997), no. 2, 599–621. MR1450427

[15] P. Gabriel and M. Zisman, *Calculus of fractions and homotopy theory*, Ergebnisse der Mathematik und ihrer Grenzgebiete, Band 35, Springer-Verlag New York, Inc., New York, 1967. MR0210125

[16] Sergei I. Gelfand and Yuri I. Manin, *Methods of homological algebra*, 2nd ed., Springer Monographs in Mathematics, Springer-Verlag, Berlin, 2003. MR1950475

[17] A. Grothendieck, *Éléments de géométrie algébrique. III. Étude cohomologique des faisceaux cohérents. I*, Inst. Hautes Études Sci. Publ. Math. **11** (1961), 167. MR217085

[18] Robin Hartshorne, *Residues and duality*, Lecture notes of a seminar on the work of A. Grothendieck, given at Harvard 1963/64. With an appendix by P. Deligne. Lecture Notes in Mathematics, No. 20, Springer-Verlag, Berlin-New York, 1966. MR0222093

[19] Luc Illusie, *Existence de résolutions globales*, Théorie des intersections et théorème de Riemann-Roch, Springer-Verlag, Berlin, 1971, Séminaire de Géométrie Algébrique du Bois-Marie 1966–1967 (SGA 6, Exposé II), pp. 160–221. Lecture Notes in Mathematics, Vol. 225.

[20] Luc Illusie, *Généralités sur les conditions de finitude dans les catégories dérivées*, Théorie des intersections et théorème de Riemann-Roch, Springer-Verlag, Berlin, 1971, Séminaire de Géométrie Algébrique du Bois-Marie 1966–1967 (SGA 6, Exposé I), pp. 78–159. Lecture Notes in Mathematics, Vol. 225.

[21] Srikanth B. Iyengar and Ryo Takahashi, *Annihilation of cohomology and strong generation of module categories*, Int. Math. Res. Not. IMRN **2** (2016), 499–535, DOI 10.1093/imrn/rnv136. MR3493424

[22] Masaki Kashiwara and Pierre Schapira, *Categories and sheaves*, Grundlehren der Mathematischen Wissenschaften [Fundamental Principles of Mathematical Sciences], vol. 332, Springer-Verlag, Berlin, 2006. MR2182076

[23] Bernhard Keller, Daniel Murfet, and Michel Van den Bergh, *On two examples by Iyama and Yoshino*, Compos. Math. **147** (2011), no. 2, 591–612, DOI 10.1112/S0010437X10004902. MR2776613

[24] Bernhard Keller and Michel Van den Bergh, *On two examples by Iyama and Yoshino*, (e-print `arXiv:0803.0720v1`).

[25] G. M. Kelly, *Chain maps inducing zero homology maps*, Proc. Cambridge Philos. Soc. **61** (1965), 847–854, DOI 10.1017/s0305004100039207. MR188273

[26] F. William Lawvere, *Metric spaces, generalized logic, and closed categories* (English, with Italian summary), Rend. Sem. Mat. Fis. Milano **43** (1973), 135–166 (1974), DOI 10.1007/BF02924844. MR352214

[27] F. William Lawvere, *Metric spaces, generalized logic, and closed categories* (English, with Italian summary), Rend. Sem. Mat. Fis. Milano **43** (1973), 135–166 (1974), DOI 10.1007/BF02924844. MR352214

[28] Tom Leinster, *The magnitude of metric spaces*, Doc. Math. **18** (2013), 857–905, DOI 10.1007/s00500-013-1105-8. MR3084566

[29] Joseph Lipman and Amnon Neeman, *Quasi-perfect scheme-maps and boundedness of the twisted inverse image functor*, Illinois J. Math. **51** (2007), no. 1, 209–236. MR2346195

[30] Valery A. Lunts, *Categorical resolution of singularities*, J. Algebra **323** (2010), no. 10, 2977–3003, DOI 10.1016/j.jalgebra.2009.12.023. MR2609187

[31] Amnon Neeman, *The categories $\mathcal{T}^c$ and $\mathcal{T}^b_c$ determine each other*, `arXiv:1806.06471`.

[32] Amnon Neeman, *The category $\left[\mathcal{T}^c\right]^{\mathrm{op}}$ as functors on $\mathcal{T}^b_c$*, `arXiv:1806.05777`.

[33] Amnon Neeman, *Strong generators in $\mathbf{D}^{\mathrm{perf}}(X)$ and $\mathbf{D}^b_{\mathrm{coh}}(X)$*, Annals of Mathematics, `arXiv:1703.04484`.

[34] Amnon Neeman, *Triangulated categories with a single compact generator and a Brown representability theorem*, `arXiv:1804.02240`.

[35] Amnon Neeman, *Metrics on triangulated categories*, J. Pure Appl. Algebra **224** (2020), no. 4, 106206, 13, DOI 10.1016/j.jpaa.2019.106206. MR4021920

[36] Dmitri Orlov, *Remarks on generators and dimensions of triangulated categories* (English, with English and Russian summaries), Mosc. Math. J. **9** (2009), no. 1, 153–159, back matter, DOI 10.17323/1609-4514-2009-9-1-143-149. MR2567400

[37] Dmitri Orlov, *Smooth and proper noncommutative schemes and gluing of DG categories*, Adv. Math. **302** (2016), 59–105, DOI 10.1016/j.aim.2016.07.014. MR3545926

[38] Jeremy Rickard, *Morita theory for derived categories*, J. London Math. Soc. (2) **39** (1989), no. 3, 436–456, DOI 10.1112/jlms/s2-39.3.436. MR1002456

[39] Raphaël Rouquier, *Dimensions of triangulated categories*, J. K-Theory **1** (2008), no. 2, 193–256, DOI 10.1017/is007011012jkt010. MR2434186

[40] Ross Street, *Homotopy classification of filtered complexes*, J. Austral. Math. Soc. **15** (1973), 298–318. MR0340380

[41] R. W. Thomason and Thomas Trobaugh, *Higher algebraic K-theory of schemes and of derived categories*, The Grothendieck Festschrift, Vol. III, Progr. Math., vol. 88, Birkhäuser Boston, Boston, MA, 1990, pp. 247–435, DOI 10.1007/978-0-8176-4576-2_10. MR1106918

[42] Jean-Louis Verdier, *Des catégories dérivées des catégories abéliennes* (French, with French summary), Astérisque **239** (1996), xii+253 pp. (1997). With a preface by Luc Illusie; Edited and with a note by Georges Maltsiniotis. MR1453167

[43] Charles A. Weibel, *An introduction to homological algebra*, Cambridge Studies in Advanced Mathematics, vol. 38, Cambridge University Press, Cambridge, 1994. MR1269324

Centre for Mathematics and its Applications, Mathematical Sciences Institute, Building 145, The Australian National University, Canberra, ACT 2601, Australia

Email address: `Amnon.Neeman@anu.edu.au`

Contemporary Mathematics
Volume **769**, 2021
https://doi.org/10.1090/conm/769/15417

Methods of constructive category theory

Sebastian Posur

ABSTRACT. We give an introduction to constructive category theory by answering two guiding computational questions. The first question is: how do we compute the set of all natural transformations between two finitely presented functors like Ext and Tor over a commutative coherent ring R? We give an answer by introducing category constructors that enable us to build up a category which is both suited for performing explicit calculations and equivalent to the category of all finitely presented functors. The second question is: how do we determine the differentials on the pages of a spectral sequence associated to a filtered cochain complex only in terms of operations directly provided by the axioms of an abelian category? Its answer relies on a constructive method for performing diagram chases based on a calculus of relations within an arbitrary abelian category.

CONTENTS

2020 *Mathematics Subject Classification.* Primary 18E10, 18E05, 18A25, 18E25.
Key words and phrases. Constructive category theory, finitely presented functors, diagram chases.
The author was supported by Deutsche Forschungsgemeinschaft (DFG) grant SFB-TRR 195: *Symbolic Tools in Mathematics and their Application.*

Introduction

Basic algorithms in computer algebra provide answers for basic mathematical questions. The Gaussian algorithm computes solutions of a given linear system over a field k. The Euclidean algorithm computes the gcd of elements in an Euclidean domain. Buchberger's algorithm [**Buc06**] computes Gröbner bases of (homogeneous) ideals $I = \langle f_1, \dots, f_r \rangle$ in a (graded) polynomial ring $R := k[x_1, \dots, x_n]$ for $r, n \in \mathbb{N}$, allowing us to answer many basic questions[1]: when do two representatives of elements in the residue class ring $k[x_1, \dots, x_n]/I$ define the same element? How to find a finite set of generators of the ideal $k[x_1, \dots, x_m] \cap I$ for $m < n$? How to find a generating set for the syzygies of the given generators of I? Generalizations of Buchberger's algorithm [**Gre99**] provide answers to similar questions for some non-commutative rings, like finite dimensional quotients of path algebras.

In this article, we demonstrate a strategy that uses these basic algorithms as building blocks for answering a more high-level mathematical question:

(1) How do we compute the set of all natural transformations between two finitely presented functors over a commutative coherent[2] ring R?

Examples of finitely presented functors over such rings are given by $\mathrm{Ext}^i(M, -)$ and $\mathrm{Tor}_i(M, -)$ for a finitely presented R-module M and $i \in \mathbb{N}_0$.

The first section of this article is dedicated to answering this question. The main idea is to use a constructive formulation of category theory. We regard a category $\mathbf{A}$ as a computational entity on whose objects and morphisms we can operate by algorithms. For example, composition of morphisms is an algorithm that takes two morphisms $\alpha : A \to B$, $\beta : B \to C$ as input and outputs a new morphism $\alpha \cdot \beta : A \to C$. Equality of morphisms is an algorithm that takes two morphisms $\alpha : A \to B$, $\alpha' : A \to B$, and outputs `true` if α and α' are equal, `false` otherwise.

The basic algorithms of computer algebra can now be used to render concrete instances of categories computable in the above sense. For example, if we regard the quotient ring $R = k[x_1, \dots, x_n]/I$ as a category with a single object whose morphisms are given by the elements of R and composition by ring multiplication, then deciding equality of morphisms is the same as deciding equality of ring elements, which is algorithmically realized by Buchberger's algorithm. As another example, Gaussian elimination serves as an algorithm to realize the computation of kernels in a computational model of finite dimensional vector spaces.

Once reinterpreted in purely categorical terms, we can forget about the internal functioning of the basic algorithms and start building up algorithms that solely rely on category theory specific notions. In this way, we will be able to answer the more

[1] For learning how to answer these questions computationally, we refer the reader to [**GP02**].

[2] A commutative ring is coherent if kernels of R-module homomorphisms between finitely generated free R-modules are themselves finitely generated.

high-level mathematical question stated above, i.e., we end up with an algorithmic strategy for computing sets of natural transformations between finitely presented functors.

Moreover, once we get used to the idea of using purely categorical notions as building blocks of our algorithms, we can ask further questions that are founded on this idea:

(2) How do we construct morphisms that are claimed to exist by homological algebra, like the differentials on the pages of a spectral sequence associated to a filtered cochain complex, only in terms of operations directly provided by the axioms of an abelian category, like computing kernels or cokernels?

The second section of this article deals with this second question and its answer relies on the introduction of the concept of *generalized morphisms* [**Bar09**]. They provide a key tool for a constructive treatment of homological algebra that let us compute with spectral sequences in the end.

Our constructive treatment of category theory has been implemented within a software project called CAP [**GSP18**], which consists of a collection of `GAP` [**GAP18**] packages. To reveal the feasibility of a direct computer implementation of all the outlined ideas within this article, we make use of a more constructive language of mathematics (see, for example, [**MRR88**]). Concretely, this means that we make an intuitive use of terms like *data types* and *algorithms* instead of sets and functions, and treat the notion of equality between elements of data types as an extra datum that has to be provided by an explicit algorithm (instead of being inherently available like in the case of sets). Since this more constructive language encompasses classical mathematics, all given constructions and theorems are also valid classically.

We assume a classical understanding of basic notions in category theory: categories, functors, natural transformations, and equivalences of categories.

1. Category constructors

In this section, we make use of the concept of category constructors in order to build up a category equivalent to the category of finitely presented functors. Simply put, a category constructor is an operation that produces a category from some given input:

$$\boxed{\text{some input}} \xrightarrow{\text{category constructor}} \boxed{\text{a category}}$$

For example, we can regard a ring R as a single object category $\mathcal{C}(R)$ whose morphisms are given by the elements of R and composition is given by ring multiplication. This defines a category constructor:

$$\boxed{R \text{ a ring}} \xrightarrow{\mathcal{C}(-)} \boxed{R \text{ regarded as a category}}$$

The input of a category constructor can of course itself consist of a category, for example in the case of taking the opposite category:

$$\boxed{\mathbf{A} \text{ a category}} \xrightarrow{(-)^{\mathrm{op}}} \boxed{\mathbf{A}^{\mathrm{op}} := \text{opposite of } \mathbf{A}}$$

Other important examples of category constructors introduced in this section are:

- the additive closure $\mathbf{A} \mapsto \mathbf{A}^{\oplus}$, see Subsection 1.3, which turns an **Ab**-category $\mathbf{A}$ into an additive one,
- the Freyd category $\mathbf{A} \mapsto \mathcal{A}(\mathbf{A})$, see Subsection 1.5, which equips an additive category $\mathbf{A}$ with cokernels.

An iterative application of category constructors can lead to intriguing results. Let R be a commutative coherent ring, R-fpmod the category of finitely presented R-modules, and $\mathrm{fp}(R\text{-fpmod}, \mathbf{Ab})$ the category of all finitely presented functors $R\text{-fpmod} \to \mathbf{Ab}$ (where $\mathbf{Ab}$ denotes the category of abelian groups), i.e., functors that arise as cokernels of representable functors. Triggering a cascade of category constructors yields an equivalence

$$\mathrm{fp}(R\text{-fpmod}, \mathbf{Ab}) \simeq \mathcal{A}(\mathcal{A}(\mathcal{C}(R)^{\oplus})^{\mathrm{op}}).$$

Thus, knowing how to compute homomorphism sets within $\mathcal{A}(\mathcal{A}(\mathcal{C}(R)^{\oplus})^{\mathrm{op}})$ allows us to compute homomorphism sets between finitely presented functors.

In order to carry out this plan, we start at the lowest level $\mathcal{C}(R)$ of this cascade and analyze how algorithms at the current level give rise to algorithms on the next level until we end up with algorithms for dealing with the top level.

1.1. Computable categories. As a very first step we need to introduce categories from a constructive point of view. We will see that it is worthwhile to pay special attention to the classically trivial notion of *equality of morphisms*.

DEFINITION 1.1. A **category** $\mathbf{A}$ consists of the following data:

(1) A data type $\mathrm{Obj}_{\mathbf{A}}$ (**objects**).
(2) Depending on $A, B \in \mathrm{Obj}_{\mathbf{A}}$, a data type $\mathrm{Hom}_{\mathbf{A}}(A, B)$ (**morphisms**), each equipped with an equivalence relation $=$ (**equality**).
(3) An algorithm that computes for given $A, B, C \in \mathrm{Obj}_{\mathbf{A}}$, $\alpha \in \mathrm{Hom}_{\mathbf{A}}(A, B)$, and $\beta \in \mathrm{Hom}_{\mathbf{A}}(B, C)$ a morphism $\alpha \cdot \beta \in \mathrm{Hom}_{\mathbf{A}}(A, C)$ (**composition**) and which respects the equality equivalence relation. For $D \in \mathrm{Obj}_{\mathbf{A}}$ and $\gamma \in \mathrm{Hom}_{\mathbf{A}}(C, D)$, we require

$$(\alpha \cdot \beta) \cdot \gamma = \alpha \cdot (\beta \cdot \gamma) \text{ (\textbf{associativity})}.$$

(4) An algorithm that constructs for given $A \in \mathrm{Obj}_{\mathbf{A}}$ a morphism $\mathrm{id}_A \in \mathrm{Hom}_{\mathbf{A}}(A, A)$ (**identities**). For $B, C \in \mathrm{Obj}_{\mathbf{A}}$, $\beta \in \mathrm{Hom}_{\mathbf{A}}(B, A)$, $\gamma \in \mathrm{Hom}_{\mathbf{A}}(A, C)$, we require

$$\beta \cdot \mathrm{id}_A = \beta \quad \text{and} \quad \mathrm{id}_A \cdot \gamma = \gamma.$$

We give several examples of categories that will quickly lead us into the realm of computationally undecidable problems.

EXAMPLE 1.2. Every monoid $(M, 1, \cdot)$ gives rise to a category $\mathcal{C}(M)$, consisting of a single object $*$, whose morphisms $* \xrightarrow{m} *$ are given by the elements $m \in M$. Composition is induced by multiplication in M, the identity is given by $1 \in M$. Equality of morphisms is simply equality of elements.

EXAMPLE 1.3. Let Σ be a finite alphabet, say $\Sigma := \{a, b, c, d, e\}$. All words built up from Σ, i.e., the elements of the free monoid $\mathrm{Free}(\Sigma)$ on Σ, together with concatenation of words form the morphisms of the single object category $\mathcal{C}(\mathrm{Free}(\Sigma))$ with the empty word as the identity. In this example, equality of morphisms is given by comparing words letter by letter.

EXAMPLE 1.4. We may alter the notion of equality in the previous Example 1.3 without altering the other defining data. For a given finite set $R \subseteq \mathrm{Free}(\Sigma) \times \mathrm{Free}(\Sigma)$, we may choose the equality in our category as the monoid equivalence relation generated by R, i.e., the smallest equivalence relation containing R that is also a submonoid of $\mathrm{Free}(\Sigma) \times \mathrm{Free}(\Sigma)$. For example, we could choose the monoid equivalence relation generated by

$$\begin{aligned} ac &= ca \\ bc &= cb \\ ce &= eca \\ ad &= da \\ bd &= db \\ de &= edb \\ cca &= ccae. \end{aligned}$$

We can still perform compositions as in $\mathcal{C}(\mathrm{Free}(\Sigma))$, but the question of deciding whether two morphisms are equal w.r.t. the concrete monoid equivalence relation above is computationally unsolvable [**Col86**].

The previous example highlights the enormous importance of equality in a constructive setup and motivates the next definition which singles out those categories for which the classically trivial proposition

$$\forall \alpha, \beta \in \mathrm{Hom}_{\mathbf{A}}(A, B) : \alpha = \beta \vee \alpha \neq \beta$$

can be realized algorithmically.

DEFINITION 1.5. A category $\mathbf{A}$ is called **computable** if we have an algorithm that decides for given $A, B \in \mathrm{Obj}_{\mathbf{A}}$, $\alpha, \beta \in \mathrm{Hom}_{\mathbf{A}}(A, B)$ whether $\alpha = \beta$ or $\alpha \neq \beta$.

EXAMPLE 1.6. The category associated to the free monoid as described in Example 1.3 is computable if we can decide equality of elements in the given alphabet Σ.

The following example generalizes the example of a free monoid to "a free monoid with multiple objects", i.e., a free category.

EXAMPLE 1.7 (Free categories). A quiver Q is a directed graph (with finitely many vertices and edges) that is allowed to contain loops and multiple edges. Edges α in Q are usually called arrows and are depicted by

$$a \xrightarrow{\alpha} b,$$

and we set $\mathrm{Source}(\alpha) := a$ and $\mathrm{Range}(\alpha) := b$. Arrows α and β are called composable if

$$\mathrm{Range}(\alpha) = \mathrm{Source}(\beta).$$

Given two nodes $a, b \in Q$, a path from a to b is a finite sequence of arrows $\alpha_1, \dots, \alpha_n$ such that any two consecutive arrows are composable, and $\mathrm{Source}(\alpha_1) = a$, $\mathrm{Range}(\alpha_n) = b$. If $a = b$, we allow $n = 0$ and call it the empty path.

Taking the set of nodes in Q as our objects, and taking paths from a to b as our morphisms, we can create a category whose composition is given by concatenation of paths. The empty paths now come in handy as identities. Equality for morphisms is given by comparing two paths arrow-wise. If we start with the quiver that contains a single node denoted by $*$ and five loops

then this category recovers Example 1.3. Free categories are computable whenever we can decide equality of arrows.

1.2. Ab-categories. We are mostly interested in categories that admit additional structure. An **Ab**-category is a category **A** for which all homomorphism data types come equipped with the structure of abelian groups such that this structure is compatible with the composition in **A**. We spell this out explicitly.

DEFINITION 1.8. An **Ab-category** is a category **A** for which we have:

(1) An algorithm that computes for given $A, B \in \mathrm{Obj}_{\mathbf{A}}$, $\alpha, \beta \in \mathrm{Hom}_{\mathbf{A}}(A, B)$ a morphism $\alpha + \beta \in \mathrm{Hom}_{\mathbf{A}}(A, B)$ (**addition**).
(2) An algorithm that constructs for given $A, B \in \mathrm{Obj}_{\mathbf{A}}$ a morphism $0 \in \mathrm{Hom}_{\mathbf{A}}(A, B)$ (**zero morphism**).
(3) An algorithm that computes for given $A, B \in \mathrm{Obj}_{\mathbf{A}}$, $\alpha \in \mathrm{Hom}_{\mathbf{A}}(A, B)$ a morphism $-\alpha \in \mathrm{Hom}_{\mathbf{A}}(A, B)$ (**additive inverse**).
(4) For all $A, B \in \mathrm{Obj}_{\mathbf{A}}$, the given data turn $\mathrm{Hom}_{\mathbf{A}}(A, B)$ into an abelian group.
(5) Composition with morphisms both from left and right in **A** becomes a bilinear map.

EXAMPLE 1.9 (Rings as categories). Analogous to Example 1.2, every ring R gives rise to an **Ab**-category $\mathcal{C}(R)$, i.e., we identify ring multiplication with composition[3].

Analogous to Example 1.4, every two-sided ideal $I \subseteq R$ lets us alter the notion of equality in this category by considering two morphisms as equal if and only if they are equal as elements in the quotient ring R/I. We denote this category by $\mathcal{C}(R, I)$ and remark that it is equivalent to $\mathcal{C}(R/I)$.

EXAMPLE 1.10. Let k be a field. Let us consider the **Ab**-category $\mathcal{C}(R, I)$ associated to the commutative polynomial ring $R := k[x_1, \dots, x_n]$ for $n \in \mathbb{N}$ and an ideal $I \subseteq R$ generated by finitely many given elements. It is a great triumph of computer algebra that this category is indeed computable provided we can decide equality in k. The decidability test for equality in R/I can then be executed using the theory of Gröbner bases. For a detailed account on Gröbner bases, see, e.g., [**CLO92**].

To a graded ring, we can attach an **Ab**-category having more than a single object.

EXAMPLE 1.11 (Graded rings as categories). A $\mathbb{Z}$-graded ring is a ring R together with a decomposition

$$R = \bigoplus_{d \in \mathbb{Z}} R_d$$

[3]Note that we defined composition in a category as *pre*composition, and not as *post*composition. It is common to regard a ring R as a category with postcomposition being identified with ring multiplication, and also to use the symbol R in order to refer to that category. Thus, our category $\mathcal{C}(R)$ equals the category R^{op}.

into a direct sum of abelian groups such that $R_d \cdot R_e \subseteq R_{d+e}$ for all $d, e \in \mathbb{Z}$. We can form an **Ab**-category out of these data as follows:

(1) Objects are given by the elements of $\mathbb{Z}$.
(2) For $a, b \in \mathbb{Z}$, homomorphisms from a to b are given by elements in R_{b-a}.
(3) Composition is multiplication in R.

We denote this category by $\mathcal{C}(\bigoplus_{d\in\mathbb{Z}} R_d)$. For a graded ideal $I \subseteq R$, we alter the notion of equality analogously to Example 1.9 in order to obtain an **Ab**-category $\mathcal{C}(\bigoplus_{d\in\mathbb{Z}} R_d, I)$.

EXAMPLE 1.12 (Path algebras as categories). We can linearize Example 1.7 as follows. Let k be a commutative ring. Again, we take as objects the nodes in our quiver Q, but now, we allow as morphisms from a node a to a node b any formal k-linear combination of paths from a to b. Extending concatenation of paths k-bilinearly, we obtain in this way an **Ab**-category that we denote by $\mathcal{C}(k, Q)$.

Note that the morphisms from a to b in $\mathcal{C}(k, Q)$ identify with all elements in the path algebra $k[Q]$ that start at a and end at b. In this sense, our category only stores the uniform elements of $k[Q]$. Ideals $I \subseteq k[Q]$ generated by uniform elements let us alter the notion of equality analogously to Example 1.9, and we denote the corresponding category by $\mathcal{C}(k, Q, I)$.

1.3. Additive closure. In this subsection, we want to introduce a categorical concept that grasps the idea of forming matrices whose entries consist of morphisms in an underlying **Ab**-category.

DEFINITION 1.13. An **additive category** is an **Ab**-category $\mathbf{A}$ for which we have:

(1) An algorithm that computes for a given finite (possibly empty) list of objects $A_1, \dots, A_n$ in $\mathrm{Obj}_{\mathbf{A}}$ (for $n \in \mathbb{N}_0$) an object $\bigoplus_{i=1}^n A_i \in \mathrm{Obj}_{\mathbf{A}}$ (**direct sum**). If we are additionally given an integer $j \in \{1 \dots n\}$, we furthermore have algorithms for computing morphisms $\pi_j \in \mathrm{Hom}_{\mathbf{A}}(\bigoplus_{i=1}^n A_i, A_j)$ (**direct sum projection**) and $\iota_j : \mathrm{Hom}_{\mathbf{A}}(A_j, \bigoplus_{i=1}^n A_i)$ (**direct sum injection**).
(2) The identities
   - $\sum_{i=1}^n \pi_i \cdot \iota_i = \mathrm{id}_{\bigoplus_{i=1}^n A_i}$,
   - $\iota_i \cdot \pi_i = \mathrm{id}_{A_i}$,
   - $\iota_i \cdot \pi_j = 0$,

   hold for all $i, j = 1, \dots, n$, $i \neq j$.

Direct sums in $\mathbf{A}$ and matrices having morphisms in $\mathbf{A}$ as its entries are closely linked as follows: given a morphism

$$\bigoplus_{i=1}^m A_i \xrightarrow{\alpha} \bigoplus_{j=1}^n B_j$$

between direct sums in $\mathbf{A}$, we can form the matrix of morphisms

$$\Big(A_s \xrightarrow{\iota_s} \bigoplus_{i=1}^m A_i \xrightarrow{\alpha} \bigoplus_{j=1}^n B_j \xrightarrow{\pi_t} B_t\Big)_{st}.$$

Conversely, any matrix of morphisms

$$\big(A_s \xrightarrow{\alpha_{st}} B_t\big)_{\substack{s=1,\dots,m \\ t=1,\dots,n}}$$

defines a morphism

$$\sum_{\substack{s=1,\dots,m\\t=1,\dots,n}} (\pi_s \cdot \alpha \cdot \iota_t) : \bigoplus_{i=1}^{m} A_i \longrightarrow \bigoplus_{j=1}^{n} B_j.$$

Both constructions are mutually inverse thanks to the equational identities 1.13.(2) that hold for direct sums.

If an **Ab**-category **A** does not yet admit direct sums, it is easy to construct its additive closure by employing exactly the philosophy of thinking of morphisms between direct sums as matrices. We will now show this construction as an example of a *category constructor*.

CONSTRUCTION 1.14. Let **A** be an **Ab**-category. We construct its **additive closure** $\mathbf{A}^{\oplus}$ as follows: an object in $\mathbf{A}^{\oplus}$ is given by an integer $m \geqslant 0$ and a list

$$(A_1, \dots, A_m)$$

of objects $A_i \in \mathbf{A}$ for $i = 1, \dots, m$. We think of this list as formally representing the object

$$\bigoplus_{i=1}^{m} A_i.$$

A morphism from one such list $(A_1, \dots, A_m)$ to another $(B_1, \dots, B_n)$ is given by a matrix

$$\begin{pmatrix} \alpha_{11} & \cdots & \alpha_{1n} \\ \vdots & \ddots & \vdots \\ \alpha_{m1} & \cdots & \alpha_{mn} \end{pmatrix}$$

consisting of morphisms $\alpha_{ij} : A_i \longrightarrow B_j$ in **A**. Now, composition can be defined by the usual formula for matrix multiplication, and matrices with identity morphisms on the diagonal and zero morphisms off-diagonal serve as identities in this category. Equality for morphisms is checked entrywise.

REMARK 1.15. $\mathbf{A}^{\oplus}$ is computable if and only if **A** is.

It is quite easy to check that $\mathbf{A}^{\oplus}$ is indeed an additive category. Futhermore, we can always view **A** as a subcategory of $\mathbf{A}^{\oplus}$ by identifying an object in $A \in \mathbf{A}$ as a list with a single element (A). The empty list $()$ defines a zero object in $\mathbf{A}^{\oplus}$, i.e., an object whose identity morphism equals the zero morphism.

EXAMPLE 1.16. If k is a field, then the objects in $\mathcal{C}(k)^{\oplus}$ (see Example 1.9) are simply given by natural numbers $\mathbb{N}_0$, and a morphism from $m \in \mathbb{N}_0$ to $n \in \mathbb{N}_0$ is an $m \times n$ matrix with entries in k.

The map $m \mapsto k^{1\times m}$ and the identification of elements in $k^{m\times n}$ with k-linear maps $k^{1\times m} \longrightarrow k^{1\times n}$ gives rise to an equivalence of categories between $\mathcal{C}(k)^{\oplus}$ and the category of all finite dimensional k-vector spaces. We set

$$\mathrm{Rows}_k := \mathcal{C}(k)^{\oplus},$$

since we think of the objects $n \in \mathbb{N}_0$ as the vector spaces $k^{1\times n}$ of rows. From a computational point of view, Rows_k often serves as a workhorse: due to the power of Gaussian elimination, whenever we can reduce a problem in another category to linear algebra, we can try and solve it within Rows_k.

EXAMPLE 1.17. More generally, if R is a ring, then objects in $\mathcal{C}(R)^{\oplus}$ identify with row modules $R^{1\times n}$ for $n \in \mathbb{N}_0$, and every R-module homomorphism $R^{1\times m} \longrightarrow R^{1\times n}$ is given by a matrix in $R^{m\times n}$. But since not every R-module is free in general, $\mathcal{C}(R)^{\oplus}$ is only equivalent to a subcategory of the category of all finitely generated R-modules. We set

$$\mathrm{Rows}_R := \mathcal{C}(R)^{\oplus}.$$

If $\mathbf{A}$ has more than just a single object, then compositionality of morphisms in $\mathbf{A}^{\oplus}$ relies on more than just matching numbers of columns and rows.

EXAMPLE 1.18. If we take the additive closure of the category $\mathcal{C}(\bigoplus_{d\in\mathbb{Z}} R_d)$ introduced in Example 1.11, then we get a category whose objects can be seen as finite lists of integers. A morphism from such a list $(m_1, \dots, m_s)$ to another list $(n_1, \dots, n_t)$ with $s, t \in \mathbb{N}_0$ is given by a matrix

$$(a_{ij})_{\substack{i=1,\dots,s\\ j=1,\dots,t}}$$

with homogeneous entries in R whose degrees satisfy

$$\deg(a_{ij}) + m_i = n_j \tag{1}$$

whenever $a_{ij} \neq 0$.

As an example, let k be a field and $R = k[x, y]$ be the $\mathbb{Z}$-graded polynomial ring with $\deg(x) = \deg(y) = 1$. Then

$$(0,1) \xrightarrow{\begin{pmatrix} xy \\ x+y \end{pmatrix}} (2)$$

is an example of a morphism in $\mathcal{C}(\bigoplus_{d\in\mathbb{Z}} R_d)$. Note that the matrix alone does not determine the source and range of this morphism, since, for example

$$(-1,0) \xrightarrow{\begin{pmatrix} xy \\ x+y \end{pmatrix}} (1)$$

is also a valid example of a morphism. If we fix the matrix and the source/range in the first example and forget its range/source

$$(0,1) \xrightarrow{\begin{pmatrix} xy \\ x+y \end{pmatrix}} (?) \qquad\qquad (?,?) \xrightarrow{\begin{pmatrix} xy \\ x+y \end{pmatrix}} (2)$$

then Equation 1 makes it possible to reconstruct the missing information. However, such a reconstruction is not possible in general: the $s \times t$ zero matrix defines a valid morphism between any two objects $(m_1, \dots, m_s)$ and $(n_1, \dots, n_t)$.

EXAMPLE 1.19. Similarly, taking the additive closure of the category $\mathcal{C}(k, Q)$ introduced in Example 1.12, we get a category whose objects are finite lists of nodes in Q, and morphisms from a list $(v_1, \dots, v_s)$ to $(w_1, \dots, w_s)$ are matrices

$$(a_{ij})_{\substack{i=1,\dots,s\\ j=1,\dots,t}}$$

whose entries consist of uniform elements in the path algebra $k[Q]$, where a_{ij} is either zero or starts at v_i and ends at w_j.

1.4. Homomorphism structures. The question of how to describe the homomorphisms between two objects "as a whole" is just as important as the decidability problem of equality for two individual morphisms. Classically, one could restrict the attention to so-called locally small categories, which are categories $\mathbf{A}$ in which the members of the family $\mathrm{Hom}_{\mathbf{A}}(A, B)$ can all be interpreted as objects in **Set**, the category of sets. This enables us to view Hom as a functor

$$\mathrm{Hom} : \mathbf{A}^{\mathrm{op}} \times \mathbf{A} \to \mathbf{Set}.$$

For our constructive approach, we will simply generalize this point of view and axiomatize those features that we need from a Hom-functor to make computational use of it. But before we do this, we state the definition of a functor within our constructive setup.

DEFINITION 1.20. A **functor** F between two categories $\mathbf{A}$ and $\mathbf{B}$ consists of the following data:

(1) An algorithm that computes for given $A \in \mathrm{Obj}_{\mathbf{A}}$ an object $F(A) \in \mathrm{Obj}_{\mathbf{B}}$.
(2) An algorithm that computes for given $A, B \in \mathrm{Obj}_{\mathbf{A}}$, $\alpha \in \mathrm{Hom}_{\mathbf{C}}(A, B)$ a morphism $F(\alpha) \in \mathrm{Hom}_{\mathbf{B}}(F(A), F(B))$. This algorithm needs to be compatible with the notion of equality for morphisms.
(3) For $A \in \mathrm{Obj}_{\mathbf{C}}$, $F(\mathrm{id}_A) = \mathrm{id}_{F(A)}$.
(4) For $A, B, C \in \mathrm{Obj}_{\mathbf{C}}$, $\alpha \in \mathrm{Hom}_{\mathbf{A}}(A, B)$, $\beta \in \mathrm{Hom}_{\mathbf{A}}(B, C)$, we have
$$F(\alpha \cdot \beta) = F(\alpha) \cdot F(\beta).$$

REMARK 1.21. Note that since we did not impose a notion of equality on the data type $\mathrm{Obj}_{\mathbf{A}}$, it is not meaningful to declare the operation of F on objects to be compatible with equality like we did in the case of morphisms.

DEFINITION 1.22. Let $\mathbf{A}$, $\mathbf{B}$ be categories. A **B-homomorphism structure** for $\mathbf{A}$ consists of the following data:

(1) An object $1 \in \mathbf{B}$ called the **distinguished object**.
(2) A functor $H : \mathbf{A}^{\mathrm{op}} \times \mathbf{A} \to \mathbf{B}$.
(3) A bijection $\nu : \mathrm{Hom}_{\mathbf{A}}(A, B) \xrightarrow{\sim} \mathrm{Hom}_{\mathbf{B}}(1, H(A, B))$ natural in $A, B \in \mathbf{A}$, i.e,
$$\nu(\alpha \cdot X \cdot \beta) = \nu(X) \cdot H(\alpha, \beta)$$
for all composable triples of morphisms α, X, β.

Moreover, if we are in the context of **Ab**-categories, we also impose the condition that H is a bilinear functor, i.e., acts linearly on morphisms in each component.

EXAMPLE 1.23. Let k be a field. We are going to describe a homomorphism structure for Rows_k (see Example 1.16) that is inspired by the fact that Rows_k is equivalent to the category of finite dimensional k-vector spaces and that linear maps between two given finite dimensional vector spaces form themselves a finite dimensional vector space.

In the language of homomorphism structures, we can construct a Rows_k-homomorphism structure for Rows_k. We define a functor H on objects (which are simply elements in $\mathbb{N}_0$) by multiplication of natural numbers, and on morphisms (which are matrices) by

$$H(\alpha, \beta) := \alpha^{\mathrm{tr}} \otimes \beta,$$

where $(-)^{\mathrm{tr}}$ is transposition and $\otimes$ denotes the Kronecker product. As a distinguished object, we take the natural number $1 \in \mathbb{N}_0$. Now, for given $m, n \in \mathbb{N}_0$, any morphism from 1 to mn, i.e., any row vector $(a_i)_{i=1,\ldots mn}$, can be interpreted as an $m \times n$ matrix by "line-breaking" after each n-entries. Conversely, every $m \times n$ matrix can be converted to such a row by simply concatenating all rows. Thus, we have found a natural way to transfer "vectors" of mn, i.e., morphisms $1 \to mn$, into morphisms $m \to n$ in Rows_k.

So, note that it is not the object $mn \in \mathrm{Rows}_k$ alone that encodes $\mathrm{Hom}_{\mathrm{Rows}_k}(m, n)$, but it is the object mn in the context of a homomorphisms structure that allows us to *interpret* it as an encoding of homomorphisms from m to n.

Next, we describe homomorphism structures for special cases of the **Ab**-categories given in Examples 1.9, 1.11, 1.12.

EXAMPLE 1.24. Let R be a *commutative* ring. We can construct a $\mathcal{C}(R)$-homomorphism structure for $\mathcal{C}(R)$ (see Example 1.9) as follows: the operation

$$H : \mathcal{C}(R)^{\mathrm{op}} \times \mathcal{C}(R) \longrightarrow \mathcal{C}(R) : (* \xleftarrow{a} *, * \xrightarrow{b} *) \mapsto (* \xrightarrow{a \cdot b} *)$$

defines a bilinear functor due to the commutativity of R. For the distinguished object, we have no other choice but to take the unique object $*$ in $\mathcal{C}(R)$. Finally, ν can be chosen as the identity on $\mathrm{Hom}_{\mathcal{C}(R)}(*, *)$.

EXAMPLE 1.25. Let k be a field and let R be a $\mathbb{Z}$-graded k-algebra. If every R_d is of finite k-dimension with bases $\{r_d^1, \ldots, r_d^{\dim_k(R_d)}\}$, then we may write for every $a, b, c \in \mathbb{Z}$ and $r \in R_c$, $s \in R_{b-(a+c)}$ the k-linear operator

$$R_a \longrightarrow R_b : x \mapsto r \cdot x \cdot s$$

in terms of the given bases in order to obtain matrices $M_{a,b,r,s}$. This enables us to describe for $\mathcal{C}(\bigoplus_{d \in \mathbb{Z}} R_d)$ (see Example 1.11) a Rows_k-homomorphism structure with

$$H(a, b) := \dim_k(R_{b-a})$$

for $a, b \in \mathbb{Z}$, and for $a', b' \in \mathbb{Z}$, $r \in R_{a-a'}$, $s \in R_{b'-b}$,

$$H(a \xleftarrow{r} a', b \xrightarrow{s} b') := M_{(b-a),(b'-a'),r,s}.$$

The distinguished object is $1 \in \mathrm{Rows}_k$, and $\nu_{a,b}$ computes for an element $r \in R_{b-a}$ its list of coefficients w.r.t. the basis $\{r_{b-a}^1, \ldots, r_{b-a}^{\dim_k(R_{b-a})}\}$.

Moreover, if R is commutative (but the R_d not necessarily finite dimensional), we could also construct a different homomorphism structure for $\mathcal{C}(\bigoplus_{d \in \mathbb{Z}} R_d)$, namely a $\mathcal{C}(\bigoplus_{d \in \mathbb{Z}} R_d)$-homomorphism structure with

$$H(a, b) := b - a$$

for $a, b \in \mathbb{Z}$ and for $a', b' \in \mathbb{Z}$, $r \in R_{a-a'}$, $s \in R_{b'-b}$,

$$H(a \xleftarrow{r} a', b \xrightarrow{s} b') := (b - a) \xrightarrow{r \cdot s} (b' - a').$$

This time, the distinguished object is $0 \in \mathcal{C}(\bigoplus_{d \in \mathbb{Z}} R_d)$, and ν given by the identity

$$\mathrm{Hom}_{\mathcal{C}(\bigoplus_{d \in \mathbb{Z}} R_d)}(a, b) = R_{b-a} = \mathrm{Hom}_{\mathcal{C}(\bigoplus_{d \in \mathbb{Z}} R_d)}(0, b - a).$$

So, we see that it is neither necessarily the case that **B** is equivalent to **A**, nor that there is only a single homomorphism structure for a given category **A**.

Example 1.26. Let k be a field and Q be a quiver. If Q is acyclic, then the homomorphisms in $\mathcal{C}(k,Q)$ from a vertex v to a vertex w form a finite dimensional k-vector space. Similarly to Example 1.25, this allows us to create an Rows_k-homomorphism structure for $\mathcal{C}(k,Q)$ with

$$H(v,w) := \text{number of paths from } v \text{ to } w.$$

It is natural to ask how a structure that we have given to a category may transfer to a category obtained by a category constructor. We can indeed transfer homomorphism structures to the additive closure.

Construction 1.27. Let $\mathbf{A}$ be an $\mathbf{Ab}$-category and $\mathbf{B}$ be an additive category. Let furthermore $(H,1,\nu)$ be a $\mathbf{B}$-homomorphism structure for $\mathbf{A}$. Then we can extend $(H,1,\nu)$ to a $\mathbf{B}$-homomorphisms structure $(H^{\oplus},1,\nu^{\oplus})$ for $\mathbf{A}^{\oplus}$ by extending bilinearly

$$H^{\oplus}\left((B_j)_j \xleftarrow{(\alpha_{ij})_{ij}} (A_i)_i, (C_s)_s \xrightarrow{(\beta_{st})_{st}} (D_t)_t\right) := \bigoplus_{j,s} H(B_j,C_s) \xrightarrow{\left(H(\alpha_{ij},\beta_{st})\right)_{(js)(it)}} \bigoplus_{i,t} H(A_i,D_t).$$

The natural isomorphism $\nu^{\oplus}$ is defined via the composition of natural isomorphisms

$$\begin{aligned}\mathrm{Hom}_{\mathbf{A}^{\oplus}}((B_j)_j,(C_s)_s) &\simeq \bigoplus_{j,s} \mathrm{Hom}_{\mathbf{A}^{\oplus}}(B_j,C_s)\\ &\simeq \bigoplus_{j,s} \mathrm{Hom}_{\mathbf{B}}(1,H(B_j,C_s))\\ &\simeq \mathrm{Hom}_{\mathbf{B}}\left(1,H^{\oplus}((B_j)_j,(C_s)_s)\right).\end{aligned}$$

Remark 1.28. We can also use Construction 1.27 in the case when $\mathbf{B}$ is an $\mathbf{Ab}$-category that is not necessarily additive by first applying the full embedding $\mathbf{B} \hookrightarrow \mathbf{B}^{\oplus}$ in order to obtain a $\mathbf{B}^{\oplus}$-homomorphism structure for $\mathbf{A}$, and then proceed as described.

Example 1.29. Let k be a field. Let H denote the $\mathcal{C}(k)$-homomorphism structure of $\mathcal{C}(k)$ described in Example 1.24. Applying Construction 1.27 to H (via Remark 1.28) yields exactly the Rows_k-homomorphism structure of $\mathrm{Rows}_k = \mathcal{C}(k)^{\oplus}$ that we described in Example 1.23.

1.5. Freyd category. In this subsection, we introduce a further category constructor: the Freyd category [**Fre66, Bel00**]. Freyd categories provide a unified approach to categories of finitely presented modules, finitely presented graded modules, and finitely presented functors.

Let R be a ring. Recall that a (left) R-module M is called **finitely presented** if there exist $a,b \in \mathbb{N}_0$ and an exact sequence

$$0 \longleftarrow M \longleftarrow R^{1\times a} \xleftarrow{\rho_M} R^{1\times b},$$

which is called a **presentation** of M. Since ρ_M is induced by a matrix with rows $r_1,\dots,r_b \in R^{1\times a}$, being finitely presented means nothing but the existence of an isomorphism

$$M \simeq R^{1\times a}/\langle r_1,\dots,r_b\rangle.$$

Thus, we may think of a presentation as a way to store finitely many relations $r_1,\dots,r_b$ that we would like to impose on an free module $R^{1\times a}$. Let N be another finitely presented module with presentation $\rho_N : R^{1\times b'} \to R^{1\times a'}$. By the comparison theorem [**Wei94**], we can lift any morphism $\mu : M \to N$ to a commutative diagram

$$\begin{array}{ccccccc}
0 & \longleftarrow & M & \longleftarrow & R^{1\times a} & \xleftarrow{\rho_M} & R^{1\times b} \\
 & & \downarrow{\scriptstyle \mu} & & \downarrow & & \downarrow \\
0 & \longleftarrow & N & \longleftarrow & R^{1\times a'} & \xleftarrow{\rho_N} & R^{1\times b'}
\end{array}$$

and conversely, any commutative diagram

$$\begin{array}{ccc}
R^{1\times a} & \xleftarrow{\rho_M} & R^{1\times b} \\
\downarrow & & \downarrow \\
R^{1\times a'} & \xleftarrow{\rho_N} & R^{1\times b'}
\end{array}$$

induces a morphism $\mu : M \to N$. Moreover, such a μ is zero if and only if we have a commutative diagram with exact rows

$$\begin{array}{ccccccc}
0 & \longleftarrow & M & \longleftarrow & R^{1\times a} & \xleftarrow{\rho_M} & R^{1\times b} \\
 & & \downarrow{\scriptstyle \mu} & & \downarrow & \searrow & \\
0 & \longleftarrow & N & \longleftarrow & R^{1\times a'} & \xleftarrow{\rho_N} & R^{1\times b'}.
\end{array}$$

It follows that computing with finitely presented modules and their homomorphisms can be replaced by computing with presentations (which are nothing but morphisms in the additive category Rows_R, see Example 1.17), and commutative squares involving presentations (which are simply commutative squares within Rows_R) considered up to an equivalence relation. The concept of a Freyd category formalizes this calculus with Rows_R being replaced by an arbitrary additive category $\mathbf{A}$.

CONSTRUCTION 1.30 (Freyd categories). Let $\mathbf{A}$ be an additive category. We create $\mathcal{A}(\mathbf{A})$, the so-called **Freyd category** of $\mathbf{A}$. Its objects consist of morphisms

$$(A \xleftarrow{\rho_A} R_A)$$

in $\mathbf{A}$. We think of such morphisms as formally representing the cokernel of ρ_A. Note that neither R_A nor ρ_A do formally depend on A, however, we like to decorate these objects with A as an index and think of them as an encoding for "relations" imposed on A. A morphism between two objects in $\mathcal{A}(\mathbf{A})$, i.e., $(A \xleftarrow{\rho_A} R_A)$ to $(B \xleftarrow{\rho_B} R_B)$, is given by a morphism

$$\alpha : A \longrightarrow B$$

such that $\exists \rho_\alpha : R_A \longrightarrow R_B$ making the diagram

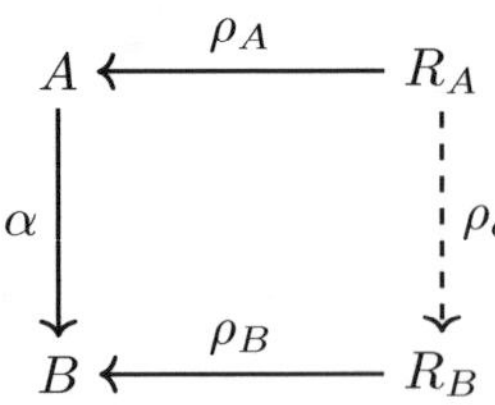

commutative. The equality of two morphisms $A \xrightarrow{\alpha} B$, $A \xrightarrow{\alpha'} B$ from $(A \xleftarrow{\rho_A} R_A)$ to $(B \xleftarrow{\rho_B} R_B)$ is defined by the existence of a λ (called **witness for α and α' being equal**) rendering the diagram

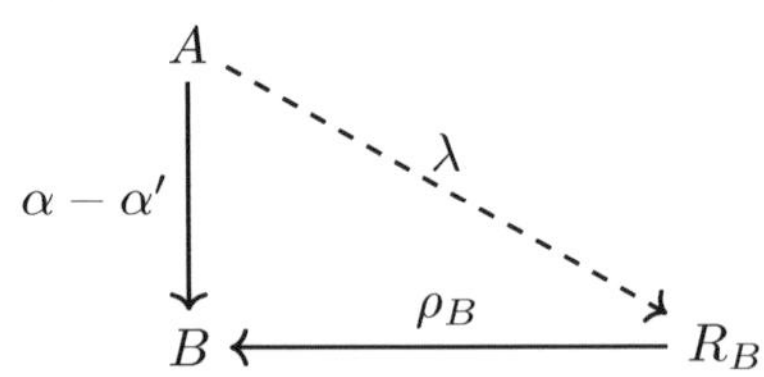

commutative. Composition and identity morphisms are inherited from $\mathbf{A}$. It is easy to check that the notion of equality for morphisms yields an equivalence relation compatible with composition and identities.

REMARK 1.31. Two commutative squares

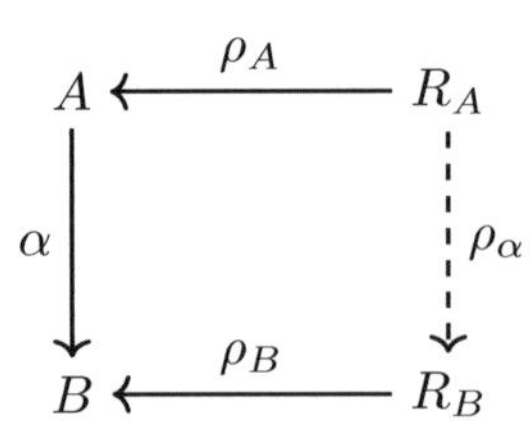

and

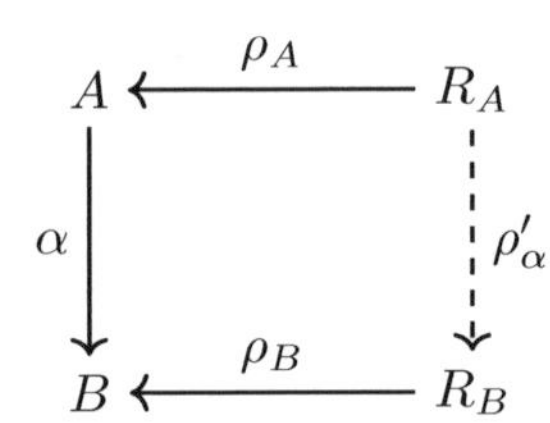

are equal as morphisms in $\mathcal{A}(\mathbf{A})$ with $0 : A \longrightarrow R_B$ as a witness, which is why we depict the arrows corresponding to $\rho_\alpha, \rho'_\alpha$ with a dashed line: they merely need to exist, but do not otherwise contribute to the actual morphism.

If R-fpmod denotes the category of finitely presented (left) R-modules, then the discussion in the beginning of this subsection can be summarized by the existence of an equivalence

$$R\text{-fpmod} \simeq \mathcal{A}(\mathrm{Rows}_R).$$

Note that the decisive feature of row modules $R^{1\times a}$ that makes this equivalence work is their projectiveness as R-modules. Thus, if we let Proj_R denote the full subcategory of the category of R-modules spanned by all finitely presented projective modules, and if $\mathbf{A}$ is any full subcategory satisfying

$$\mathrm{Rows}_R \subseteq \mathbf{A} \subseteq \mathrm{Proj}_R,$$

we still have

$$R\text{-fpmod} \simeq \mathcal{A}(\mathbf{A}).$$

If k is a field and Q a quiver, then $\mathcal{C}(k,Q)^{\oplus}$ (see Example 1.19) identifies with the full additive subcategory of the category of modules over the path algebra $k[Q]$ generated by the projectives $k[Q]e_v$, where e_v denotes the idempotent associated to the node $v \in Q$. Since this subcategory contains $k[Q]$ and thus $\mathrm{Rows}_{k[Q]}$, we obtain an equivalence

$$k[Q]\text{-fpmod} \simeq \mathcal{A}(\mathcal{C}(k,Q)^{\oplus}).$$

The discussion in this subsection neatly generalizes to finitely presented graded modules. If $R = \bigoplus_{d\in\mathbb{Z}} R_d$ is a $\mathbb{Z}$-graded ring, then $\mathcal{C}(\bigoplus_{d\in\mathbb{Z}} R_d)^{\oplus}$ (see Example 1.18) identifies with the full additive subcategory of the category of graded R-modules generated by the shifts $R(d)$ for $d \in \mathbb{Z}$, i.e., by the graded modules with graded parts $R(d)_e := R_{d+e}$ for all $e \in \mathbb{Z}$, and we again have an equivalence

$$R\text{-fpgrmod} \simeq \mathcal{A}(\mathcal{C}(\bigoplus_{d\in\mathbb{Z}} R_d)^{\oplus}),$$

with R-fpgrmod denoting the category of finitely presented graded R-modules.

Thus, the abstract study of Freyd categories enables us to study all these computational models of finitely presented modules in one go.

For an additive category $\mathbf{A}$, let $\mathrm{Hom}(\mathbf{A}^{\mathrm{op}}, \mathbf{Ab})$ denote the category of contravariant additive functors from $\mathbf{A}$ into the category of abelian groups $\mathbf{Ab}$. By Yoneda's lemma, the functor

$$\mathbf{A} \longrightarrow \mathrm{Hom}(\mathbf{A}^{\mathrm{op}}, \mathbf{Ab}) : A \mapsto (-, A)$$

is full and faithful, where $(-, A)$ denotes the contravariant Hom-functor. Thus, we can think of $\mathbf{A}$ as the full subcategory of $\mathrm{Hom}(\mathbf{A}^{\mathrm{op}}, \mathbf{Ab})$ generated by all representable functors. Again, by Yoneda's lemma, representable functors are projective objects in $\mathrm{Hom}(\mathbf{A}^{\mathrm{op}}, \mathbf{Ab})$, and a straightforward generalization of the discussion in the beginning of this subsection shows that we can identify $\mathcal{A}(\mathbf{A})$ with the full subcategory of $\mathrm{Hom}(\mathbf{A}^{\mathrm{op}}, \mathbf{Ab})$ generated by so-called *finitely presented functors*. A functor $F : \mathbf{A}^{\mathrm{op}} \longrightarrow \mathbf{Ab}$ is finitely presented if there exists $A, B \in \mathbf{A}$ and $\alpha : A \to B$ and an exact sequence

$$0 \longleftarrow F \longleftarrow (-, B) \xleftarrow{(-,\alpha)} (-, A)$$

in $\mathrm{Hom}(\mathbf{A}^{\mathrm{op}}, \mathbf{Ab})$, i.e., F arises as the cokernel of a morphism between representable functors. Analogously, one defines finitely presented covariant functors on $\mathbf{A}$, and the category of all such functors is equivalent to $\mathcal{A}(\mathbf{A}^{\mathrm{op}})$.

EXAMPLE 1.32. If $\mathbf{A}$ is an abelian category with enough projectives and $A \in \mathbf{A}$, then

$$\mathrm{Ext}^i(A, -) : \mathbf{A} \to \mathbf{Ab}$$

if finitely presented for all $i \geqslant 0$ [**Aus66**]. For example, in order to write $\mathrm{Ext}^1(A, -)$ as an object in $\mathcal{A}(\mathbf{A}^{\mathrm{op}})$, take any short exact sequence

$$0 \longleftarrow A \longleftarrow P \longleftarrow \Omega^1(A) \longleftarrow 0$$

with P projective. Then the morphism $(\Omega^1(A) \longrightarrow P)$ considered as an object in $\mathcal{A}(\mathbf{A}^{\mathrm{op}})$ corresponds to $\mathrm{Ext}^1(A, -)$. For higher Exts, we need to compute more steps of a projective resolution of A.

We have seen in this subsection that if we start with a ring R and consider it as a single object category $\mathcal{C}(R)$, then we can apply a cascade of category constructors

$$\mathcal{A}(\mathcal{A}(\mathcal{C}(R)^{\oplus})^{\mathrm{op}})$$

and end up with a category equivalent to finitely presented functors on finitely presented modules over R. Thus, the question of how to compute with finitely presented functors now reduces to the understanding of how to compute with Freyd categories.

1.6. Computing with Freyd categories. We explain how to perform several explicit constructions within Freyd categories, like computing cokernels, kernels, lifts along monomorphisms, and homomorphism structures. For details about the correctness of these constructions, we refer the reader to [**Pos21**].

1.6.1. *Equality of morphisms.* Being computable for $\mathbf{A}$ does by no means imply computability of $\mathcal{A}(\mathbf{A})$. We specify the decisive algorithmic feature of $\mathbf{A}$ that turns $\mathcal{A}(\mathbf{A})$ into a computable category.

DEFINITION 1.33. We say a category $\mathbf{A}$ has **decidable lifts** if we have an algorithm that takes as an input a cospan

$$A \xrightarrow{\alpha} B \xleftarrow{\gamma} C$$

and either outputs a lift $\lambda : A \to C$ rendering the diagram

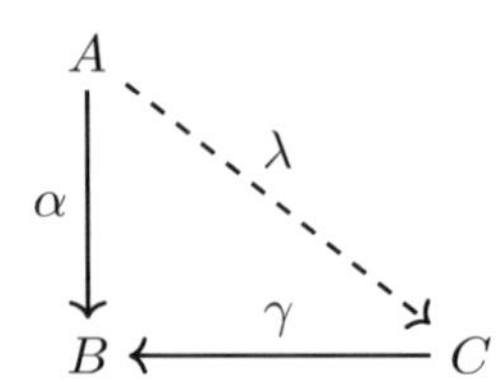

commutative, or disproves the existence of such a lift.

Clearly, whenever an additive category $\mathbf{A}$ has decidable lifts, we are able to decide equality in $\mathcal{A}(\mathbf{A})$.

EXAMPLE 1.34. Let k be a field with decidable equality of elements. Then, the category Rows_k has decidable lifts: a cospan in Rows_k is nothing but a pair of matrices α, γ over k having the same number of columns, and we can decide whether there exists a matrix λ over k such that $\lambda \cdot \gamma = \alpha$ using Gaussian elimination.

EXAMPLE 1.35. The following class of examples is vital for constructive algebraic geometry. Let k be a field with decidable equality of elements. For

$$R = k[x_1, \dots, x_n]/I,$$

where $I \subseteq k[x_1, \dots, x_n]$ is an ideal, Gröbner basis techniques imply that Rows_R has decidable lifts. Moreover, if $\mathfrak{p} \subseteq k[x_1, \dots, x_n]/I$ is a prime ideal, then for the localization

$$R = (k[x_1, \dots, x_n]/I)_{\mathfrak{p}},$$

Rows_R has decidable lifts. A general algorithm proving this fact can be found in [**Pos18**]. Computing lifts in more specialized cases of such rings are treated for example in [**BLH11**] or [**GP02**].

We can employ homomorphism structures for making lifts decidable.

LEMMA 1.36. *Let $\mathbf{A}$ have a $\mathbf{B}$-homomorphism structure $(H, \nu, 1)$. For a given cospan $A \xrightarrow{\alpha} B \xleftarrow{\gamma} C$ in $\mathbf{A}$, there exists a lift*

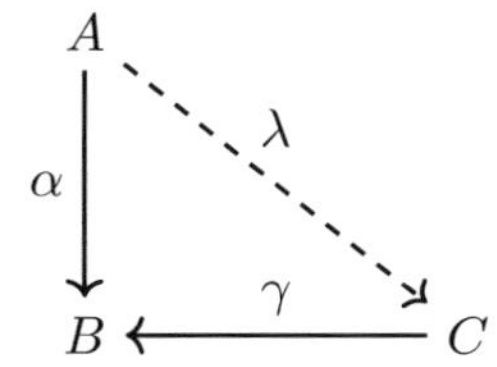

in $\mathbf{A}$ *if and only if there exists a lift*

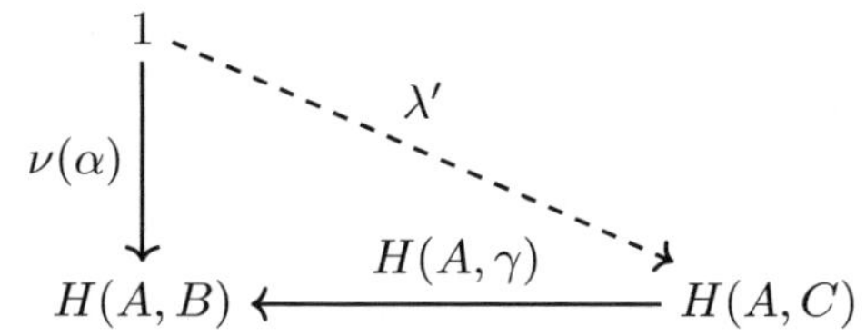

in $\mathbf{B}$. *In other words, we can decide lifts in* $\mathbf{A}$ *whenever we can decide lifts in* $\mathbf{B}$.

PROOF. It is easy to see that

$$\nu : \mathrm{Hom}_{\mathbf{A}}(A, C) \longrightarrow \mathrm{Hom}(1, H(A, C))$$

induces a bijection between lifts of the former system and lifts of the latter, since, by naturality, we have

$$\nu(\alpha) = \nu(\lambda \cdot \gamma) = \nu(\lambda) \cdot H(A, \gamma).$$

□

EXAMPLE 1.37. Let k be a field with decidable equality and let Q be an acyclic quiver. Then the Rows_k-homomorphism structure of $\mathcal{C}(k, Q)^{\oplus}$ described in Example 1.26, the statement in Lemma 1.36, and the decidability of lifts in Rows_k (Example 1.34) imply the decidability of lifts in $\mathcal{C}(k, Q)^{\oplus}$.

The same holds for $\mathbb{Z}$-graded k-algebras $R = \bigoplus_{d \in \mathbb{Z}} R_d$ with finite dimensional degree-parts, see Example 1.25.

1.6.2. *Cokernels.* Just as the additive closure turns an **Ab**-category into an additive one, Freyd categories endow additive categories with cokernels.

DEFINITION 1.38. Let $\mathbf{A}$ be an additive category. Given $A, B \in \mathrm{Obj}_{\mathbf{A}}$, $\alpha \in \mathrm{Hom}_{\mathbf{A}}(A, B)$, a cokernel of α consists of the following data:

(1) An object $\mathrm{CokernelObject}(\alpha)$ (**cokernel object**), also denoted by $\mathrm{coker}(\alpha)$, and a morphism

$$\mathrm{CokernelProjection}(\alpha) \in \mathrm{Hom}_{\mathbf{A}}(B, \mathrm{CokernelObject}(\alpha)) \quad \textbf{(cokernel projection)}$$

such that $\alpha \cdot \mathrm{CokernelProjection}(\alpha) = 0$.

(2) An algorithm that computes for given $T \in \mathrm{Obj}_{\mathbf{A}}$, $\tau \in \mathrm{Hom}_{\mathbf{A}}(B, T)$ such that $\alpha \cdot \tau = 0$ a morphism

$$\mathrm{CokernelColift}(\alpha, \tau) \in \mathrm{Hom}_{\mathbf{A}}(\mathrm{coker}(\alpha), T) \quad \textbf{(cokernel colift)}$$

such that

$$\mathrm{CokernelProjection}(\alpha) \cdot \mathrm{CokernelColift}(\alpha, \tau) = \tau,$$

where $\mathrm{CokernelColift}(\alpha, \tau)$ is uniquely determined (up to equality of morphisms) by this property.

EXAMPLE 1.39. Let R be a ring and let $\rho : R^{1\times b} \longrightarrow R^{1\times a}$ be an R-module homomorphism. Then $\operatorname{coker}(\rho) \in R\text{-fpmod}$ is mapped to an object in $\mathcal{A}(\mathrm{Rows}_R)$ via the equivalence

$$R\text{-fpmod} \simeq \mathcal{A}(\mathrm{Rows}_R),$$

and this object is given, up to isomorphism, by the morphism ρ itself. In this sense, taking the cokernel of a morphism between two row modules is a completely formal act.

Every morphism in $\mathcal{A}(\mathbf{A})$ has a cokernel by means of the following construction, whose proof of correctness can be found in [**Pos21**, Section 3.1].

CONSTRUCTION 1.40. The following algorithm creates cokernel projections in $\mathcal{A}(\mathbf{A})$:

$$\begin{array}{ccc} A & \xleftarrow{\rho_A} & R_A \\ {\scriptstyle\alpha}\downarrow & & \downarrow{\scriptstyle\rho_\alpha} \\ B & \xleftarrow{\rho_B} & R_B \end{array} \quad \xmapsto{\text{CokernelProjection}} \quad \begin{array}{ccc} B & \xleftarrow{\rho_B} & R_B \\ {\scriptstyle\mathrm{id}_B}\downarrow & & \downarrow{\scriptstyle(\mathrm{id}_{R_B} \;\; 0)} \\ B & \xleftarrow{\begin{pmatrix}\rho_B \\ \alpha\end{pmatrix}} & R_B \oplus A \end{array}$$

Moreover, for any morphism

$$\begin{array}{ccc} B & \xleftarrow{\rho_B} & R_B \\ {\scriptstyle\tau}\downarrow & & \downarrow{\scriptstyle\rho_\tau} \\ T & \xleftarrow{\rho_T} & R_T \end{array}$$

and any witness $A \xrightarrow{\lambda} R_T$ for the composition

$$\begin{array}{ccc} A & \xleftarrow{\rho_A} & R_A \\ {\scriptstyle\alpha\cdot\tau}\downarrow & & \downarrow{\scriptstyle\rho_\alpha\cdot\rho_\tau} \\ T & \xleftarrow{\rho_T} & R_T \end{array}$$

being equal to zero in $\mathcal{A}(\mathbf{A})$, we can construct a cokernel colift:

$$\begin{array}{ccc} B & \xleftarrow{\begin{pmatrix}\rho_B \\ \alpha\end{pmatrix}} & R_B \oplus A \\ {\scriptstyle\tau}\downarrow & & \downarrow{\scriptstyle\begin{pmatrix}\rho_\tau \\ \lambda\end{pmatrix}} \\ T & \xleftarrow{\rho_T} & R_T \end{array}$$

1.6.3. *Kernels.* Unlike cokernels, kernels in $\mathcal{A}(\mathbf{A})$, if they exist, cannot be constructed formally but only with the help of additional algorithms in $\mathbf{A}$.

DEFINITION 1.41. Let $\mathbf{A}$ be an additive category. Given $A, B \in \mathrm{Obj}_{\mathbf{A}}$, $\alpha \in \mathrm{Hom}_{\mathbf{A}}(A, B)$, a **kernel** of α consists of the following data:

(1) An object KernelObject$(\alpha) \in \mathrm{Obj}_{\mathbf{A}}$ (**kernel object**), also denoted by $\ker(\alpha)$, and a morphism

$$\mathrm{KernelEmbedding}(\alpha) \in \mathrm{Hom}_{\mathbf{A}}(\mathrm{KernelObject}(\alpha), A) \quad (\textbf{kernel embedding})$$

such that KernelEmbedding$(\alpha) \cdot \alpha = 0$.

(2) An algorithm that computes for given $T \in \mathrm{Obj}_{\mathbf{A}}$, $\tau \in \mathrm{Hom}_{\mathbf{A}}(T, A)$ such that $\tau \cdot \alpha = 0$ a morphism

$$\mathrm{KernelLift}(\alpha, \tau) \in \mathrm{Hom}_{\mathbf{A}}(T, \mathrm{KernelObject}(\alpha)) \quad (\textbf{kernel lift})$$

such that

$$\mathrm{KernelLift}(\alpha, \tau) \cdot \mathrm{KernelEmbedding}(\alpha) = \tau$$

where KernelLift(α, τ) is uniquely determined (up to equality of morphisms) by this property.

REMARK 1.42. Let R be a ring. Assume that we can produce for every R-module homomorphism of the form $\rho : R^{1\times b} \longrightarrow R^{1\times a}$ another R-module homomorphism

$$\kappa : R^{1\times c} \longrightarrow R^{1\times b}$$

whose image spans the kernel of ρ as an R-module. Then, by using such a procedure twice, we are able to construct an exact sequence

$$R^{1\times a} \xleftarrow{\ \rho\ } R^{1\times b} \xleftarrow{\ \kappa\ } R^{1\times c} \xleftarrow{\ \kappa'\ } R^{1\times c'}$$

in which κ' is a finite presentation of the kernel of ρ.

Abstracting the procedure $\rho \mapsto \kappa$ from Rows_R to an arbitrary additive category $\mathbf{A}$ leads to the notion of a weak kernel, which is defined exactly like a kernel, but we drop the uniqueness assumption of the kernel lift.

DEFINITION 1.43. Let $\mathbf{A}$ be an additive category. Given $A, B \in \mathrm{Obj}_{\mathbf{A}}$, $\alpha \in \mathrm{Hom}_{\mathbf{A}}(A, B)$, a **weak kernel** of α consists of the following data:

(1) An object WeakKernel$(\alpha) \in \mathrm{Obj}_{\mathbf{A}}$ (**weak kernel object**) and a morphism

$$\mathrm{WeakKernelEmbedding}(\alpha) \in \mathrm{Hom}_{\mathbf{A}}(\mathrm{WeakKernel}(\alpha), A) \quad (\textbf{weak kernel embedding})$$

such that WeakKernelEmbedding$(\alpha) \cdot \alpha = 0$.

(2) An algorithm that computes for given $T \in \mathrm{Obj}_{\mathbf{A}}$, $\tau \in \mathrm{Hom}_{\mathbf{A}}(T, A)$ such that $\tau \cdot \alpha = 0$ a morphism

$$\mathrm{WeakKernelLift}(\alpha, \tau) \in \mathrm{Hom}_{\mathbf{A}}(T, \mathrm{WeakKernel}(\alpha)) \quad (\textbf{weak kernel lift})$$

such that

$$\mathrm{WeakKernelLift}(\alpha, \tau) \cdot \mathrm{WeakKernelEmbedding}(\alpha) = \tau.$$

EXAMPLE 1.44. We unravel the definition of a weak kernel in the concrete case where R is a ring and $\mathbf{A} = \mathrm{Rows}_R$. So, given a matrix $R^{1\times b} \xrightarrow{\rho} R^{1\times a}$, i.e., a morphism in Rows_R, a weak kernel of ρ consists of

(1) an object $R^{1\times c}$,
(2) a matrix $R^{1\times c} \xrightarrow{\kappa} R^{1\times b}$ such that $\kappa \cdot \rho = 0$,
(3) and for every other matrix $R^{1\times t} \xrightarrow{\tau} R^{1\times b}$ such that $\tau \cdot \rho = 0$, we can find a lift $R^{1\times t} \xrightarrow{u(\tau)} R^{1\times c}$ making the diagram

$$\begin{array}{ccccc} R^{1\times c} & & & & \\ \uparrow {\scriptstyle u(\tau)} & \searrow^{\kappa} & & & \\ & & R^{1\times b} & \xrightarrow{\rho} & R^{1\times a} \\ & \nearrow_{\tau} & & & \\ R^{1\times t} & & & & \end{array}$$

commutative. In matrix terms, this means that the rows of κ have to span the row kernel (also called **syzygies**) of ρ, since we can express every collection of rows τ lying in the row kernel of ρ as a linear combination (given by $u(\tau)$) of the rows in κ.

But since these linear combinations do not have to be uniquely determined, we deal with *weak* kernels here. Thus, the existence of weak kernels in Rows_R is equivalent to finding a finite generating system for row kernels of matrices over R. A ring for which row kernels are finitely generated is called **(left-)coherent**.

REMARK 1.45. Algorithms to compute syzygies in Rows_R mainly rely on the theory of Gröbner bases. For the cases of quotients of commutative polynomial rings (both graded and non-graded), see, e.g., [**GP02**]. For non-commutative cases (including finite dimensional quotients of path algebras), see, e.g., [**Gre99**].

Our goal is to describe kernels in $\mathcal{A}(\mathbf{A})$ with the help of weak kernels in $\mathbf{A}$. In order to be able to do so, we need the construction of weak pullbacks from weak kernels.

DEFINITION 1.46. Let $\mathbf{A}$ be an additive category. Given a cospan $A \xrightarrow{\alpha} B \xleftarrow{\gamma} C$ in $\mathbf{A}$, a **weak pullback** consists of the following data:

(1) An object $\mathrm{WeakPullback}(\alpha,\gamma) \in \mathbf{A}$.
(2) Morphisms

$${}_{\alpha}\begin{bmatrix}1\\0\end{bmatrix}_{\gamma} : \mathrm{WeakPullback}(\alpha,\gamma) \to A$$

and

$${}_{\alpha}\begin{bmatrix}0\\1\end{bmatrix}_{\gamma} : \mathrm{WeakPullback}(\alpha,\gamma) \to C$$

such that

$${}_{\alpha}\begin{bmatrix}1\\0\end{bmatrix}_{\gamma} \cdot \alpha = {}_{\alpha}\begin{bmatrix}0\\1\end{bmatrix}_{\gamma} \cdot \gamma.$$

(3) An algorithm that computes for $T \in \mathbf{A}$ and morphisms $p : T \to A$, $q : T \to C$ with $p \cdot \alpha = q \cdot \gamma$ a morphism

$${}_{\alpha}\begin{bmatrix}p & q\end{bmatrix}_{\gamma} : T \to \mathrm{WeakPullback}(\alpha,\gamma)$$

satisfying

$$p = {}_{\alpha}\begin{bmatrix}p & q\end{bmatrix}_{\gamma} \cdot {}_{\alpha}\begin{bmatrix}1\\0\end{bmatrix}_{\gamma} \quad \text{and} \quad q = {}_{\alpha}\begin{bmatrix}p & q\end{bmatrix}_{\gamma} \cdot {}_{\alpha}\begin{bmatrix}0\\1\end{bmatrix}_{\gamma}.$$

REMARK 1.47. The only difference between pullbacks and weak pullbacks lies in the uniqueness of the induced morphism, which is missing in the case of weak pullbacks.

CONSTRUCTION 1.48. We show how to construct weak pullbacks from weak kernels in an additive category $\mathbf{A}$. Let

$$\begin{array}{ccc} & & C \\ & & \downarrow{\scriptstyle\gamma} \\ A & \xrightarrow[\alpha]{} & B \end{array}$$

be a cospan. We define the **diagonal difference**

$$\delta := \begin{pmatrix} \alpha \\ -\gamma \end{pmatrix} : A \oplus C \to B.$$

Then, we may set

(1) the weak pullback object

$$\mathrm{WeakPullback}(\alpha, \gamma) := \mathrm{WeakKernel}(\delta),$$

(2) the first weak pullback projection

$$_{\alpha}\begin{bmatrix} 1 \\ 0 \end{bmatrix}_{\gamma} := \mathrm{WeakPullback}(\alpha, \gamma) \xrightarrow{\mathrm{WeakKernelEmbedding}(\delta)} A \oplus C \xrightarrow{\begin{pmatrix} 1 \\ 0 \end{pmatrix}} A,$$

(3) the second weak pullback projection

$$_{\alpha}\begin{bmatrix} 0 \\ 1 \end{bmatrix}_{\gamma} := \mathrm{WeakPullback}(\alpha, \gamma) \xrightarrow{\mathrm{WeakKernelEmbedding}(\delta)} A \oplus C \xrightarrow{\begin{pmatrix} 0 \\ 1 \end{pmatrix}} C.$$

Moreover, for any pair $p : T \to A$, $q : T \to C$ such that $p \cdot \alpha = q \cdot \gamma$, we set

(4) the morphism into the weak pullback

$$_{\alpha}\begin{bmatrix} p & q \end{bmatrix}_{\gamma} := \mathrm{WeakKernelLift}\left(\delta, \begin{pmatrix} p & q \end{pmatrix}\right) \begin{array}{ccc} \mathrm{WeakPullback}(\alpha, \gamma) & & \\ \uparrow & \searrow & \\ \vdots & & A \oplus C \xrightarrow{\ \delta\ } B. \\ T & \nearrow_{\begin{pmatrix} p & q \end{pmatrix}} & \end{array}$$

Correctness of the construction. The equation $p \cdot \alpha = q \cdot \gamma$ is equivalent to $\begin{pmatrix} p & q \end{pmatrix} \cdot \delta = 0$. $\square$

Example 1.49. Let R be a ring. Computing the weak pullback of two morphisms in Rows_R, i.e., of two matrices α, γ over R having the same number of columns, amounts to computing the syzygies of the stacked matrix

$$\begin{pmatrix} \alpha \\ -\gamma \end{pmatrix}.$$

Construction 1.50 (Kernels in Freyd categories). Let $\mathbf{A}$ be an additive category in which we can compute weak kernels. By Construction 1.48, this means that we are able to construct weak pullbacks. We will use these for the construction of kernels in the Freyd category. Given a morphism

$$\begin{array}{ccc} A & \xleftarrow{\rho_A} & R_A \\ {\scriptstyle \alpha}\downarrow & & \vdots\, {\scriptstyle \rho_\alpha} \\ B & \xleftarrow{\rho_B} & R_B \end{array}$$

in $\mathcal{A}(\mathbf{A})$. Generalizing the idea given in Remark 1.42, we can construct its kernel object and kernel embedding as

$$\begin{array}{ccc} \mathrm{WeakPullback}(\rho_B, \alpha) & \xleftarrow{{}_{\kappa}\begin{bmatrix}1\\0\end{bmatrix}_{\alpha}} & \mathrm{WeakPullback}(\kappa, \rho_A) \\ \kappa := {}_{\rho_B}\begin{bmatrix}0\\1\end{bmatrix}_{\alpha} \downarrow & & \vdots\, {}_{\kappa}\begin{bmatrix}0\\1\end{bmatrix}_{\alpha} \\ A & \xleftarrow{\rho_A} & R_A \end{array}$$

If we have a test morphism

$$\begin{array}{ccc} T & \xleftarrow{\rho_T} & R_T \\ {\scriptstyle \tau}\downarrow & & \vdots\, {\scriptstyle \rho_\tau} \\ A & \xleftarrow{\rho_A} & R_A \end{array}$$

whose composition with our first morphism yields zero in $\mathcal{A}(\mathbf{A})$, i.e., there exists a lift

$$\begin{array}{ccc} T & & \\ {\scriptstyle \tau \cdot \alpha}\downarrow & \searrow^{\sigma} & \\ B & \xleftarrow{\rho_B} & R_B, \end{array}$$

then we can construct the kernel lift

$$\begin{array}{ccc} T & \xleftarrow{\rho_A} & R_T \\ {}_{\rho_B}\begin{bmatrix}\sigma & \tau\end{bmatrix}_{\alpha} \downarrow & & \vdots\, {}_{\kappa}\begin{bmatrix} {}_{\rho_B}\begin{bmatrix}\sigma & \tau\end{bmatrix}_{\alpha} & \rho_\tau \end{bmatrix}_{\rho_A} \\ \mathrm{WeakPullback}(\rho_B, \alpha) & \xleftarrow{{}_{\kappa}\begin{bmatrix}1\\0\end{bmatrix}_{\alpha}} & \mathrm{WeakPullback}(\kappa, \rho_A) \end{array}$$

CORRECTNESS OF THE CONSTRUCTION. See [**Pos21**, Section 3.2]. □

1.6.4. *The abelian case.* Knowing how to construct kernels and cokernels in Freyd categories allows us to construct pullbacks and pushouts: for pullbacks, we can proceed analogously to Construction 1.48. For pushouts, we can proceed dually.

The construction of kernels in $\mathcal{A}(\mathbf{A})$ relies on having weak kernels in $\mathbf{A}$. However, even more can be computed once $\mathbf{A}$ has weak kernels:

THEOREM 1.51 ([**Fre66**]). *$\mathcal{A}(\mathbf{A})$ is abelian if and only if $\mathbf{A}$ has weak kernels.*

Here is the definition of an abelian category as it can be found in textbooks like [**Wei94**]: an abelian category is an additive category $\mathbf{A}$ with kernels and cokernels such that

(1) every mono is the kernel of its cokernel,
(2) every epi is the cokernel of its kernel.

Let us unravel these new requirements from an algorithmic point of view. The first statement tells us that whenever we are given a monomorphism $\alpha \in \mathrm{Hom}_{\mathbf{A}}(A, B)$, it should have the same categorical properties as the kernel embedding of the morphism CokernelProjection(α). Since we are able to compute kernel lifts for a given kernel embedding, we have to be able to compute such lifts for α as well. Thus, an algorithmic rereading of the first statement is given as follows: an abelian category comes equipped with an algorithm that computes for a given monomorphism $\alpha \in \mathrm{Hom}_{\mathbf{A}}(A, B)$ and given morphism $\tau \in \mathrm{Hom}_{\mathbf{A}}(T, B)$ such that $\tau \cdot \mathrm{CokernelProjection}(\alpha) = 0$ the **lift along a monomorphism** $u \in \mathrm{Hom}_{\mathbf{A}}(T, A)$ (i.e., $u \cdot \alpha = \tau$).

Dually, the second statement can be rephrased as: an abelian category comes equipped with an algorithm that computes for a given epimorphism $\alpha \in \mathrm{Hom}_{\mathbf{A}}(A, B)$ and given morphism $\tau \in \mathrm{Hom}_{\mathbf{A}}(A, T)$ such that KernelEmbedding $(\alpha) \cdot \tau = 0$ the **colift along an epimorphism** $u \in \mathrm{Hom}_{\mathbf{A}}(B, T)$ (i.e., $\alpha \cdot u = \tau$).

We will show how to compute lifts along monomorphisms in $\mathcal{A}(\mathbf{A})$.

REMARK 1.52. Suppose given a monomorphism

$$\begin{array}{ccc} A & \xleftarrow{\ \rho_A\ } & R_A \\ {\scriptstyle\alpha}\downarrow & & \downarrow{\scriptstyle\rho_\alpha} \\ B & \xleftarrow{\ \rho_B\ } & R_B \end{array}$$

in $\mathcal{A}(\mathbf{A})$. Then its kernel embedding (see Construction 1.50)

$$\begin{array}{ccc} \mathrm{WeakPullback}(\rho_B, \alpha) & \xleftarrow{\ {}_{\kappa}\begin{bmatrix}1\\0\end{bmatrix}_{\alpha}\ } & \mathrm{WeakPullback}(\kappa, \rho_A) \\ {\scriptstyle \kappa := {}_{\rho_B}\begin{bmatrix}0\\1\end{bmatrix}_{\alpha}}\downarrow & & \downarrow{\scriptstyle {}_{\kappa}\begin{bmatrix}0\\1\end{bmatrix}_{\alpha}} \\ A & \xleftarrow{\ \rho_A\ } & R_A \end{array}$$

is zero in $\mathcal{A}(\mathbf{A})$. We call a witness for this kernel embedding being zero, which is nothing but a lift

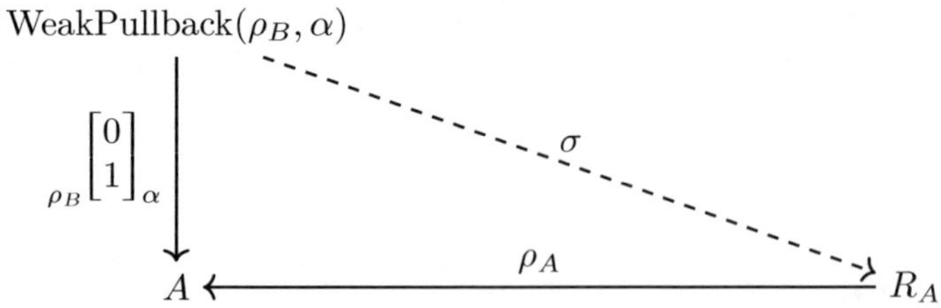

a **witness for being a monomorphism** of our original morphism.

CONSTRUCTION 1.53 (Lift along monomorphism in Freyd categories). Let

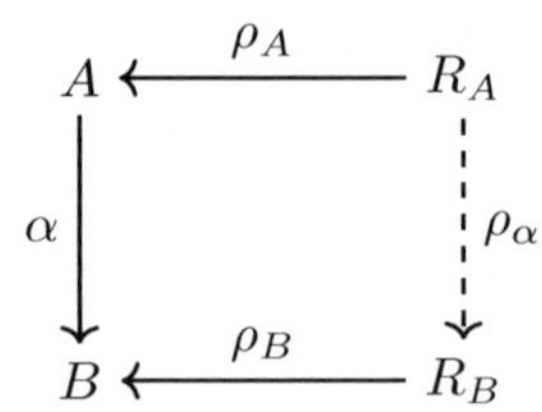

be a monomorphism in $\mathcal{A}(\mathbf{A})$ together with a witness for being a monomorphism (see Remark 1.52)

$$\sigma : \mathrm{WeakPullback}(\rho_B, \alpha) \longrightarrow R_A.$$

Moreover, let

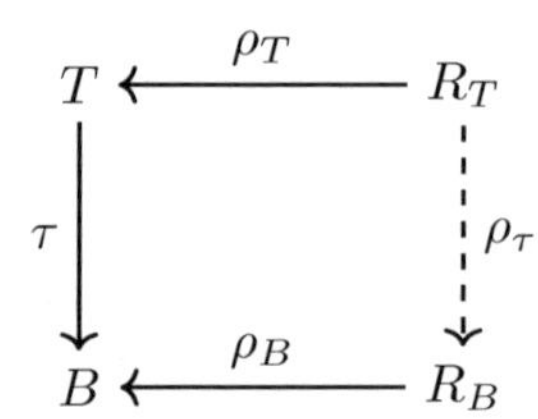

be a test morphism, i.e., a morphism in $\mathcal{A}(\mathbf{A})$ whose composition with the cokernel projection

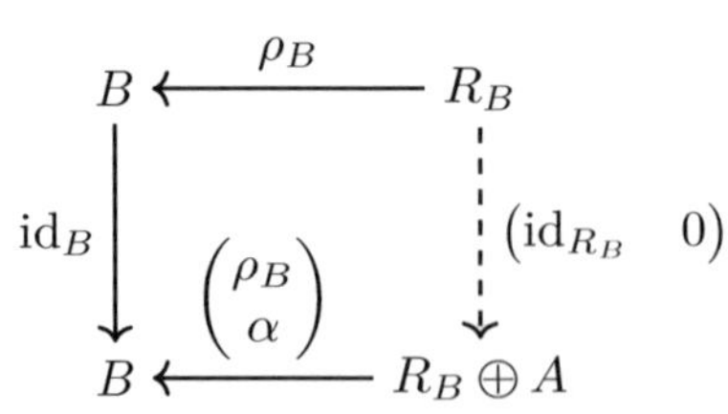

of our monomorphism yields zero, which, in turn, is witnessed by a lift

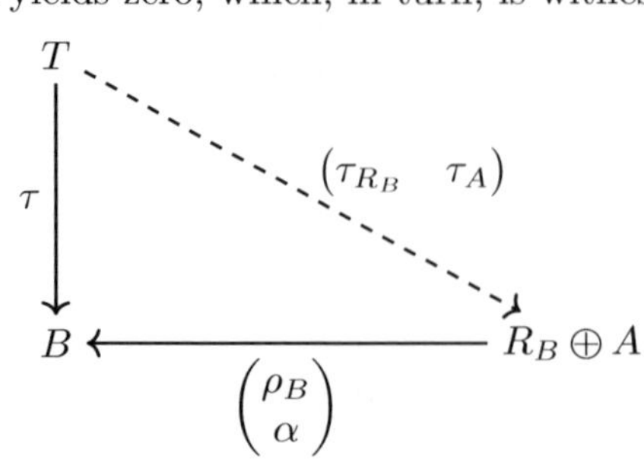

Then, we can construct the lift along monomorphism as

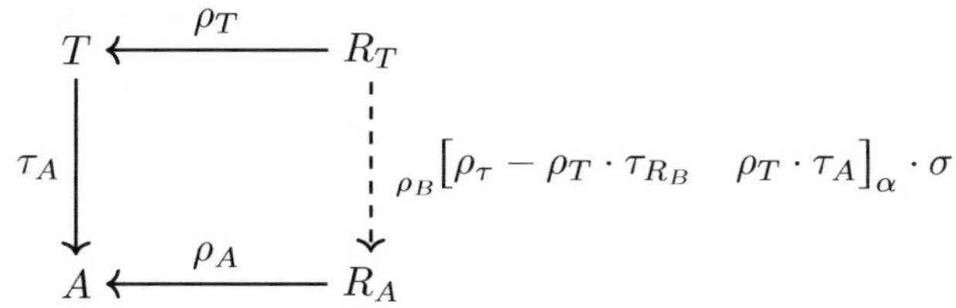

Correctness of the construction. See [**Pos21**, Section 3.3]. □

How to proceed for colifts along epimorphisms can be seen in [**Pos21**, Section 3.4].

1.6.5. *Homomorphisms.* We end this first section with a discussion of how to compute sets of homomorphisms in Freyd categories, since this enables us, among other things, to compute sets of natural transformations between finitely presented functors.

Let $\mathbf{A}$ be an additive category and let $(A \xleftarrow{\rho_A} R_A)$ and $(B \xleftarrow{\rho_B} R_B)$ be objects in $\mathcal{A}(\mathbf{A})$. Recall that a morphism between these two objects

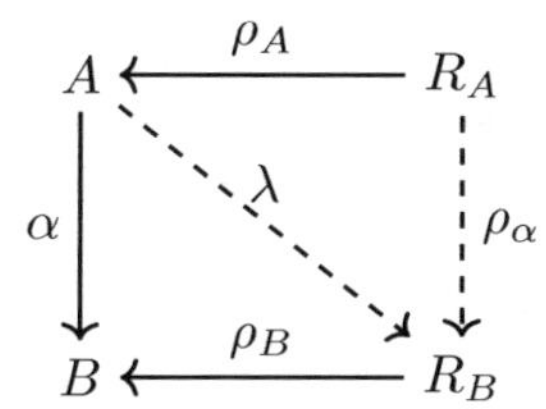

consists of an element $\alpha \in \mathrm{Hom}_{\mathbf{A}}(A, B)$ considered up to addition with an element of the form $\lambda \cdot \rho_B$ such that there exists ρ_α with $\rho_A \cdot \alpha = \rho_\alpha \cdot \rho_B$. In other words, the abelian group

$$\mathcal{H} := \mathrm{Hom}_{\mathcal{A}(\mathbf{A})}\left((A \xleftarrow{\rho_A} R_A), (B \xleftarrow{\rho_B} R_B)\right)$$

is given by a certain subquotient of the abelian group $\mathrm{Hom}_{\mathbf{A}}(A, B)$ that fits into the following commutative diagram of abelian groups with exact rows and columns:

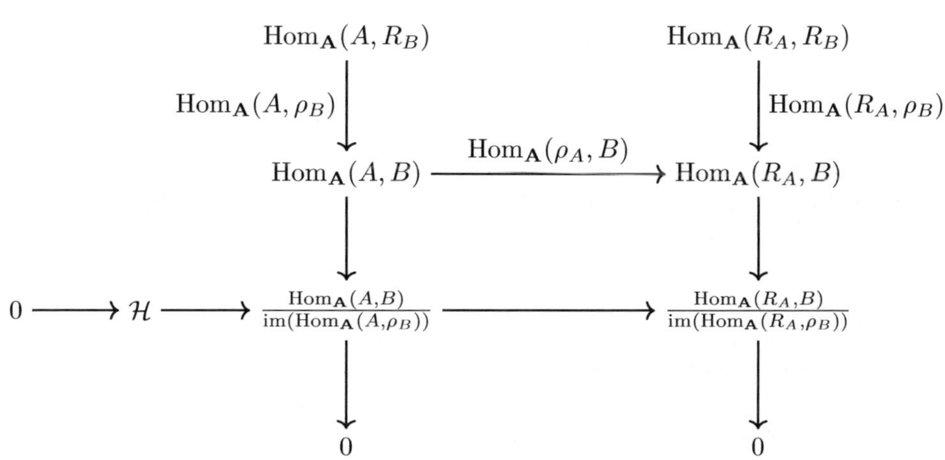

Figure 1. $\mathcal{H}$ as a subquotient of abelian groups.

Now, assume that $\mathbf{A}$ has a $\mathbf{B}$-homomorphism structure $(H, 1, \nu)$, where $\mathbf{B}$ is an abelian category. Then, inspired by the diagram of abelian groups above, we may construct a diagram with exact rows and columns in $\mathbf{B}$:

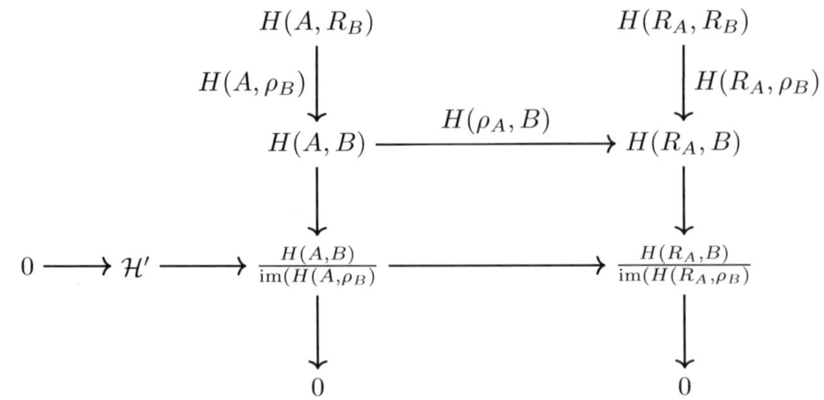

FIGURE 2. Constructing a homomorphism structure for Freyd categories.

If $1 \in \mathbf{B}$ is a projective object, then $\mathrm{Hom}_{\mathbf{B}}(1, -)$ is exact. Applying $\mathrm{Hom}_{\mathbf{B}}(1, -)$ to the diagram in Figure 2 recovers the diagram of abelian groups depicted in Figure 1. But this means

$$\mathrm{Hom}_{\mathbf{B}}(1, \mathcal{H}') \simeq \mathcal{H} \simeq \mathrm{Hom}_{\mathcal{A}(\mathbf{A})}\left((A \xleftarrow{\rho_A} R_A), (B \xleftarrow{\rho_B} R_B)\right).$$

In other words, we used the $\mathbf{B}$-homomorphism structure on $\mathbf{A}$ to define a $\mathbf{B}$-homomorphism structure on $\mathcal{A}(\mathbf{B})$ (for more details, see [**Pos21**, Section 6.2]).

1.7. Computing natural transformations. As an application of the abstract algorithms that allow us to compute within Freyd categories, we show how to compute sets of natural transformations between finitely presented functors. Within this subsection, R denotes a commutative coherent ring.

CONSTRUCTION 1.54. Recall from Subsection 1.5 that the cascade of category constructors

$$\mathcal{A}(\mathcal{A}(\mathcal{C}(R)^{\oplus})^{\mathrm{op}})$$

defines a category equivalent to finitely presented functors on the category of finitely presented modules over R. We use the findings of the previous subsections to define an $\mathcal{A}(\mathcal{C}(R)^{\oplus})$-homomorphism structure for this category.

(1) By Example 1.24, $\mathcal{C}(R)$ has a $\mathcal{C}(R)$-homomorphism structure.
(2) By Construction 1.27 and Remark 1.28, we can extend this to a $\mathcal{C}(R)^{\oplus}$-homomorphism structure for $\mathcal{C}(R)^{\oplus}$.
(3) By applying the natural embedding $\mathcal{C}(R)^{\oplus} \longrightarrow \mathcal{A}(\mathcal{C}(R)^{\oplus})$, the category $\mathcal{C}(R)^{\oplus}$ has an $\mathcal{A}(\mathcal{C}(R)^{\oplus})$-homomorphism structure.
(4) Since R is coherent, $\mathcal{A}(\mathcal{C}(R)^{\oplus})$ is abelian and the distinguished object of the homomorphism structure, corresponding to R, is projective. Thus, by the findings of Subsubsection 1.6.5, we obtain an $\mathcal{A}(\mathcal{C}(R)^{\oplus})$-homomorphism structure for $\mathcal{A}(\mathcal{C}(R)^{\oplus})$.

(5) If an additive category $\mathbf{A}$ has a $\mathbf{B}$-homomorphism structure, then $\mathbf{A}^{\mathrm{op}}$ has a $\mathbf{B}$-homomorphism structure as well. In particular, $\mathcal{A}(\mathcal{C}(R)^{\oplus})^{\mathrm{op}}$ has a $\mathcal{A}(\mathcal{C}(R)^{\oplus})$-homomorphism structure.
(6) Last, we apply the findings of Subsubsection 1.6.5 again and arrive at the desired $\mathcal{A}(\mathcal{C}(R)^{\oplus})$-homomorphism structure for $\mathcal{A}(\mathcal{A}(\mathcal{C}(R)^{\oplus})^{\mathrm{op}})$.

We demonstrate how the algorithm for the computation of homomorphisms that results from Construction 1.54 is carried out concretely. For simplifying the notation we use the equivalence $\mathcal{A}(\mathcal{C}(R)^{\oplus}) \simeq R\text{-fpmod}$, but keep in mind that computing kernels, cokernels, and homomorphisms for R-fpmod can all be carried out by means of the results in Subsection 1.5 on Freyd categories. We start with a simple example.

EXAMPLE 1.55. Given the functors $\mathrm{Hom}_{\mathbb{Z}}(\mathbb{Z}/2\mathbb{Z},-)$ and $\mathrm{Ext}^1_{\mathbb{Z}}(\mathbb{Z}/2\mathbb{Z},-)$, we want to confirm computationally

$$\mathrm{Hom}\left(\mathrm{Hom}_{\mathbb{Z}}(\mathbb{Z}/2\mathbb{Z},-),\mathrm{Ext}^1_{\mathbb{Z}}(\mathbb{Z}/2\mathbb{Z},-)\right) \simeq \mathrm{Ext}^1_{\mathbb{Z}}(\mathbb{Z}/2\mathbb{Z},\mathbb{Z}/2\mathbb{Z}) \simeq \mathbb{Z}/2\mathbb{Z}.$$

The functor $\mathrm{Hom}_{\mathbb{Z}}(\mathbb{Z}/2\mathbb{Z},-)$ considered as an object in $\mathcal{A}(\mathbb{Z}\text{-fpmod}^{\mathrm{op}})$ is given by

$$\mathbb{Z}/2\mathbb{Z} \longrightarrow 0.$$

The functor $\mathrm{Ext}^1_{\mathbb{Z}}(\mathbb{Z}/2\mathbb{Z},-)$ considered as an object in $\mathcal{A}(\mathbb{Z}\text{-fpmod}^{\mathrm{op}})$ is given by

$$\mathbb{Z} \xrightarrow{\;2\;} \mathbb{Z},$$

see Example 1.32. Now, plugging these data into the diagram in Figure 2 and computing the cokernels, the induced morphism, and the kernel, we end up with the diagram

$$\begin{array}{ccccccc}
 & & & & \mathbb{Z}/2\mathbb{Z} \simeq H(\mathbb{Z},\mathbb{Z}/2\mathbb{Z}) & & 0 \simeq H(\mathbb{Z},0) \\
 & & & & {\scriptstyle 2}\downarrow & & \downarrow \\
 & & & & \mathbb{Z}/2\mathbb{Z} \simeq H(\mathbb{Z},\mathbb{Z}/2\mathbb{Z}) & \longrightarrow & 0 \simeq H(\mathbb{Z},0) \\
 & & & & \downarrow & & \downarrow \\
0 & \longrightarrow & \boxed{\mathbb{Z}/2\mathbb{Z}} & \longrightarrow & \mathbb{Z}/2\mathbb{Z} & \longrightarrow & 0 \\
 & & & & \downarrow & & \downarrow \\
 & & & & 0 & & 0
\end{array}$$

where we find our desired result inside the box.

Let M be a finitely presented R-module. In order to provide more complicated examples, we show how to represent the functor $(M \otimes -)$ in $\mathcal{A}(R\text{-fpmod}^{\mathrm{op}})$, see also [**Aus66**, Lemma 6.1]. Let

$$0 \longleftarrow M \longleftarrow R^{1\times a} \xleftarrow{\;\rho_M\;} R^{1\times b}$$

be a presentation of M. The right exactness of the tensor product yields an exact sequence of functors

$$0 \longleftarrow (M \otimes -) \longleftarrow (R^{1\times a} \otimes -) \xleftarrow{\;\rho_M \otimes -\;} (R^{1\times b} \otimes -)$$

where $\otimes$ is taken over R. For any free module $R^{1\times c}$ where $c \in \mathbb{N}_0$, there are isomorphisms

$$R^{1\times c} \otimes N \simeq N^{1\times c} \simeq \mathrm{Hom}_R(R^{1\times c}, N)$$

natural in $N \in R\text{-fpmod}$. Applied to the exact sequence above yields the presentation

$$0 \longleftarrow (M \otimes -) \longleftarrow (R^{1\times a}, -) \xleftarrow{(\rho_M^{\mathrm{tr}}, -)} (R^{1\times b}, -).$$

Thus, $(M \otimes -)$ is given as an object in $\mathcal{A}(R\text{-fpmod}^{\mathrm{op}})$ by

$$R^{1\times a} \xrightarrow{\rho_M^{\mathrm{tr}}} R^{1\times b}.$$

EXAMPLE 1.56. Let $R := \mathbb{Q}[x,y]$ and let

$$M := R^{1\times 2}/\langle (x \quad y) \rangle.$$

We wish to compute

$$\mathrm{Hom}\left((M \otimes_R -), \mathrm{Ext}^1(M, -)\right).$$

As seen above, the functor $(M \otimes_R -)$ considered as an object in $\mathcal{A}(R\text{-fpmod}^{\mathrm{op}})$ is given by

$$R^{1\times 2} \xrightarrow{\begin{pmatrix} x \\ y \end{pmatrix}} R^{1\times 1}$$

and the functor $\mathrm{Ext}^1_R(M, -)$ considered as an object in $\mathcal{A}(R\text{-fpmod}^{\mathrm{op}})$ is given by

$$R^{1\times 1} \xrightarrow{(x \quad y)} R^{1\times 2}.$$

Again, we use the diagram in Figure 2 for our computation

$$\begin{array}{ccccc}
 & & R^{2\times 2} & & R^{2\times 1} \\
 & & \downarrow {\scriptstyle (A \mapsto (x \quad y)\, A)} & {\scriptstyle (w \mapsto w \begin{pmatrix} x \\ y \end{pmatrix})} & \downarrow {\scriptstyle (v \mapsto (x \quad y)\, v)} \\
 & & R^{1\times 2} & \longrightarrow & R \\
 & & \downarrow & & \downarrow \\
0 \longrightarrow & \boxed{(R/\langle x,y\rangle)^{1\times 2}} & \rightarrow (R/\langle x,y\rangle)^{1\times 2} & \xrightarrow{0} & R/\langle x,y\rangle \\
 & & \downarrow & & \downarrow \\
 & & 0 & & 0
\end{array}$$

from which we conclude

$$\mathrm{Hom}\left((M \otimes_R -), \mathrm{Ext}^1(M, -)\right) \simeq (R/\langle x,y\rangle)^{1\times 2}.$$

Last, the functors $\mathrm{Tor}_i(M, -)$ for $i > 0$ are also finitely presented and can thus be represented as objects in $\mathcal{A}(R\text{-fpmod}^{\mathrm{op}})$, see also [**Pre09**, Theorem 10.2.35]. For $\mathrm{Tor}_1(M, -)$, let

$$0 \longleftarrow M \xleftarrow{\epsilon} R^{1\times a} \xleftarrow{\iota} R^{1\times b} \xleftarrow{\rho} R^{1\times c}$$

be an exact sequence, and set

$$\Omega^1(M) := \ker(\epsilon) \simeq \operatorname{im}(\iota) \simeq \operatorname{coker}(\rho).$$

We have an isomorphism

$$\operatorname{Tor}_1(M,N) \simeq \ker\left(\Omega^1(M) \otimes N \to R^{1\times a} \otimes N\right)$$

natural in $N \in R$-fpmod, which means that $\operatorname{Tor}_1(M,-)$ can be computed as the kernel of

$$(\Omega^1(M) \otimes -) \to (R^{1\times a} \otimes -). \tag{2}$$

Thus, all we need to do is to translate this natural transformation to a morphism in $\mathcal{A}(R\text{-fpmod}^{\mathrm{op}})$ and take its kernel. Lifting the embedding $\Omega^1(M) \longrightarrow R^{1\times a}$ to presentations is simply given by the following commutative diagram with exact rows:

$$\begin{array}{ccccccc}
0 & \longleftarrow & \Omega^1(M) & \longleftarrow & R^{1\times b} & \overset{\rho}{\longleftarrow} & R^{1\times c} \\
 & & \downarrow & & \downarrow{\scriptstyle \iota} & & \downarrow \\
0 & \longleftarrow & R^{1\times a} & \longleftarrow & R^{1\times a} & \longleftarrow & 0
\end{array}$$

The transposition of its right square is our desired representation of (2) in $\mathcal{A}(R\text{-fpmod}^{\mathrm{op}})$:

$$\begin{array}{c}
(\Omega^1(M) \otimes -) \\
\downarrow \\
(R^{1\times a} \otimes -)
\end{array}
\quad \text{corresponds to} \quad
\begin{array}{ccc}
R^{1\times b} & \xrightarrow{\rho^{\mathrm{tr}}} & R^{1\times c} \\
{\scriptstyle \iota^{\mathrm{tr}}}\uparrow & & \uparrow \\
R^{1\times a} & \xrightarrow{0} & 0
\end{array}$$

For the construction of its kernel, we apply Construction 1.50 with $\mathbf{A} = R\text{-fpmod}^{\mathrm{op}}$. Since pullbacks in abelian categories are in particular weak pullbacks, and since pullbacks and pushouts are dual concepts, we end up with

$$\begin{array}{c}
\operatorname{Tor}_1(M,-) \\
\downarrow \\
(\Omega^1(M) \otimes -) \\
\downarrow \\
(R^{1\times a} \otimes -)
\end{array}
\quad \text{corresponds to} \quad
\begin{array}{ccc}
\operatorname{coker}(\iota^{\mathrm{tr}}) & \longrightarrow & \operatorname{coker}(\iota^{\mathrm{tr}}) \amalg_{R^{1\times b}} R^{1\times c} \\
\uparrow & & \uparrow \\
R^{1\times b} & \xrightarrow{\rho^{\mathrm{tr}}} & R^{1\times c} \\
{\scriptstyle \iota^{\mathrm{tr}}}\uparrow & & \uparrow \\
R^{1\times a} & \xrightarrow{0} & 0
\end{array}$$

where $\operatorname{coker}(\iota^{\mathrm{tr}}) \amalg_{R^{1\times b}} R^{1\times c}$ denotes the pushout of the cokernel projection $R^{1\times b} \to \operatorname{coker}(\iota^{\mathrm{tr}})$ and ρ^{tr}. For higher Tors, we simply need to replace $\Omega^1(M)$ with a higher syzygy object.

EXAMPLE 1.57. We set $R := \mathbb{Q}[x,y]$ and again take a look at the module

$$M := R^{1\times 2}/\langle (x \quad y) \rangle.$$

This time, we wish to compute

$$\mathrm{Hom}\left(\mathrm{Tor}_1(M,-),\mathrm{Ext}^1(M,-)\right).$$

Again, the functor $\mathrm{Ext}^1_R(M,-)$ considered as an object in $\mathcal{A}(R\text{-fpmod}^{\mathrm{op}})$ is given by

$$R^{1\times 1} \xrightarrow{\begin{pmatrix} x & y \end{pmatrix}} R^{1\times 2}.$$

Using the description preceding this example, we see that $\mathrm{Tor}_1(M,-)$ considered as an object in $\mathcal{A}(R\text{-fpmod}^{\mathrm{op}})$ is given by

$$R^{1\times 1}/\langle x,y\rangle \longrightarrow 0.$$

Again, we use the diagram in Figure 2 for our computation

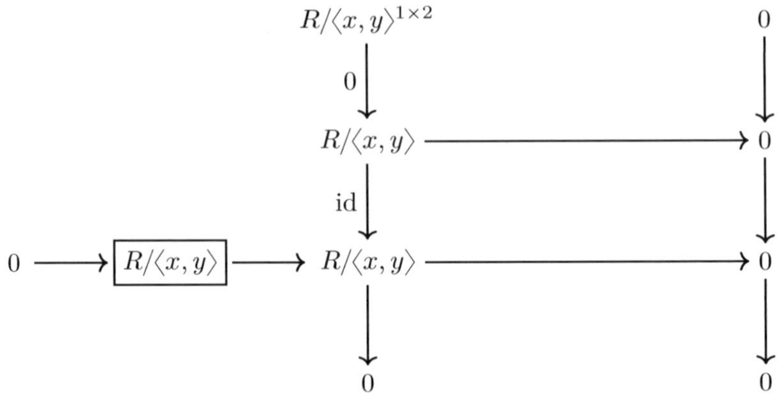

from which we conclude

$$\mathrm{Hom}\left(\mathrm{Tor}_1(M,-),\mathrm{Ext}^1(M,-)\right) \simeq R/\langle x,y\rangle.$$

2. Constructive diagram chases

Diagram chases are a powerful tool used in homological algebra for proving the existence of morphisms situated in some diagram of prescribed shape. In this section, we will demonstrate how to perform diagram chases constructively. The main idea is to employ a calculus that replaces the morphisms in an abelian category $\mathbf{A}$ with a more flexible notion, yielding a new category $\mathbf{G}(\mathbf{A})$, analogous to the replacement of functions in the category of sets with relations. This idea has first been pursued in an axiomatic way by Brinkmann and Puppe in [**BP69**] and [**Pup62**], and rendered into an explicit calculus by Hilton in [**Hil66**]. A calculus of relations in so-called regular categories, which are more general than abelian categories, was given by Johnstone [**Joh02**].

The first algorithmic usage of this calculus in the context of spectral sequence computations is due to Barakat in [**Bar09**]. Here, the term *generalized morphism* is coined for morphisms in $\mathbf{G}(\mathbf{A})$ and we will follow this convention. Other appropriate terms would be: relations, correspondences, or pseudo morphisms[4].

The presented material follows closely the presentation of generalized morphisms given in [**Pos17**], especially Subsections 2.2 and 2.3.

[4]Suggested by Jean Michel.

2.1. Additive relations. We start with the following diagram with exact rows in the category of abelian groups **Ab**:

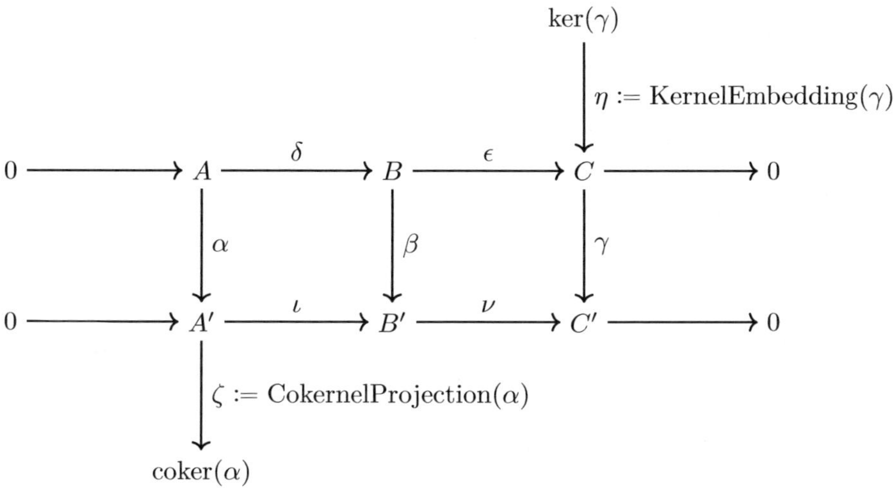

The famous snake lemma claims the existence of a morphism

$$\partial : \ker(\gamma) \longrightarrow \operatorname{coker}(\alpha)$$

fitting into an exact sequence

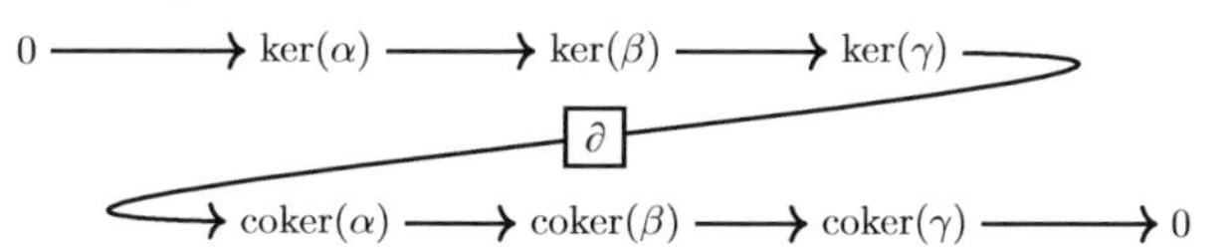

We will focus on the existence part of this lemma. A description of ∂ can be given on the level of elements:

(1) Start with an element $c \in \ker(\gamma)$.
(2) Regard it as an element $c \in C$.
(3) **Choose** an element $b \in \epsilon^{-1}(\{c\})$.
(4) Map b via β and obtain $b' := \beta(b) \in B'$.
(5) Find the uniquely determined element $a' \in \iota^{-1}(\{b'\})$.
(6) Consider the residue class of a' in $\operatorname{coker}(\alpha)$.

It is quite easy to prove that each of these steps can actually be carried out and that the resulting map

$$\ker(\gamma) \to \operatorname{coker}(\alpha) : c \mapsto a' + \operatorname{im}(\alpha)$$

is a group homomorphism independent of the choice made in step (3).

A common approach to prove the existence of ∂ not only in the category of abelian groups but in every abelian category is to use embedding theorems [**Fre64**]. Such theorems reduce constructions in a small abelian category to the case of categories of modules where one can happily perform element-wise constructions like the one we did above.

We are going to follow a more computer-friendly approach that will enable us to construct ∂ only using operations within our given abelian category and without passing to an ambient module category. To see how this goal can be achieved, let us take a look at the most crucial step within the construction of ∂ in the category of abelian groups above, namely step (3). It is highly uncanonical to choose just any preimage of c, and in fact, every choice is just as good as every other choice. A possible way to overcome this problem is by not making any choice at all, but to work with the whole preimage $\epsilon^{-1}(\{c\})$ instead. Following this idea, the steps in the construction of ∂ above can be reformulated as follows:

(1) Start with an element $c \in \ker(\gamma)$.
(2) Regard it as an element $c \in C$.
(3) Construct the whole preimage $b := \epsilon^{-1}(\{c\}) \subseteq B$.
(4) Construct the image $b' := \beta(b) \subseteq B'$.
(5) Construct the whole preimage $a' := \iota^{-1}(\{b'\}) \subseteq A'$.
(6) Construct the image of a' under the cokernel projection: $\{x + \mathrm{im}(\alpha) \mid x \in a'\}$. It will consist of a single element.

We got rid of the uncanonical step in this set of instructions and all we do is to take images and fibers of *sets of elements* instead of single elements. One possible way to formulate these new instructions in a more categorical way is given by replacing the notion of a group homomorphism by the notion of an additive relation.

DEFINITION 2.1. An **additive relation** from an abelian group A to an abelian group B is given by a subgroup $f \subseteq A \times B$.

EXAMPLE 2.2. Every abelian group homomorphism $\alpha : A \to B$ in **Ab** defines via its graph an additive relation

$$[\alpha] := \{(a, b) \mid \alpha(a) = b\} \subseteq A \times B.$$

EXAMPLE 2.3. If $f \subseteq A \times B$ is an additive relation, then so is its **pseudo-inverse**

$$f^{-1} := \{(b, a) \mid (a, b) \in f\} \subseteq B \times A.$$

Additive relations $f \subseteq A \times B$ and $g \subseteq B \times C$ can be composed via

$$f \cdot g := \{(a, c) \mid \exists b \in B : (a, b) \in f, (b, c) \in g\} \subseteq A \times C.$$

This composition turns abelian groups and additive relations into a category $\mathrm{Rel}(\mathbf{Ab})$ with graphs of the identity group homomorphisms as its identities. Mapping a group homomorphism to its graph lets us think of **Ab** as a non-full subcategory of $\mathrm{Rel}(\mathbf{Ab})$.

Our reformulated set of instructions for computing ∂ can now conveniently be written as a simple composition of relations:

$$[\partial] = [\eta] \cdot [\epsilon]^{-1} \cdot [\beta] \cdot [\iota]^{-1} \cdot [\zeta].$$

To sum it up, it can be said that performing constructions in **Ab** via diagram chases boils down to calculations in $\mathrm{Rel}(\mathbf{Ab})$. Thus, it is our goal to find a calculus for working with relations in an arbitrary abelian category **A**.

2.2. Category of generalized morphisms. From now on, we denote by **A** an arbitrary abelian category. Given two objects $A, B \in \mathbf{A}$, a **span** S (from A to B) is simply given by an object $C \in \mathbf{A}$ together with a pair of morphisms $(A \xleftarrow{\alpha} C, C \xrightarrow{\beta} B)$. We depict a span as

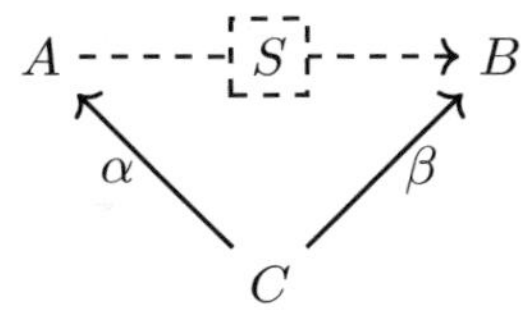

or as

$$A \xleftarrow{\alpha} C \xrightarrow{\beta} B.$$

Note that we included a direction within our definition of a span in the sense that swapping the order of the pair of morphisms defines a different span (from B to A).

DEFINITION 2.4. The **category of spans of A**, denoted by $\mathrm{Span}(\mathbf{A})$, is defined by the following data:

(1) Objects are given by $\mathrm{Obj}_{\mathbf{A}}$.
(2) Morphisms from A to B are spans from A to B.
(3) Two spans $(A \xleftarrow{\alpha} C \xrightarrow{\beta} B)$ and $(A \xleftarrow{\alpha'} C' \xrightarrow{\beta'} B)$ are considered to be **equal as spans** if there exists an isomorphism $\iota : C \longrightarrow C'$ compatible with the spans, i.e., such that $\alpha = \iota \cdot \alpha'$ and $\beta = \iota \cdot \beta'$.
(4) The identity of A is given by $(A \xleftarrow{\mathrm{id}} A \xrightarrow{\mathrm{id}} A)$, where id denotes the identity of A regarded as an object in $\mathbf{A}$.
(5) Composition of $(A \xleftarrow{\alpha} D \xrightarrow{\beta} B)$ and $(B \xleftarrow{\gamma} E \xrightarrow{\delta} C)$ is given by the outer span in the following diagram:

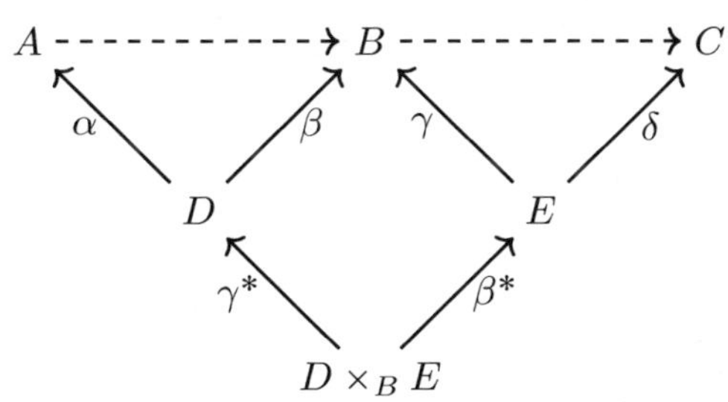

We have to check compatibility of composition and identities with our notion of equality for spans.

LEMMA 2.5.

(1) *The identity in* $\mathrm{Span}(\mathbf{A})$ *acts like a unit up to equality of spans.*
(2) *Composition of morphisms in* $\mathrm{Span}(\mathbf{A})$ *is associative up to equality of spans.*

PROOF. For the first assertion, let $(A \xleftarrow{\alpha} D \xrightarrow{\beta} B)$ be a span. Composition with the identity $(B \xleftarrow{\mathrm{id}} B \xrightarrow{\mathrm{id}} B)$ from the right yields the diagram

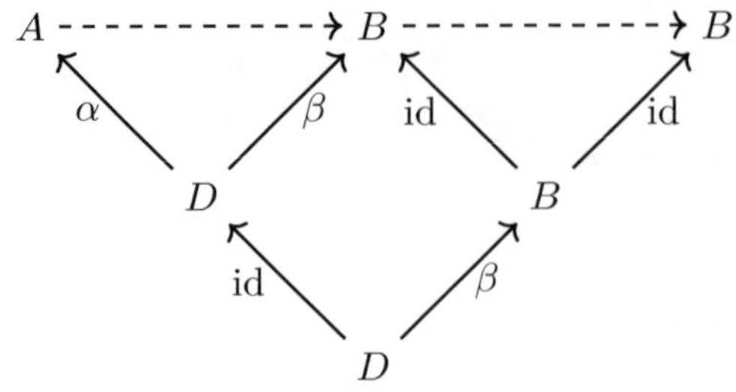

This proves that the identity is a right unit. An analogous argument shows that it is also a left unit. For the second assertion, consider the following diagram of consecutive pullbacks:

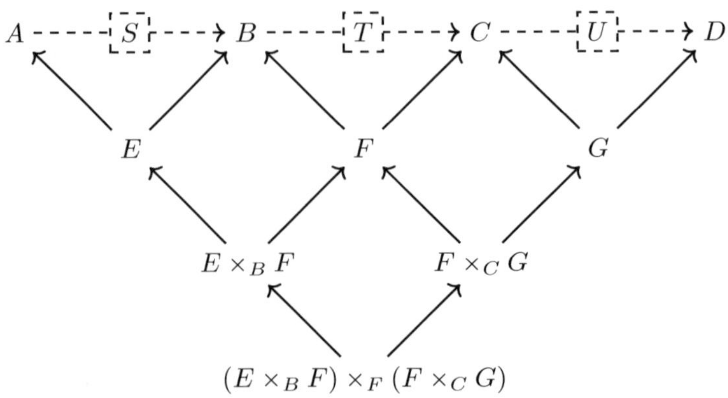

By transitivity of pullbacks, the rectangles with vertices $E, B, F \times_C G, (E \times_B F) \times_F (F \times_C G)$ and $C, G, E \times_B F, (E \times_B F) \times_F (F \times_C G)$ are also pullback squares. But this means that the outer span of the above diagram is isomorphic to both $S \cdot (T \cdot U)$ and $(S \cdot T) \cdot U$. $\square$

Definition 2.6. Given a span $(A \xleftarrow{\alpha} C \xrightarrow{\beta} B)$, we define its **associated relation** as the image of the morphism

$$(\alpha, \beta) : C \longrightarrow A \oplus B.$$

In particular, the associated relation of a span is a subobject of $A \oplus B$.

Definition 2.7. We say two spans from A to B are **stably equivalent** if their associated relations are equal as subobjects of $A \oplus B$.

Remark 2.8. Being stably equivalent is coarser than being equal as spans.

Lemma 2.9. *Let $\epsilon : D \twoheadrightarrow C$ be an epimorphism in $\mathbf{A}$. Every span of the form*

$$(A \xleftarrow{\alpha} C \xrightarrow{\beta} B)$$

is stably equivalent to the outer span in the diagram given by composition with ϵ:

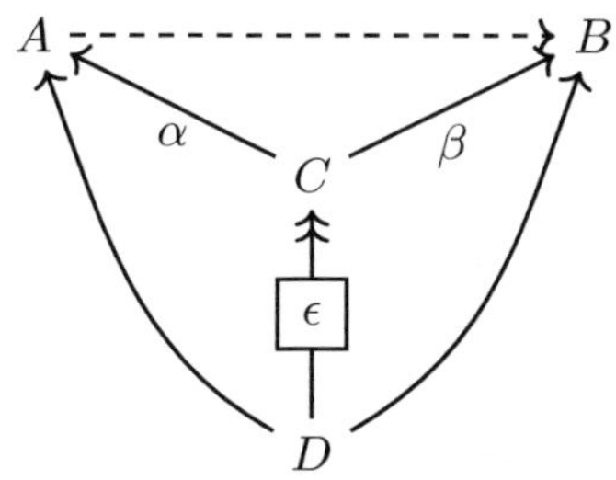

PROOF. We have $(\epsilon \cdot \alpha, \epsilon \cdot \beta) = \epsilon \cdot (\alpha, \beta)$, and in an abelian category, the image is not affected by epimorphisms. Thus, $\mathrm{im}\left((\epsilon \cdot \alpha, \epsilon \cdot \beta)\right) = \mathrm{im}\left((\alpha, \beta)\right)$. □

THEOREM 2.10. *Being stably equivalent defines a congruence on* Span(**A**).

PROOF. Let $S = (A \longleftarrow D \longrightarrow B)$ be a span and let $(\zeta, \eta) : I \hookrightarrow B \oplus C$ be a monomorphism. Let $T = (B \longleftarrow E \longrightarrow C)$ be a span obtained by composing ζ, η with an epimorphism $\epsilon : E \twoheadrightarrow I$. By transitivity of the pullback, we get $S \cdot T$ as the outer span in the following diagram:

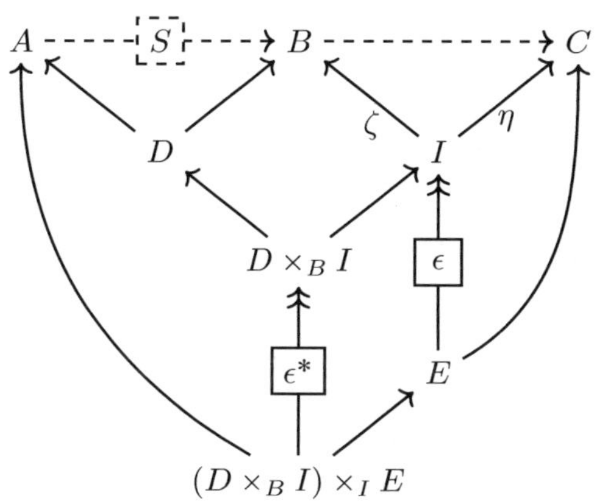

In an abelian category the pullback of an epimorphism yields an epimorphism. Thus, ϵ^* is an epimorphism. Now, we apply Lemma 2.9 to see that the stable equivalence class of $S \cdot T$ only depends on (ζ, η), which is the associated relation of T. Thus, if T and T' have the same associated relation, i.e., are stably equivalent, then so are $S \cdot T$ and $S \cdot T'$. By the symmetry of the situation, a similar statement holds for stably equivalent S, S' and compositions $S \cdot T$, $S' \cdot T$. This shows the claim. □

Due to Theorem 2.10, we can now define the generalized morphism category.

DEFINITION 2.11. Let **A** be an abelian category. The quotient category of Span(**A**) modulo stable equivalences is called the **generalized morphism category of A**, and denoted by **G**(**A**). Concretely, it consists of the following data:

(1) Objects are given by $\mathrm{Obj}_{\mathbf{A}}$.
(2) Morphisms from A to B are spans from A to B.
(3) Two spans are considered to be **equal as generalized morphisms** if and only if they are stably equivalent.
(4) Identity and composition are given as in Definition 2.4.

We call a span from A to B a **generalized morphism** when we regard it as a morphism in $\mathbf{G}(\mathbf{A})$.

2.3. Computation rules. We will see that computing within $\mathbf{G}(\mathbf{A})$ boils down to computing compositions of morphisms and pseudo-inverses of morphisms in $\mathbf{A}$. Every morphism $\alpha : A \to B$ in $\mathbf{A}$ gives rise to a morphism

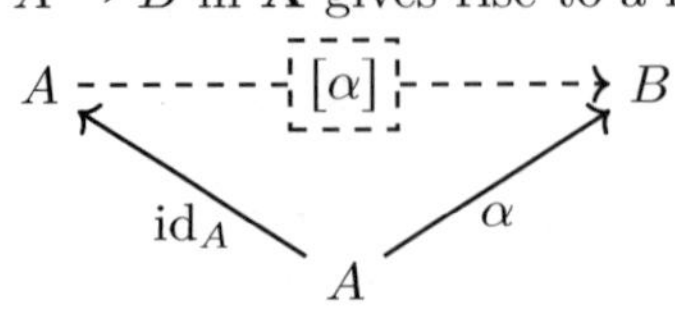

in $\mathbf{G}(\mathbf{A})$. Since the pullback of the identity can again be chosen as the identity, we actually have a functor

$$[-] : \mathbf{A} \longrightarrow \mathbf{G}(\mathbf{A}).$$

Moreover, assume that we have $[\alpha] = [\alpha']$ for a given pair $\alpha, \alpha' : A \to B$. Since the morphisms $(1, \alpha) : A \longrightarrow A \oplus B$ and $(1, \alpha') : A \longrightarrow A \oplus B$ are monos, it follows that $\alpha = \alpha'$. Thus, our functor $[-]$ is faithful, and we can regard $\mathbf{A}$ as a subcategory of $\mathbf{G}(\mathbf{A})$. Any morphism in $\mathbf{G}(\mathbf{A})$ which is equal to a morphism of the form $[\alpha]$ for $\alpha \in \mathbf{A}$ is called **honest**.

The most prominent feature of $\mathbf{G}(\mathbf{A})$ is the operation of taking pseudo-inverses.

DEFINITION 2.12. For a span $S = (A \xleftarrow{\alpha} C \xrightarrow{\beta} B)$ from A to B, we call the span $(B \xleftarrow{\beta} C \xrightarrow{\alpha} A)$ from B to A its **pseudo-inverse** and denote it by S^{-1}.

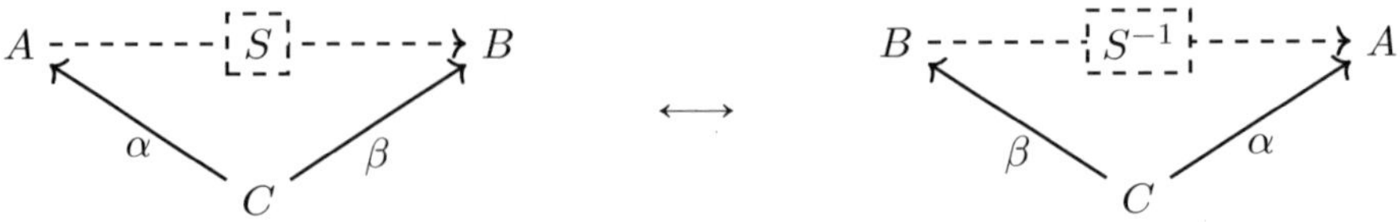

REMARK 2.13. Taking pseudo-inverses is compatible with stable equivalences. Thus, it defines an equivalence of categories

$$(-)^{-1} : \mathbf{G}(\mathbf{A})^{\mathrm{op}} \to \mathbf{G}(\mathbf{A}).$$

Now, we show that we may represent every generalized morphism as a composition of a pseudo-inverse of an honest morphism with another honest morphism.

LEMMA 2.14. *Every span* $(A \xleftarrow{\alpha} C \xrightarrow{\beta} B)$ *is equal to* $[\alpha]^{-1} \cdot [\beta]$ *as generalized morphisms.*

PROOF. A square consisting of identities is a pullback square. Thus, we have an equation of generalized morphisms (even as spans):

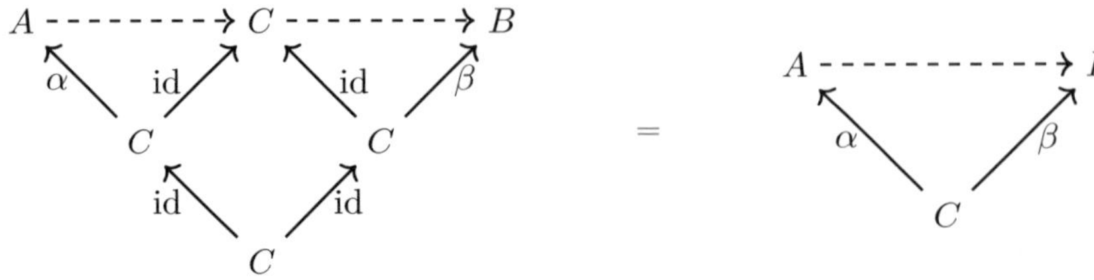

$\square$

THEOREM 2.15. *Given a mono ι in $\mathbf{A}$, then $[\iota]$ is split in $\mathbf{G}(\mathbf{A})$ with its pseudo-inverse as a retraction. Dually, given an epi ϵ in $\mathbf{A}$, then $[\epsilon]$ is split in $\mathbf{G}(\mathbf{A})$ with its pseudo-inverse as a section.*

PROOF. The composition of $[\iota]$ with $[\iota]^{-1}$ yields the diagram

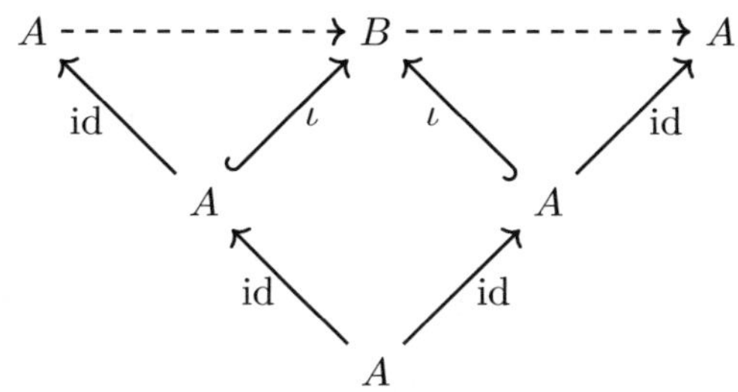

The dual statement can be proved analogously. □

COROLLARY 2.16. *Given a commutative diagram*

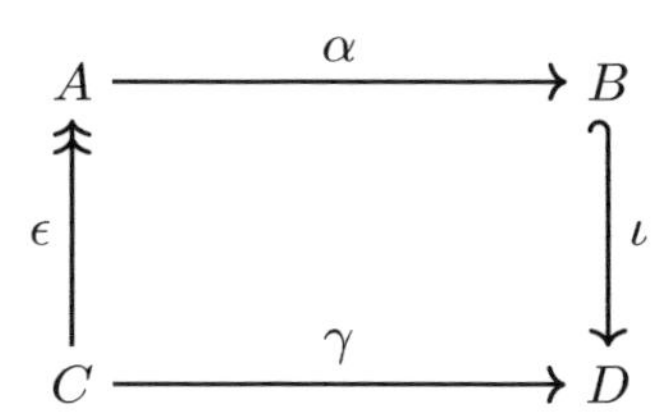

in $\mathbf{A}$ with ϵ epi and ι mono, we get a commutative diagram

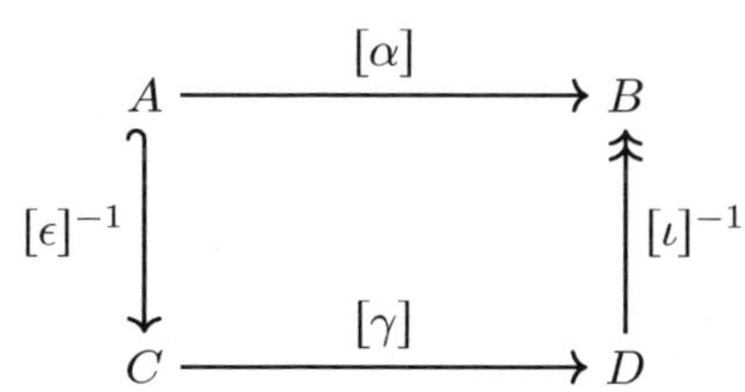

in $\mathbf{G}(\mathbf{A})$, i.e., the equation

$$[\alpha] = [\epsilon]^{-1} \cdot [\gamma] \cdot [\iota]^{-1}$$

holds.

PROOF. We simply multiply the equation

$$[\epsilon] \cdot [\alpha] \cdot [\iota] = [\gamma]$$

from the left with $[\epsilon]^{-1}$ and from the right with $[\iota]^{-1}$. Then we apply Theorem 2.15. □

THEOREM 2.17. *Given a pullback diagram*

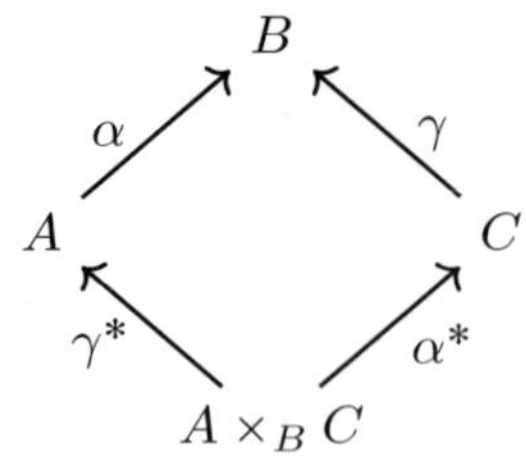

the ***pullback computation rule***

$$[\alpha] \cdot [\gamma]^{-1} = [\gamma^*]^{-1} \cdot [\alpha^*]$$

holds. Dually, given a pushout square

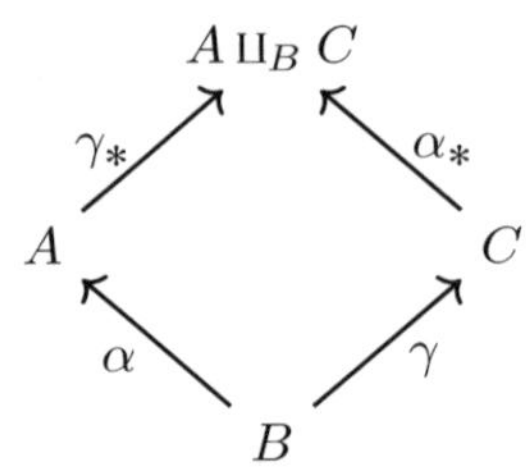

the ***pushout computation rule***

$$[\alpha]^{-1} \cdot [\gamma] = [\gamma_*] \cdot [\alpha_*]^{-1}$$

holds.

Proof. From the diagram

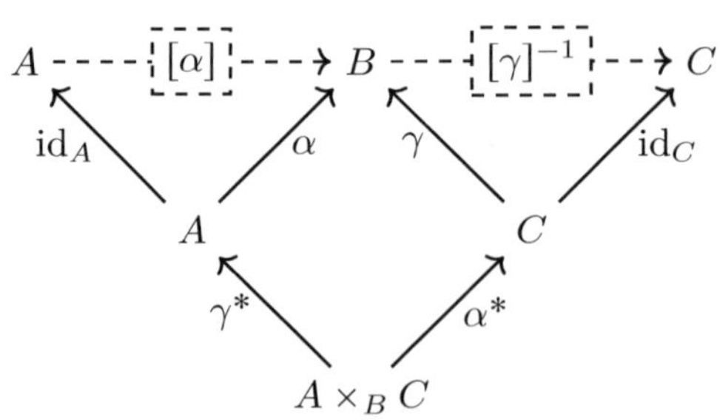

and Lemma 2.14, we get the pullback computation rule.

Next, we consider the situation for the pushout computation rule. Let

$$\alpha_*^* : A \times_{A \amalg_B C} C \to A$$

and

$$\gamma_*^* : A \times_{A \amalg_B C} C \to C$$

be the pullback projections of γ_*, α_*:

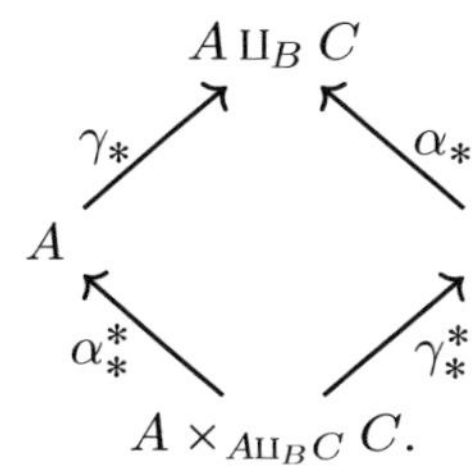

By the pullback computation rule, we have

$$[\gamma_*] \cdot [\alpha_*]^{-1} = [\alpha_*^*]^{-1} \cdot [\gamma_*^*].$$

But taking pushout followed by taking pullback yields a monomorphism

$$(\alpha_*^*, \gamma_*^*) : A \times_{A \amalg_B C} C \longrightarrow A \oplus C$$

which identifies with the image embedding of the morphism

$$(\alpha, \gamma) : B \longrightarrow A \oplus C,$$

since images in abelian categories are defined as the kernel embeddings of cokernel projections. It follows that

$$[\alpha_*^*]^{-1} \cdot [\gamma_*^*] = [\alpha]^{-1} \cdot [\gamma].$$

□

2.4. Cohomology. Generalized morphisms are a convenient tool to write down closed formulas for morphisms whose existence is induced by some prescribed diagram. We demonstrate this principle by means of a standard example in homological algebra, namely the induced morphism on cohomology.

THEOREM 2.18. *Suppose given a commutative diagram in* **A** *of the following form:*

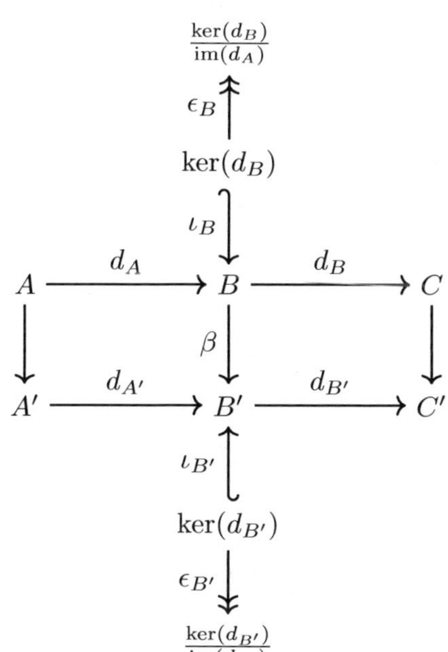

where we have $\operatorname{im}(d_A) \subseteq \ker(d_B)$, $\operatorname{im}(d_{A'}) \subseteq \ker(d_{B'})$, *and* $\iota_B, \iota_{B'}$ *are the kernel embeddings, and* ϵ_B, $\epsilon_{B'}$ *are the natural projections. Then the induced morphism on cohomologies*

$$\frac{\ker(d_B)}{\operatorname{im}(d_A)} \longrightarrow \frac{\ker(d_{B'})}{\operatorname{im}(d_{A'})}$$

is given by the following composition of generalized morphisms:

$$[\epsilon_B]^{-1} \cdot [\iota_B] \cdot [\beta] \cdot [\iota_{B'}]^{-1} \cdot [\epsilon_{B'}].$$

PROOF. The induced morphism on cohomologies is constructed by the cokernel functor applied to the commutative square

$$\begin{array}{ccc} \operatorname{im}(d_A) & \hookrightarrow & \ker(d_B) \\ \downarrow & & \downarrow \\ \operatorname{im}(d_{A'}) & \hookrightarrow & \ker(d_{B'}) \end{array}$$

which itself is defined by restricting β. Thus, we have a commutative diagram

$$\begin{array}{ccccc} B & \xleftarrow{\iota_B} & \ker(d_B) & \xrightarrow{\epsilon_B} & \frac{\ker(d_B)}{\operatorname{im}(d_A)} \\ \beta \downarrow & & \gamma \downarrow & & \downarrow \delta \\ B' & \xleftarrow{\iota_{B'}} & \ker(d_{B'}) & \xrightarrow{\epsilon_{B'}} & \frac{\ker(d_{B'})}{\operatorname{im}(d_{A'})} \end{array}$$

where the dashed arrow δ is the induced morphism on cohomologies.

Now, since ϵ_B is an epi, by Corollary 2.16 we have

$$[\delta] = [\epsilon_B]^{-1} \cdot [\gamma] \cdot [\epsilon_{B'}].$$

Moreover, since $\iota_{B'}$ is a mono, by Corollary 2.16 we have

$$[\gamma] = [\iota_B] \cdot [\beta] \cdot [\iota_{B'}]^{-1}.$$

Substituting the latter formula in the former yields the claim. □

2.5. Snake lemma. The induced morphism in the famous snake lemma can also be constructed as a composition of the obvious generalized morphisms. For seeing this, we analyze the construction of the snake following [**ML98**] in the light of the theory of generalized morphisms.

The starting point of the snake lemma is a commutative diagram in **A** with exact rows:

$$\begin{array}{ccccccccc}
 & & & & & & \ker(\gamma) & & \\
 & & & & & & \big\downarrow{\scriptstyle \eta := \mathrm{KernelEmbedding}(\gamma)} & & \\
0 & \longrightarrow & A & \xrightarrow{\ \delta\ } & B & \xrightarrow{\ \epsilon\ } & C & \longrightarrow & 0 \\
 & & \big\downarrow{\scriptstyle\alpha} & & \big\downarrow{\scriptstyle\beta} & & \big\downarrow{\scriptstyle\gamma} & & \\
0 & \longrightarrow & A' & \xrightarrow{\ \iota\ } & B' & \xrightarrow{\ \nu\ } & C' & \longrightarrow & 0 \\
 & & \big\downarrow{\scriptstyle \zeta := \mathrm{CokernelProjection}(\alpha)} & & & & & & \\
 & & \mathrm{coker}(\alpha) & & & & & &
\end{array}$$

In [**ML98**], Mac Lane constructs the snake morphism

$$\delta : \ker(\gamma) \longrightarrow \mathrm{coker}(\alpha)$$

by first computing the pullback

$$\begin{array}{ccc}
\ker(\gamma) \times_C B & \xrightarrow{\ \epsilon^*\ } & \ker(\gamma) \\
{\scriptstyle\eta^*}\big\downarrow & & \big\downarrow{\scriptstyle\eta} \\
B & \xrightarrow{\ \epsilon\ } & C
\end{array}$$

and pushout

$$\begin{array}{ccc}
A' & \xrightarrow{\ \iota\ } & B' \\
{\scriptstyle\zeta}\big\downarrow & & \big\downarrow{\scriptstyle\zeta_*} \\
\mathrm{coker}(\alpha) & \xrightarrow{\ \iota_*\ } & \mathrm{coker}(\alpha) \amalg_{A'} B'
\end{array}$$

and second proving the existence of a unique morphism δ rendering the diagram

$$\begin{array}{ccc}
\ker(\gamma) & \overset{\delta}{\dashrightarrow} & \mathrm{coker}(\alpha) \\
{\scriptstyle\epsilon^*}\big\uparrow & & \big\downarrow{\scriptstyle\iota_*} \\
\ker(\gamma) \times_C B & \xrightarrow{\ \eta^* \cdot \beta \cdot \zeta_*\ } & \mathrm{coker}(\alpha) \amalg_{A'} B'
\end{array}$$

commutative.

Analyzing this process in the light of generalized morphisms, the first step of taking the pullback/pushout can be interpreted as rewriting the generalized morphisms

$$[\eta] \cdot [\epsilon]^{-1} = [\epsilon^*]^{-1} \cdot [\eta^*] \tag{3}$$

and

$$[\iota]^{-1} \cdot [\zeta] = [\zeta_*] \cdot [\iota_*]^{-1} \tag{4}$$

employing the pullback/pushout computation rule. From Corollary 2.16, we know that we can produce δ as the composition

$$[\delta] = [\epsilon^*]^{-1} \cdot [\eta^*] \cdot [\beta] \cdot [\zeta_*] \cdot [\iota_*]^{-1}. \tag{5}$$

Substituting (3) and (4) in (5), the equation

$$[\delta] = [\eta] \cdot [\epsilon]^{-1} \cdot [\beta] \cdot [\iota]^{-1} \cdot [\zeta]$$

follows, which is nothing but straightforwardly following the arrows regardless of their direction from $\ker(\gamma)$ to $\mathrm{coker}(\alpha)$:

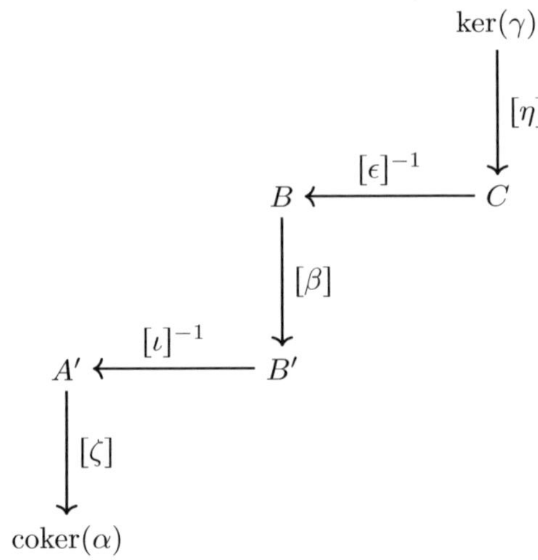

REMARK 2.19. This is not a proof of the snake lemma, but a way to construct the connecting homomorphism once we know it exists. For a proof of the snake lemma using the language of generalized morphisms, see [**Pos17**, Lemma II.2.1]

2.6. Generalized homomorphism theorem. To any morphism $\alpha : A \longrightarrow B$ in an abelian category $\mathbf{A}$, we can associate two canonical subobjects: its image $\mathrm{im}(\alpha)$ and its kernel $\ker(\alpha)$. The homomorphism theorem states that, using these canonical subobjects, we get a commutative diagram

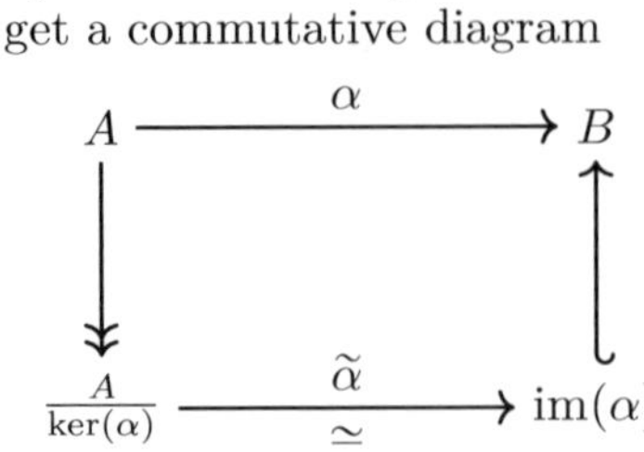

Given a generalized morphism 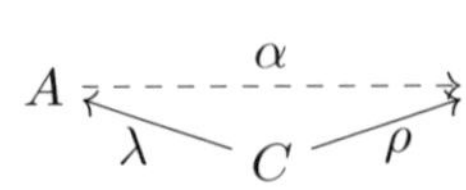 , we have four canonical subobjects:

- **Domain:**
  $\mathrm{dom}(\alpha) := \mathrm{im}(\lambda) \subseteq A$

- **Generalized kernel:**
  $\mathrm{gker}(\alpha) := \lambda(\ker(\rho)) \subseteq A$
- **Generalized image:**
  $\mathrm{gim}(\alpha) := \mathrm{im}(\rho) \subseteq B$
- **Defect:**
  $\mathrm{def}(\alpha) := \rho(\ker(\lambda)) \subseteq B$

We claim that a generalized homomorphism theorem holds, namely, the existence of a commutative diagram

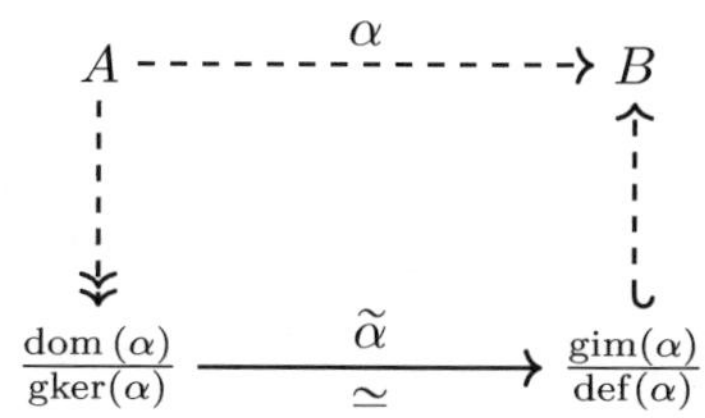

The two vertical arrows are simply given by the generalized subquotient projection

$$A \hookleftarrow \mathrm{dom}(\alpha) \twoheadrightarrow \frac{\mathrm{dom}(\alpha)}{\mathrm{gker}(\alpha)},$$

which is an epimorphism in $\mathbf{G}(\mathbf{A})$ by Theorem 2.15, and the generalized subquotient injection

$$\frac{\mathrm{gim}(\alpha)}{\mathrm{def}(\alpha)} \twoheadleftarrow \mathrm{gim}(\alpha) \hookrightarrow B,$$

which is a monomorphism in $\mathbf{G}(\mathbf{A})$ also by Theorem 2.15.

The validity of the generalized homomorphism theorem can be easily extracted from the following commutative diagram and from the pushout computation rule:

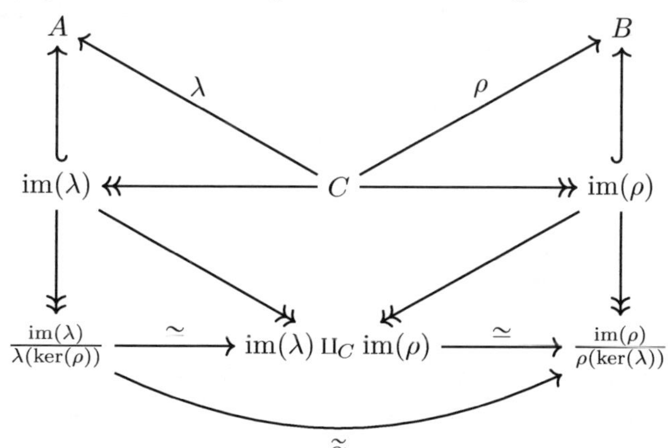

2.7. Computing spectral sequences. This subsection serves as an introduction to spectral sequences. We use generalized morphisms as a fundamental tool in our explanation. This has two advantages:

(1) The main idea behind spectral sequences becomes quite transparent when you already have generalized morphisms available as a tool.
(2) Instead of mere existence theorems, we will get explicit formulas for all the differentials within a spectral sequence.

Let **A** be an abelian category. A spectral sequence is a lot of data that can naturally be associated to a given filtered cochain complex, i.e., a cochain complex

$$\cdots \longrightarrow M^i \xrightarrow{\partial^i} M^{i+1} \xrightarrow{\partial^{i+1}} M^{i+2} \xrightarrow{\partial^{i+2}} M^{i+3} \longrightarrow \cdots$$

in which each object M^i is equipped with a chain of subobjects

$$M^i \supseteq \cdots \supseteq F^j M^i \supseteq F^{j+1} M^i \supseteq F^{j+2} M^i \supseteq \ldots$$

compatible with the differentials, i.e., ∂^i restricts to a morphism

$$F^j \partial^i : F^j M^i \longrightarrow F^j M^{i+1}$$

for every $i, j \in \mathbb{Z}$. To simplify our explanation, we will concentrate on a finite excerpt of such a filtered cochain complex, and denote it as follows:

$$\cdots \longrightarrow A \xrightarrow{\partial^A} B \xrightarrow{\partial^B} C \xrightarrow{\partial^C} D \longrightarrow \cdots$$

with chain of subobjects

$$A \supseteq \cdots \supseteq A^j \supseteq A^{j+1} \supseteq A^{j+2} \supseteq \ldots$$

and likewise for B, C, and D. The restrictions of the differentials to the j-th subobjects are denoted by adding an extra index, e.g., $\partial^{A,j} : A^j \longrightarrow B^j$.

For every $j \in \mathbb{Z}$, we can restrict our filtered cochain complex to its j-th graded part and again obtain a cochain complex:

$$\cdots \longrightarrow \frac{A^j}{A^{j+1}} \xrightarrow{\overline{\partial^{A,j}}} \frac{B^j}{B^{j+1}} \xrightarrow{\overline{\partial^{B,j}}} \frac{C^j}{C^{j+1}} \xrightarrow{\overline{\partial^{C,j}}} \frac{D^j}{D^{j+1}} \longrightarrow \cdots$$

It is the common convention to arrange this $\mathbb{Z}$-indexed family of cochain complexes between the graded parts as follows:

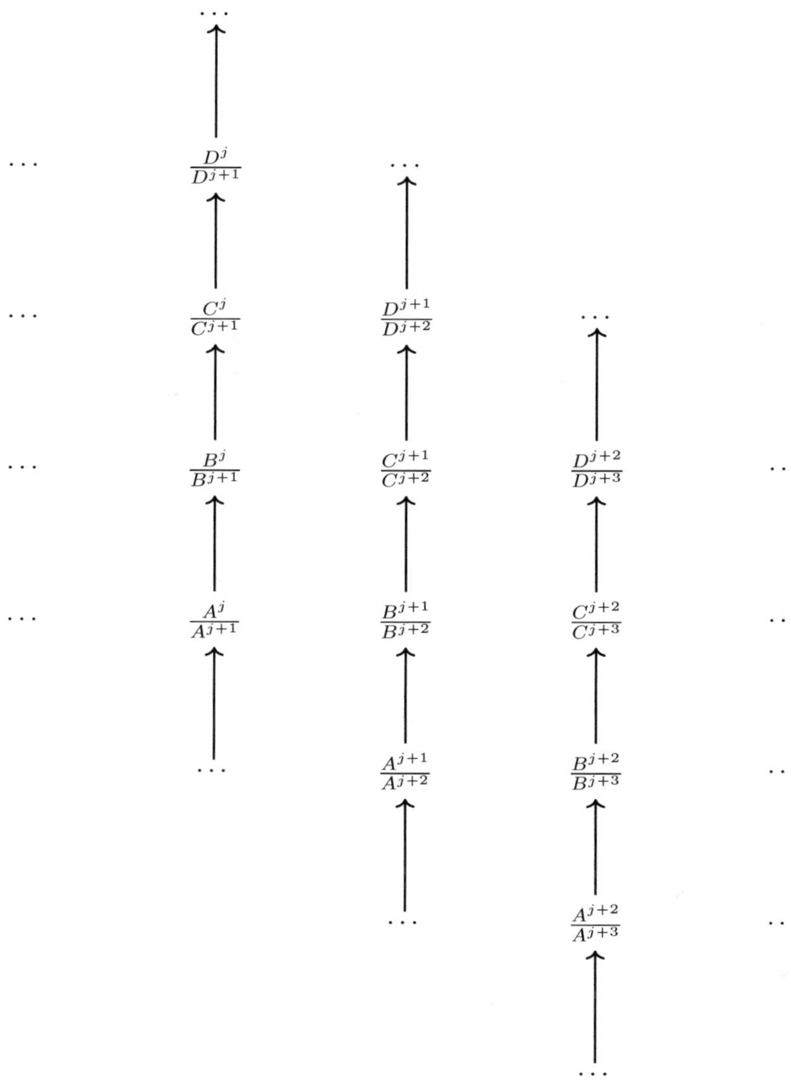

Let us take a closer look at the induced differentials $\overline{\partial^{A,j}}$. They fit into a commutative diagram

$$\begin{array}{ccc} A & \xrightarrow{\partial^{A}} & B \\ {\scriptstyle \iota^{A,j}}\uparrow & & \uparrow{\scriptstyle \iota^{B,j}} \\ A^j & \xrightarrow{\partial^{A,j}} & B^j \\ {\scriptstyle \epsilon^{A,j}}\downarrow & & \downarrow{\scriptstyle \epsilon^{B,j}} \\ \frac{A^j}{A^{j+1}} & \xrightarrow{\overline{\partial^{A,j}}} & \frac{B^j}{B^{j+1}} \end{array}$$

which shows, using Corollary 2.16, that we may express $\overline{\partial^{A,j}}$ as a composition of generalized morphisms, following the outer path from $\frac{A^j}{A^{j+1}}$ to $\frac{B^j}{B^{j+1}}$ in the diagram above:

$$[\overline{\partial^{A,j}}] = [\epsilon^{A,j}]^{-1} \cdot [\iota^{A,j}] \cdot [\partial^A] \cdot [\iota^{B,j}]^{-1} \cdot [\epsilon^{B,j}].$$

To simplify this expression, let us introduce

$$\mathrm{emb}^{A,j} := [\epsilon^{A,j}]^{-1} \cdot [\iota^{A,j}] : \frac{A^j}{A^{j+1}} \dashrightarrow A$$

as notation for the **generalized subquotient embedding** and

$$\mathrm{proj}^{B,j} := [\iota^{B,j}]^{-1} \cdot [\epsilon^{B,j}] : B \dashrightarrow \frac{B^j}{B^{j+1}}$$

as notation for the **generalized subquotient projection**. Then, the induced morphism between graded parts is literally given by restricting $\partial^A : A \to B$ to the appropriate subquotients:

$$[\overline{\partial^{A,j}}] = \mathrm{emb}^{A,j} \cdot [\partial^A] \cdot \mathrm{proj}^{B,j}.$$

Now, the main idea behind spectral sequences is that too much information is lost when we only focus on restrictions of ∂^A to subquotients of the same index j, and thus, we should try and see what happens if we increase the index of the projection by 1:

$$\partial_1^{A,j} := \mathrm{emb}^{A,j} \cdot [\partial^A] \cdot \mathrm{proj}^{B,j+1}.$$

In general, we cannot expect this generalized morphism to be honest anymore and so we depict it with a dashed arrow

$$\partial_1^{A,j} : \frac{A^j}{A^{j+1}} \dashrightarrow \frac{B^{j+1}}{B^{j+2}}.$$

We can assemble these generalized differentials within a structure that we would like to call a *generalized* cochain complex:

(6)
$$\cdots \dashrightarrow \frac{A^j}{A^{j+1}} \overset{\partial_1^{A,j}}{\dashrightarrow} \frac{B^{j+1}}{B^{j+2}} \overset{\partial_1^{B,j+1}}{\dashrightarrow} \frac{C^{j+2}}{C^{j+3}} \overset{\partial_1^{C,j+2}}{\dashrightarrow} \frac{D^{j+3}}{D^{j+4}} \dashrightarrow \cdots$$

DEFINITION 2.20. We define a **generalized cochain complex** to be a $\mathbb{Z}$-indexed family of objects M^i together with a $\mathbb{Z}$-indexed family of *generalized* morphisms

$$\partial^i : M^i \dashrightarrow M^{i+1}$$

such that

$$\mathrm{gim}(\partial^i) \subseteq \mathrm{gker}(\partial^{i+1}).$$

We show that two consecutive morphisms in (6), e.g., $\partial_1^{A,j}$ and $\partial_1^{B,j+1}$, satisfy

(7)
$$\mathrm{gim}(\partial_1^{A,j}) \subseteq \mathrm{gker}(\partial_1^{B,j+1}).$$

Indeed, we can calculate

$$\mathrm{gim}(\partial_1^{A,j}) = \frac{(\partial^A(A^j) \cap B^{j+1}) + B^{j+2}}{B^{j+2}}$$

and

$$\mathrm{gker}(\partial_1^{B,j+1}) = \frac{((\partial^B)^{-1}(C^{j+3}) \cap B^{j+1}) + B^{j+2}}{B^{j+2}}$$

where we use standard notation for dealing with subobjects in abelian categories, i.e., $\cap$ and $(-)^{-1}$ are shorthand for the corresponding pullbacks, and $+$ for the join of subobjects. Since

$$\partial^A(A^j) \subseteq \mathrm{im}(\partial^A) \subseteq \ker(\partial^B) = (\partial^B)^{-1}(0) \subseteq (\partial^B)^{-1}(C^{j+3})$$

we really get our desired inclusion (7). Thus, (6) forms a generalized cochain complex.

The whole collection of generalized cochain complexes that we get in this way may be depicted as follows:

$$\begin{array}{ccccccccc}
 & & \cdots & & & & & & \\
\cdots & \dashrightarrow & \frac{D^j}{D^{j+1}} & \dashrightarrow & \cdots & & & & \\
\cdots & \dashrightarrow & \frac{C^j}{C^{j+1}} & \dashrightarrow & \frac{D^{j+1}}{D^{j+2}} & \dashrightarrow & \cdots & & \\
\cdots & \dashrightarrow & \frac{B^j}{B^{j+1}} & \dashrightarrow & \frac{C^{j+1}}{C^{j+2}} & \dashrightarrow & \frac{D^{j+2}}{D^{j+3}} & \dashrightarrow & \cdots \\
\cdots & \dashrightarrow & \frac{A^j}{A^{j+1}} & \dashrightarrow & \frac{B^{j+1}}{B^{j+2}} & \dashrightarrow & \frac{C^{j+2}}{C^{j+3}} & \dashrightarrow & \cdots \\
 & & \cdots & \dashrightarrow & \frac{A^{j+1}}{A^{j+2}} & \dashrightarrow & \frac{B^{j+2}}{B^{j+3}} & \dashrightarrow & \cdots \\
 & & & & \cdots & \dashrightarrow & \frac{A^{j+2}}{A^{j+3}} & \dashrightarrow & \cdots \\
 & & & & & & \cdots & &
\end{array}$$

Increasing the index of the projection by 2 would yield the following picture (again of generalized cochain complexes):

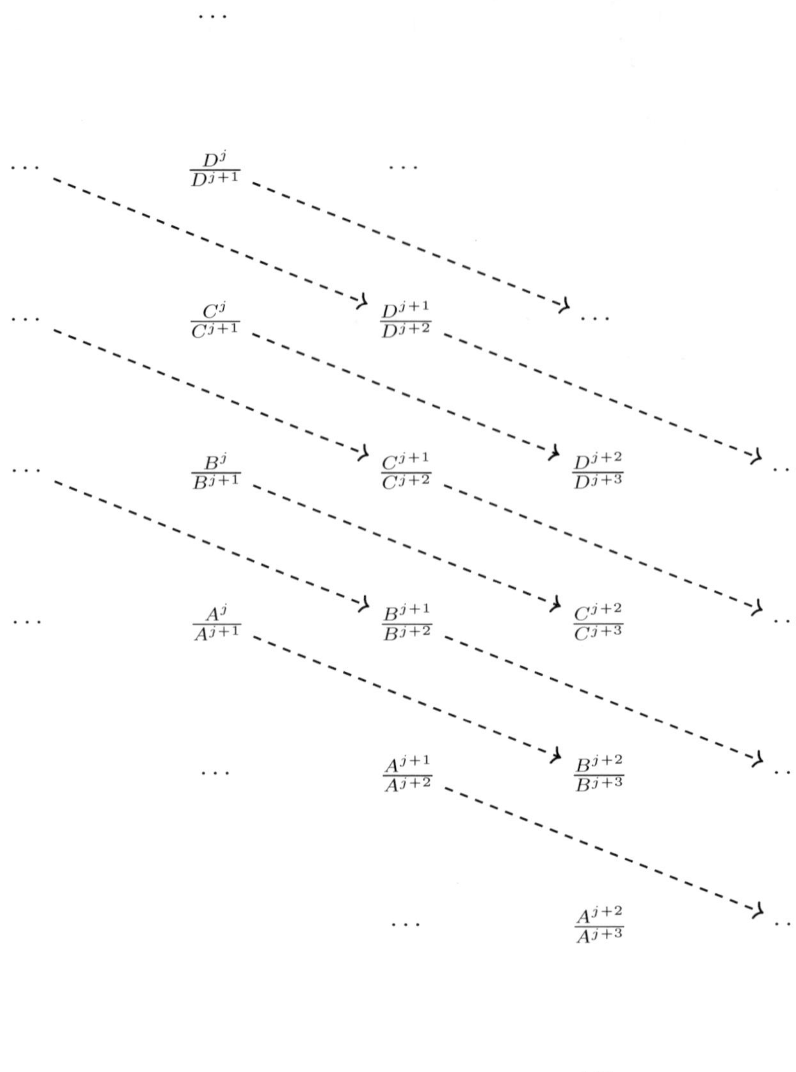

It follows that we are able to construct for every integer $i \geqslant 0$, and not only for the case $i = 0$, a $\mathbb{Z}$-indexed family of *generalized* cochain complexes

$$\cdots \dashrightarrow \frac{A^j}{A^{j+1}} \overset{\partial_i^{A,j}}{\dashrightarrow} \frac{B^{j+i}}{B^{j+i+1}} \overset{\partial_i^{B,j+i}}{\dashrightarrow} \frac{C^{j+2i}}{C^{j+2i+1}} \overset{\partial_i^{C,j+2i}}{\dashrightarrow} \frac{D^{j+3i}}{D^{j+3i+1}} \dashrightarrow \cdots \tag{8}$$

Next, we will see how to produce from a generalized cochain complex an ordinary cochain complex having honest differentials. Applying this process to the just created generalized cochain complexes will then yield our desired spectral sequence.

So, let

$$\cdots \dashrightarrow M^i \overset{\partial^i}{\dashrightarrow} M^{i+1} \overset{\partial^{i+1}}{\dashrightarrow} M^{i+2} \overset{\partial^{i+2}}{\dashrightarrow} M^{i+3} \dashrightarrow \cdots$$

be an arbitrary generalized cochain complex. Since we have

$$\operatorname{gim}(\partial^i) \subseteq \operatorname{gker}(\partial^{i+1}),$$

we also have

$$\operatorname{def}(\partial^i) \subseteq \operatorname{gim}(\partial^i) \subseteq \operatorname{gker}(\partial^{i+1}) \subseteq \operatorname{dom}(\partial^{i+1}).$$

We apply the generalized homomorphism theorem (see Subsection 2.6) to the generalized morphisms ∂^i in order to produce honest morphisms d^i fitting in the following commutative diagram:

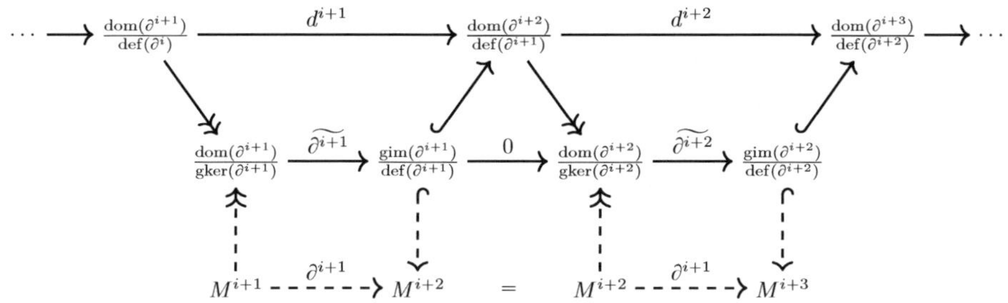

We can directly read off the equation

$$d^{i+1} \cdot d^{i+2} = 0.$$

The collection of the d^i is what we call the **associated honest cochain complex** of the generalized cochain complex given by the ∂^i. Note that the rectangles of the above diagram

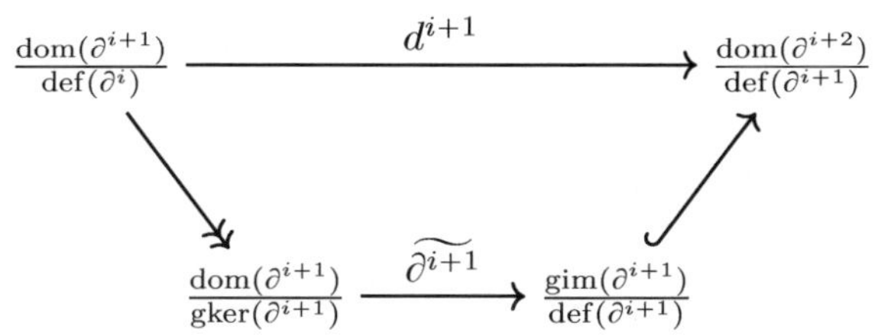

are actually decompositions of the d^{i+1} in the sense of the homomorphism theorem, since $\widetilde{\partial^{i+1}}$ is an isomorphism. But then it follows that

$$\ker(d^{i+1}) = \frac{\operatorname{gker}(\partial^{i+1})}{\operatorname{def}(\partial^i)}$$

and

$$\operatorname{im}(d^{i+1}) = \frac{\operatorname{gim}(\partial^{i+1})}{\operatorname{def}(\partial^{i+1})}.$$

In particular, we can compute the cohomologies of the associated honest cochain complex $d^\bullet$ in terms of $\partial^\bullet$:

$$\mathrm{H}^{i+1}(d^\bullet) \simeq \frac{\operatorname{gker}(\partial^{i+1})}{\operatorname{gim}(\partial^i)}. \tag{9}$$

Now, let us go back to our generalized cochain complexes (8). As we have learned in (9), computing the cohomologies of their associated honest cochain complexes boils down to the computation of generalized images and generalized kernels, for which we have:

$$\operatorname{gim}(\partial_i^{A,j}) = \frac{\left(\partial^A(A^j) \cap B^{j+i}\right) + B^{j+i+1}}{B^{j+i+1}}$$

and

$$\operatorname{gker}(\partial_i^{B,j+1}) = \frac{\left((\partial^B)^{-1}(C^{j+2i+1}) \cap B^{j+i}\right) + B^{j+i+1}}{B^{j+i+1}}.$$

Computing the remaining two canonical subobjects can be performed analogously and yields

$$\operatorname{def}(\partial_i^{A,j}) = \frac{\left(\partial^A(A^{j+1}) \cap B^{j+i}\right) + B^{j+i+1}}{B^{j+i+1}}$$

and

$$\operatorname{dom}(\partial_i^{B,j+1}) = \frac{\left((\partial^B)^{-1}(C^{j+2i}) \cap B^{j+i}\right) + B^{j+i+1}}{B^{j+i+1}}.$$

But from this, we can deduce by a simple variable substitution

$$\operatorname{gker}(\partial_i^{B,j+1}) = \operatorname{dom}(\partial_{i+1}^{B,j})$$

and

$$\operatorname{gim}(\partial_i^{A,j}) = \operatorname{def}(\partial_{i+1}^{A,j-1}).$$

In particular, we deduce

$$\frac{\operatorname{gker}(\partial_i^{B,j+1})}{\operatorname{gim}(\partial_i^{A,j})} \simeq \frac{\operatorname{dom}(\partial_{i+1}^{B,j})}{\operatorname{def}(\partial_{i+1}^{A,j-1})}.$$

Putting these information together, it follows that the cohomologies of the i-th associated honest cochain complexes determine the objects of the $(i+1)$-th associated honest cochain complexes. This is exactly the defining feature of a spectral sequence, which we are going to define now.

DEFINITION 2.21. A **cohomological spectral sequence** (starting at 0) consists of the following data: For all $p, q \in \mathbb{Z}$, $r \geqslant 0$, we have:

(1) objects $E_r^{p,q} \in \mathbf{A}$,
(2) morphisms $d_r^{p,q} : E_r^{p,q} \longrightarrow E_r^{p+r,q-(r-1)} \in \mathbf{A}$,
(3) isomorphisms $\iota_r^{p,q} : E_{r+1}^{p,q} \xrightarrow{\sim} \frac{\ker(d_r^{p,q})}{\operatorname{im}(d_r^{p-r,q+(r-1)})}$,
(4) the equation $d_r^{p,q} \cdot d_r^{p+r,q-(r-1)} = 0$ holds.

From the discussion in this subsection, it follows that if we are given a filtered cochain complex

$$\cdots \longrightarrow M^i \xrightarrow{\partial^i} M^{i+1} \xrightarrow{\partial^{i+1}} M^{i+2} \xrightarrow{\partial^{i+1}} M^{i+3} \longrightarrow \cdots$$

then we can construct a spectral sequence by first defining the auxiliary data

$$E_0^{p,q} := \frac{F^p M^{p+q}}{F^{p+1} M^{p+q}}$$

and

$$\partial_r^{p,q} \quad := \quad E_0^{p,q} \overset{\text{emb}}{\hookrightarrow} M^{p+q} \xrightarrow{\partial^{p+q}} M^{p+q+1} \overset{\text{proj}}{\twoheadrightarrow} E_0^{p+r,q-(r-1)}$$

and second constructing the data for the spectral sequence as

$$E_r^{p,q} := \frac{\operatorname{dom}(\partial_r^{p,q})}{\operatorname{def}(\partial_r^{p-r,q+(r-1)})}$$

and

$$d_r^{p,q} \quad := \quad E_r^{p,q} \overset{\text{emb}}{\hookrightarrow} M^{p+q} \xrightarrow{\partial^{p+q}} M^{p+q+1} \overset{\text{proj}}{\twoheadrightarrow} E_r^{p+r,q-(r-1)}.$$

Note that all our constructions in this subsection were formulated purely in the language of generalized morphisms. We have seen that computing with generalized morphisms only involves computations in the underlying abelian category like taking pushouts and pullbacks. It follows that we reached our second computational goal: computing the differentials on the pages of a spectral sequence associated to a filtered cochain complex only with the help of direct computations in the underlying abelian category.

References

[Aus66] Maurice Auslander, *Coherent functors*, Proc. Conf. Categorical Algebra (La Jolla, Calif., 1965), Springer, New York, 1966, pp. 189–231. MR0212070

[Bar09] Mohamed Barakat, *The homomorphism theorem and effective computations*, Habilitation thesis, Department of Mathematics, RWTH-Aachen University, April 2009.

[Bel00] Apostolos Beligiannis, *On the Freyd categories of an additive category*, Homology Homotopy Appl. **2** (2000), 147–185, DOI 10.4310/hha.2000.v2.n1.a11. MR2027559

[BLH11] Mohamed Barakat and Markus Lange-Hegermann, *An axiomatic setup for algorithmic homological algebra and an alternative approach to localization*, J. Algebra Appl. **10** (2011), no. 2, 269–293, DOI 10.1142/S0219498811004562. MR2795737

[BP69] Hans-Berndt Brinkmann and Dieter Puppe, *Abelsche und exakte Kategorien, Korrespondenzen* (German), Lecture Notes in Mathematics, Vol. 96, Springer-Verlag, Berlin-New York, 1969. MR0269713

[Buc06] Bruno Buchberger, *An algorithm for finding the basis elements of the residue class ring of a zero dimensional polynomial ideal*, J. Symbolic Comput. **41** (2006), no. 3-4, 475–511, DOI 10.1016/j.jsc.2005.09.007. Translated from the 1965 German original by Michael P. Abramson. MR2202562

[CLO92] David Cox, John Little, and Donal O'Shea, *Ideals, varieties, and algorithms*, Undergraduate Texts in Mathematics, Springer-Verlag, New York, 1992. An introduction to computational algebraic geometry and commutative algebra. MR1189133

[Col86] Donald J. Collins, *A simple presentation of a group with unsolvable word problem*, Illinois J. Math. **30** (1986), no. 2, 230–234. MR840121

[Fre64] Peter Freyd, *Abelian categories. An introduction to the theory of functors*, Harper's Series in Modern Mathematics, Harper & Row, Publishers, New York, 1964. MR0166240

[Fre66] Peter Freyd, *Representations in abelian categories*, Proc. Conf. Categorical Algebra (La Jolla, Calif., 1965), Springer, New York, 1966, pp. 95–120. MR0209333

[GAP18] The GAP Group, *GAP – Groups, Algorithms, and Programming, Version 4.9.1*, 2018, (`http://www.gap-system.org`).

[GP02] Gert-Martin Greuel and Gerhard Pfister, *A* **Singular** *introduction to commutative algebra*, Springer-Verlag, Berlin, 2002. With contributions by Olaf Bachmann, Christoph Lossen and Hans Schönemann; With 1 CD-ROM (Windows, Macintosh, and UNIX). MR1930604

[Gre99] Edward L. Green, *Noncommutative Gröbner bases, and projective resolutions.*, In: Dräxler P., Ringel C.M., Michler G.O. (eds) Computational Methods for Representations of Groups and Algebras.

[GSP18] Sebastian Gutsche, Øystein Skartsæterhagen, and Sebastian Posur, *The* `CAP` *project – Categories, Algorithms, Programming*, (`http://homalg-project.github.io/CAP_project`), 2013–2018.

[Hil66] Peter Hilton, *Correspondences and exact squares*, Proc. Conf. Categorical Algebra (La Jolla, Calif., 1965), Springer, New York, 1966, pp. 254–271. MR0204487

[Joh02] Peter T. Johnstone, *Sketches of an elephant: a topos theory compendium. Vol. 1*, Oxford Logic Guides, vol. 43, The Clarendon Press, Oxford University Press, New York, 2002. MR1953060

[ML98] Saunders Mac Lane, *Categories for the working mathematician*, 2nd ed., Graduate Texts in Mathematics, vol. 5, Springer-Verlag, New York, 1998. MR1712872

[MRR88] Ray Mines, Fred Richman, and Wim Ruitenburg, *A course in constructive algebra*, Universitext, Springer-Verlag, New York, 1988. MR919949

[Pos17] Sebastian Posur, *Constructive category theory and applications to equivariant sheaves*, Dissertation, University of Siegen, 2017, (`https://nbn-resolving.org/urn:nbn:de:hbz:467-11798`).

[Pos18] Sebastian Posur, *Linear systems over localizations of rings*, Arch. Math. (Basel) **111** (2018), no. 1, 23–32, DOI 10.1007/s00013-018-1183-z. MR3816974

[Pos21] Sebastian Posur, *A constructive approach to Freyd categories*, Appl. Categ. Structures **29** (2021), no. 1, 171–211, DOI 10.1007/s10485-020-09612-y. MR4204560

[Pre09] Mike Prest, *Purity, spectra and localisation*, Encyclopedia of Mathematics and its Applications, vol. 121, Cambridge University Press, Cambridge, 2009. MR2530988

[Pup62] Dieter Puppe, *Korrespondenzen in abelschen Kategorien* (German), Math. Ann. **148** (1962), 1–30, DOI 10.1007/BF01438388. MR141698

[Wei94] Charles A. Weibel, *An introduction to homological algebra*, Cambridge Studies in Advanced Mathematics, vol. 38, Cambridge University Press, Cambridge, 1994. MR1269324

RWTH Aachen University, 52062 Aachen, Germany
Email address: `posur@art.rwth-aachen.de`

Contemporary Mathematics
Volume **769**, 2021
https://doi.org/10.1090/conm/769/15416

The HRS tilting process and Grothendieck hearts of t-structures

Carlos E. Parra and Manuel Saorín

ABSTRACT. In this paper we revisit the problem of determining when the heart of a t-structure is a Grothendieck category, with special attention to the case of the Happel-Reiten-Smalø (HSR) t-structure in the derived category of a Grothendieck category associated to a torsion pair in the latter. We revisit the HRS tilting process deriving from it a lot of information on the HRS t-structures which have a projective generator or an injective cogenerator, and obtain several bijections between classes of pairs $(\mathcal{A}, \mathbf{t})$ consisting of an abelian category and a torsion pair in it. We use these bijections to re-prove, by different methods, a recent result of Tilting Theory and the fact that if $\mathbf{t} = (\mathcal{T}, \mathcal{F})$ is a torsion pair in a Grothendieck category $\mathcal{G}$, then the heart of the associated HRS t-structure is itself a Grothendieck category if, and only if, $\mathbf{t}$ is of finite type. We survey this last problem and recent results after its solution.

1. Introduction

The aim of this paper is twofold. On one side we want to give a summary of the main results related with the following question:

QUESTION 1.1. When is the heart of a t-structure a Grothendieck category?

We shall mainly concentrate in the route leading to the answer to the question in the case when the ambient triangulated category is the (unbounded) derived category $\mathcal{D}(\mathcal{G})$ of a Grothendieck category $\mathcal{G}$ and the t-structure is the Happel-Reiten-Smalø (HRS) t-structure in $\mathcal{D}(\mathcal{G})$ associated to a torsion pair in $\mathcal{G}$ (see Example 2.7(2)). But we include a final short section, where we briefly summarize the main results for general triangulated categories with coproducts and arbitrary t-structures. As a second goal, we want to revisit the HRS tilting process and show that it allows to prove in an easy way parts of recent results in the literature, and

---

2020 *Mathematics Subject Classification.* Primary 18E10, 18G80, 18E30, 18E40, 16B50, 16E30, 16E35.

Key words and phrases. Derived category, Grothendieck category, Happel-Reiten-Smalø t-structure, heart of a t-structure, torsion pair, t-structure.

The first named author was supported by CONICYT/FONDECYT/Iniciación/11160078. The second named author was supported by the research projects from Spanish Ministerio de Economía y Competitividad (MTM2016-77445-P) and from the Fundación 'Séneca' of Murcia (19880/GERM/15), with a part of FEDER funds.

that, with the help of a recent approach to the problem using purity, one can reprove the answer to Question 1.3 below by methods completely different to those used to get the earlier answer.

All throughout the paper, unless otherwise stated, all categories will be additive. We will mainly use two types of categories commonly studied in *Homological Algebra*, concretely *abelian categories* and *triangulated categories* (we refer to [**S**] and [**N**] for the respective definitions). The key concept for us is that of a *t-structure* in a triangulated category, introduced by Beilinson, Bernstein and Deligne [**BBD**] in their treatment of perverse sheaves. Roughly speaking a t-structure in the triangulated category $\mathcal{D}$ is a pair $\tau = (\mathcal{U}, \mathcal{W})$ of full subcategories satisfying some axioms (see Definition 2.5 for the details) which guarantee that the intersection $\mathcal{H}_\tau = \mathcal{U} \cap \mathcal{W}$ is an abelian category, commonly called the *heart* of the t-structure. This abelian category comes with a cohomological functor $H^0_\tau : \mathcal{D} \longrightarrow \mathcal{H}_\tau$. In [**BBD**] the category of perverses sheaves on a variety $\mathbb{X}$ appeared as the heart of a t-structure in $\mathcal{D}^b(\mathbb{X})$, the bounded derived category of coherent sheaves on $\mathbb{X}$.

In several modern developments of Mathematics, as Motive Theory, the homological approach to Mirror Symmetry, Modular Representation of finite groups, Representation Theory of Algebras, among others, the role of t-structures is fundamental. For this reason it is important to know when the heart of a t-structure has nice properties as an abelian category. Vaguely speaking, one would ask: When is the heart of a given t-structure a nice category?. Trying to make sense of the adjective 'nice' here, one commonly uses the following "hierarchy" among abelian categories introduced by Grothendieck [**G**]. We say that an abelian category $\mathcal{A}$ is:

(1) AB3 (resp. AB3*) when it has (arbitrary set-indexed) coproducts (resp. products);
(2) AB4 (resp. AB4*) when it is AB3 (resp. AB3*) and the coproduct functor $\coprod : [\Lambda, \mathcal{A}] \to \mathcal{A}$ (resp. product functor $\prod : [\Lambda, \mathcal{A}] \to \mathcal{A}$) is exact, for each set Λ;
(3) AB5 (resp. AB5*) when it is AB3 (resp. AB3*) and the direct limit functor $\varinjlim : [\Lambda, \mathcal{A}] \to \mathcal{A}$ (resp. inverse limit functor $\varprojlim : [\Lambda^{\mathrm{op}}, \mathcal{A}] \to \mathcal{A}$) is exact, for each directed set Λ.
(4) a Grothendieck category, when it is AB5 and has a generator or, equivalently, a set of generators.

Grothendieck categories appear quite naturally in Algebra and Geometry and their behavior is, in many aspects, similar to that of module categories over a ring (see [**S**, Chapter V]). For instance, such a category has enough injectives and every object in it has an injective envelope. Even more, by a famous theorem of Gabriel and Popescu (see [**GP**], and also [**S**, Theorem X.4.1]), such a category is always a Gabriel localisation of a module category, which roughly means that it is obtained from such a category by formally inverting some morphisms. This is the main reason why the study of when the heart of a t-structure is a Grothendieck category, i.e. Question 1.1, has deserved most of the attention, apart of the study of when it is a module category, that we barely touch in this paper. When one starts approaching the question, one quickly sees that it is hopeless unless some extra hypotheses are imposed on the ambient triangulated category $\mathcal{D}$ and/or on the t-structure τ itself. For instance, it is unavoidable to require that $\mathcal{D}$ has coproducts or, at least, to guarantee that coproducts in $\mathcal{D}$ of objects in the heart of τ always exist. On the other hand, the problem gets quite complicated if the coproduct in

$\mathcal{H}_\tau$ and the coproduct in $\mathcal{D}$ of a given family of objects in $\mathcal{H}_\tau$ do not coincide. A way of ensuring that they coincide is to require that the t-structure be *smashing*, i.e. that the co-aisle $\mathcal{W}$ of the t-structure is closed under coproducts in $\mathcal{D}$. Therefore, instead of the initial question, the following one has more hopes of being answered and has deserved a lot of attention in recent times (see Section 5):

QUESTION 1.2. Let $\mathcal{D}$ be a triangulated category with coproducts and let $\tau = (\mathcal{U}, \mathcal{W})$ be a smashing t-structrure in $\mathcal{D}$. When is the heart of τ a Grothendieck category?

Although studied historically first, the question for the HRS t-structure is a particular case of this last question. Namely, if $\mathcal{G}$ is a Grothendieck category, then its derived category $\mathcal{D}(\mathcal{G})$ is the prototypical example of a triangulated category with coproducts (and also products). When a torsion pair $\mathbf{t} = (\mathcal{T}, \mathcal{F})$ is given in $\mathcal{G}$, the associated HRS t-structure in $\mathcal{D}(\mathcal{G})$ is smashing. So restricted to this particular example, the last question is re-read as follows, and it is the main problem that we survey and re-visit in this paper:

QUESTION 1.3. Let $\mathcal{G}$ be a Grothendieck category, let $\mathbf{t} = (\mathcal{T}, \mathcal{F})$ be a torsion pair in $\mathcal{G}$ and let $\mathcal{H}_\mathbf{t}$ be the heart of the associated HRS t-structure in $\mathcal{D}(\mathcal{G})$. When is $\mathcal{H}_\mathbf{t}$ a Grothendieck category?

Let's now have a look at the new results and/or proofs of the paper. On what concerns our new look at the HRS tilting process, we give a series of results leading to a list of bijections between pairs consisting of a class of abelian categories and a class of torsion pairs in them (see Corollaries 3.16 and 4.17 for the complete list). We just point out in this introduction two of those results (see Theorem 3.9 and Corollary 3.15). The first one identifies the torsion pairs with cogenerating torsion class for which the heart is AB3 and has a projective generator (see also Theorem 3.11 for its dual).

THEOREM 1.4. *Let $\mathcal{A}$ be an abelian category and let $\mathbf{t} = (\mathcal{T}, \mathcal{F})$ be a torsion pair in $\mathcal{A}$. The following assertions are equivalent:*

(1) $\mathbf{t}$ *is a tilting torsion pair.*
(2) $\mathbf{t}$ *is a co-faithful torsion pair whose heart $\mathcal{H}_\mathbf{t}$ is an AB3 abelian category with a projective generator.*

In this case, V is 1-tilting object such that $\mathcal{T} = Gen(V)$ if, and only if, $V[0]$ is a projective generator of $\mathcal{H}_\mathbf{t}$. Moreover, an object P of $\mathcal{H}_\mathbf{t}$ is a projective generator of this latter category if, and only if, it is isomorphic to $V[0]$ for some 1-tilting object V of $\mathcal{A}$ such that $\mathcal{T} = Gen(V)$.

The next one characterizes when a co-faithful torsion pair has a heart which is a module category (see Corollary 3.15).

COROLLARY 1.5. *Let $\mathcal{A}$ be an abelian category and let $\mathbf{t} = (\mathcal{T}, \mathcal{F})$ be a torsion pair in $\mathcal{A}$. The following assertions are equivalent:*

(1) *There is a classical 1-tilting set $\mathcal{T}_0$ (resp. a classical 1-tilting object V) such that $\mathcal{T} = Gen(\mathcal{T}_0)$ (resp. $\mathcal{T} = Gen(V)$).*
(2) $\mathbf{t}$ *is a co-faithful torsion pair whose heart $\mathcal{H}_\mathbf{t}$ is equivalent to the module category over a small pre-additive category (resp. over a ring).*

One of the consequences of our new visit to the HRS tilting process is a new easy proof of the $n = 1$ case of Positselsky-Stovicek tilting-cotilting correspondence (see Corollary 3.17):

COROLLARY 1.6 (Positselski-Stovicek). *The HRS tilting process gives a one-to-one correspondence between:*

(1) *The pairs* $(\mathcal{A}, \mathbf{t})$ *consisting of an AB3* abelian category* $\mathcal{A}$ *with an injective cogenerator and a tilting torsion pair* $\mathbf{t}$ *in* $\mathcal{A}$*;*
(2) *The pairs* $(\mathcal{B}, \bar{\mathbf{t}})$ *consisting of an AB3 abelian category* $\mathcal{B}$ *with a projective generator and a cotilting torsion pair* $\bar{\mathbf{t}}$ *in* $\mathcal{B}$.

On what concerns Question 1.3, we will re-prove, by completely different methods (see Theorem 4.18), the 'only if' implication, which is the hardest one, of the following earlier result:

THEOREM 1.7. ([**PS2**, Theorem 1.2]) *Let* $\mathcal{G}$ *be a Grothendieck category, let* $\mathbf{t} = (\mathcal{T}, \mathcal{F})$ *be a torsion pair in* $\mathcal{G}$ *and let* $\mathcal{H}_{\mathbf{t}}$ *be the heart of the associated HRS t-structure in* $\mathcal{D}(\mathcal{G})$*. Then* $\mathcal{H}_{\mathbf{t}}$ *is a Grothendieck category if, and only if, the torsionfree class* $\mathcal{F}$ *is closed under taking direct limits in* $\mathcal{G}$.

The reader is referred to Section 5 for a summary of recent results concerning Question 1.2, that go beyond the HRS situation.

The organization of the paper goes as follows. In Section 2 we introduce the main concepts needed for the understanding of the paper, specially torsion pairs in abelian categories and t-structures in triangulated categories, with a look also at the HRS tilting process. It turns out that if one starts with a pair $(\mathcal{A}, \mathbf{t})$ consisting of an abelian category $\mathcal{A}$ and a torsion pair $\mathbf{t}$, then the new abelian category obtained by tilting $\mathcal{A}$ with respect to $\mathbf{t}$ need not have *Hom* sets. Corollary 2.13 gives the precise conditions to get *Hom* sets. In Section 3 we study when the heart of (the HRS t-structure associated to) a torsion pair in an abelian category has either a projective generator or an injective cogenerator. This leads naturally to quasi-(co)tilting torsion pairs (see Proposition 3.8) in abelian categories. Then expanded versions of Theorem 1.4 and its dual are proved (see Theorems 3.9 and 3.11). As a particular case, we then give an expanded version Corollary 1.5 (see Corollary 3.15) that characterizes when the heart of a co-faithful torsion pair is a module category. We end the section by giving a series of bijections induced by the HRS tilting process (see Proposition 3.16), from which Corollary 1.6 is deduced. In Section 4 we study Question 1.3. Subsections 4.1 and 4.2 are dedicated to show the milestones of the route that led to the solution of the problem in [**PS2**], i.e. to the proof of Theorem 1.7. Subsection 4.3 briefly summarizes recent results by Bazzoni, Herzog, Prihoda, Saroch and Trlifaj about the same question in the particular case when the torsion pair is tilting. We end the section by re-proving, using a recent characterization of the AB5 condition by Positselski and Stovicek (see [**Po-St**]), the fact that if the heart of a torsion pair in a Grothendieck category is itself a Grothendieck category, then the torsion pair is of finite type. The final Section 5 shows the most recent results and the present state of Question 1.2.

2. Preliminaries

In the rest of the paper, whenever $\mathcal{C}$ is an additive category and $\mathcal{X}$ is any class of objects, we shall denote by $\mathcal{X}^{\perp}$ (resp. ${}^{\perp}\mathcal{X}$) the full subcategory consisting of the objects Y such that $\mathcal{C}(X,Y)=0$ (resp. $\mathcal{C}(Y,X)=0$), for all $X\in\mathcal{X}$. When $\mathcal{X}=\{X\}$ for simplicity we will write $X^{\perp}$ (resp. ${}^{\perp}X$) instead of $\mathcal{X}^{\perp}$ (resp. ${}^{\perp}\mathcal{X}$).

Unless explicitly said otherwise, in the rest of the paper the letter $\mathcal{A}$ will denote an abelian category.

2.1. Torsion pairs.

DEFINITION 2.1. A *torsion pair* in $\mathcal{A}$ is a pair $\mathbf{t}=(\mathcal{T},\mathcal{F})$ of full subcategories satisfying the following two conditions:

1) $\mathcal{A}(T,F)=0$, for all $T\in\mathcal{T}$ and $F\in\mathcal{F}$;
2) For each object X of $\mathcal{A}$, there is an exact sequence

$$0\longrightarrow T_X\longrightarrow X\longrightarrow F_X\longrightarrow 0$$

where $T_X\in\mathcal{T}$ and $F_X\in\mathcal{F}$.

A *torsion class* in $\mathcal{A}$ is a class of objects $\mathcal{T}$ that appears as first component of a torsion pair in $\mathcal{A}$. A *torsionfree class* $\mathcal{F}$ is defined dually. Note that in a torsion pair we have $\mathcal{F}=\mathcal{T}^{\perp}$ and $\mathcal{T}={}^{\perp}\mathcal{F}$. On the other hand, in the sequence above T_X and F_X depend functorially on X, so that the assignment $X\rightsquigarrow T_X$ (resp. $X\rightsquigarrow F_X$) underlies a functor $t:\mathcal{A}\to\mathcal{T}$ (resp. $(1:t):\mathcal{A}\to\mathcal{F}$), which is right (resp. left) adjoint of the inclusion functor $\iota_{\mathcal{T}}:\mathcal{T}\hookrightarrow\mathcal{A}$ (resp. $\iota_{\mathcal{F}}:\mathcal{F}\hookrightarrow\mathcal{A}$). The composition $\iota_{\mathcal{T}}\circ t:\mathcal{A}\to\mathcal{A}$ (resp. $\iota_{\mathcal{F}}\circ(1:t):\mathcal{A}\to\mathcal{A}$), which we will still denote by t (resp. $(1:t)$), is called the *torsion radical* (resp. *torsion coradical*) associated to $\mathbf{t}$.

In particular situations, torsion and torsionfree classes are identified by the satisfaction of some closure properties. Recall that an abelian category is called *locally small* when the subobjects of any given object form a set.

PROPOSITION 2.2. *Let $\mathcal{A}$ be an abelian category and let $\mathcal{T}\subseteq\mathcal{A}$ (resp. $\mathcal{F}\subseteq\mathcal{A}$) be a full subcategory. Consider the following assertions:*

(1) *$\mathcal{T}$ (resp. $\mathcal{F}$) is a torsion (resp. torsionfree) class;*
(2) *$\mathcal{T}$ (resp. $\mathcal{F}$) is closed under taking quotients (=epimorphic images) (resp. subobjects), extensions and coproducts (resp. products), when these exist in $\mathcal{A}$.*

The implication (1) $\Longrightarrow$ (2) *holds true. When $\mathcal{A}$ is AB3 (resp. AB3*) and locally small, also* (2) $\Longrightarrow$ (1) *holds.*

PROOF. The implication (1) $\Longrightarrow$ (2) follows from the equalities $\mathcal{F}=\mathcal{T}^{\perp}$ and $\mathcal{T}={}^{\perp}\mathcal{F}$. For (2) $\Longrightarrow$ (1) see [**S**, Proposition VI.2.1]. □

Recall that, in any category $\mathcal{C}$, a class of objects $\mathcal{X}$ is called a *generating (resp. cogenerating) class* when, for each object $C\in\mathrm{Ob}(\mathcal{C})$, there is an epimorphism $X_C\twoheadrightarrow C$ (resp. monomorphism $C\rightarrowtail X_C$), for some object $X_C\in\mathcal{X}$.

We recall some particular cases of torsion pairs:

DEFINITION 2.3. Let $\mathbf{t}=(\mathcal{T},\mathcal{F})$ be a torsion pair in $\mathcal{A}$. We will say that $\mathbf{t}$ is:

(1) *faithful* (resp. *co-faithful*) when $\mathcal{F}$ (resp. $\mathcal{T}$) is a generating (resp. cogenerating) class of $\mathcal{A}$.
(2) *of finite type* when direct limits in $\mathcal{A}$ of objects in $\mathcal{F}$ exist and are in $\mathcal{F}$.

REMARK 2.4. In [**HRS**] faithful (resp. co-faithful) torsion pairs are called cotilting (resp. tilting). In this paper we separate from that terminology, reserving the term 'cotilting' (resp. 'tilting') for torsion pairs defined by 1-cotilting (resp. 1-tilting) objects (see Definition 3.4).

2.2. t-Structures. In the sequel, the letter $\mathcal{D}$ will denote a triangulated category and $?[1] : \mathcal{D} \to \mathcal{D}$ will be its suspension functor. Moreover, we put $?[0] = 1_{\mathcal{D}}$ and $?[k]$ the k-th power of $?[1]$, for each integer k. We will denote the triangles on $\mathcal{D}$ by $X \longrightarrow Y \longrightarrow Z \stackrel{+}{\longrightarrow}$, or also $X \longrightarrow Y \longrightarrow Z \longrightarrow X[1]$.

DEFINITION 2.5. Let $(\mathcal{D}, ?[1])$ be a triangulated category. A *t-structure* on $\mathcal{D}$ is a couple of full subcategories closed under direct summands $(\mathcal{U}, \mathcal{W})$ such that:

(1) $\mathcal{U}[1] \subseteq \mathcal{U}$;
(2) $\mathcal{D}(U, W[-1]) = 0$, for all $U \in \mathcal{U}$ and $W \in \mathcal{W}$;
(3) For each $X \in \mathcal{D}$, there is a distinguished triangle:
$$U_X \to X \to V_X \to U_X[1]$$
with $U_X \in \mathcal{U}$ and $V_X \in \mathcal{W}[-1]$.

In such case, the subcategory $\mathcal{U}$ is called the *aisle* of the t-structure, and $\mathcal{W}$ is called the *coaisle*. Note that in such case, we have $\mathcal{W}[-1] = \mathcal{U}^{\perp}$ and $\mathcal{U} = {}^{\perp}(\mathcal{W}[-1]) = {}^{\perp}(\mathcal{U}^{\perp})$. For this reason, we will write the t-structures using the following notation $(\mathcal{U}, \mathcal{U}^{\perp}[1])$. On the other hand, the objects U_X and V_X in the previous triangle are uniquely determined by X, up to isomorphism, so that the assignment $X \rightsquigarrow U_X$ (resp. $X \rightsquigarrow V_X$) underlies a functor $\tau_{\mathcal{U}}^{\leq} : \mathcal{D} \to \mathcal{U}$ (resp. $\tau_{\mathcal{U}}^{>} : \mathcal{D} \to \mathcal{U}^{\perp}$) which is right (resp. left) adjoint of the inclusion functor $\iota_{\mathcal{U}} : \mathcal{U} \hookrightarrow \mathcal{D}$ (resp. $\iota_{\mathcal{U}^{\perp}} : \mathcal{U}^{\perp} \hookrightarrow \mathcal{D}$). The composition $\iota_{\mathcal{U}} \circ \tau_{\mathcal{U}}^{\leq} : \mathcal{D} \to \mathcal{D}$ (resp. $\iota_{\mathcal{U}^{\perp}} \circ \tau_{\mathcal{U}}^{>} : \mathcal{D} \to \mathcal{D}$), which we will still denote by $\tau_{\mathcal{U}}^{\leq}$ (resp. $\tau_{\mathcal{U}}^{>}$) and it is called the *left truncation* (resp. *right truncation*) functor associated to the t-struture $(\mathcal{U}, \mathcal{U}^{\perp}[1])$. The full subcategory $\mathcal{H} = \mathcal{U} \cap \mathcal{U}^{\perp}[1]$ of $\mathcal{D}$ is called the *heart* of the t-structure and it is an abelian category, where the short exact sequences are the triangles in $\mathcal{D}$ having their three terms in $\mathcal{H}$. In particular, we have $\operatorname{Ext}^1_{\mathcal{H}}(M, N) = \mathcal{D}(M, N[1])$, for all objects M and N in $\mathcal{H}$. Moreover, the canonical morphism $\operatorname{Ext}^2_{\mathcal{H}}(M, N) \to \mathcal{D}(M, N[2])$ is a monomorphism in *Ab*, for all objects $M, N \in \mathcal{H}$ (see [**BBD**, Remarque 3.1.17]).

The kernel and cokernel of a morphism $f : M \to N$ on the heart $\mathcal{H}$ are computed as follows: we complete f to a triangle in $\mathcal{D}$ and consider the following diagram, where the row and column are triangles:

$$\begin{array}{ccccccc}
 & & & & \tau_{\mathcal{U}}^{\leq}(Z[-1])[1] & & \\
 & & & & \downarrow & & \\
M & \stackrel{f}{\longrightarrow} & N & \longrightarrow & Z & \stackrel{+}{\longrightarrow} & \\
 & & & & \downarrow & & \\
 & & & & \tau_{\mathcal{U}}^{>}(Z[-1])[1] & & \\
 & & & & \downarrow{\scriptstyle +} & &
\end{array}$$

From the octahedral axiom, we obtain the following triangles in $\mathcal{D}$

$$\tau_{\mathcal{U}}^{\leq}(Z[-1]) \longrightarrow M \stackrel{p_f}{\longrightarrow} I \stackrel{+}{\longrightarrow} \qquad I \stackrel{\iota_f}{\longrightarrow} N \longrightarrow \tau_{\mathcal{U}}^{>}(Z[-1])[1] \stackrel{+}{\longrightarrow} ,$$

where all terms are in $\mathcal{H}$ (with $\iota_f \circ p_f = f$). Then we have that $\mathrm{Ker}_{\mathcal{H}}(f) = \tau_{\mathcal{U}}^{\leq}(Z[-1])$ and $\mathrm{Coker}_{\mathcal{H}}(f) = \tau_{\mathcal{U}}^{>}(Z[-1])[1]$.

Recall that if $\mathcal{D}$ and $\mathcal{A}$ are a triangulated and an abelian category, respectively, then an additive functor $H : \mathcal{D} \longrightarrow \mathcal{A}$ is called *cohomological* when any triangle $X \xrightarrow{u} Y \xrightarrow{v} Z \xrightarrow{+}$ in $\mathcal{D}$ induces an exact sequence $H(X) \xrightarrow{H(u)} H(Y) \xrightarrow{H(v)} H(Z)$ in $\mathcal{A}$. The following proposition, shown by Beilinson, Bernstein and Deligne in [**BBD**], associates to each t-structure in a triangulated category an intrinsic homology theory.

PROPOSITION 2.6. *Let $(\mathcal{D}, ?[1])$ be a triangulated category. If $\sigma = (\mathcal{U}, \mathcal{U}^{\perp}[1])$ is a t-structure in $\mathcal{D}$, then the assignments $X \rightsquigarrow \tau_{\mathcal{U}}^{\leq}(\tau_{\mathcal{U}}^{>}(X[-1])[1])$ and $X \rightsquigarrow \tau_{\mathcal{U}}^{>}(\tau_{\mathcal{U}}^{\leq}(X)[-1])[1]$ define two naturally isomorphic functors from $\mathcal{D}$ to $\mathcal{H}$, which are cohomological. In the sequel we will fix a (cohomological) functor $H_{\sigma}^{0} : \mathcal{D} \to \mathcal{H}$ that is naturally isomorphic to those two functors.*

EXAMPLES 2.7. The following examples of t-structures will be of great interest in this paper.

(1) Let $\mathcal{A}$ be an abelian category for which $\mathcal{D}(\mathcal{A})$ exists i.e. $\mathcal{D}(\mathcal{A})$ has *Hom* sets. For each $m \in \mathbb{Z}$, we will denote by $\mathcal{D}^{\leq m}(\mathcal{A})$ (resp. $\mathcal{D}^{\geq m}(\mathcal{A})$) the full subcategory of $\mathcal{D}(\mathcal{A})$ consisting of the cochain complexes X such that $H^k(X) = 0$, for all $k > m$ (resp. $k < m$). Moreover, we put $\mathcal{D}^{[a,b]}(\mathcal{A}) := \mathcal{D}^{\leq b}(\mathcal{A}) \cap \mathcal{D}^{\geq a}(\mathcal{A})$ for any integers a and b. Then, the pair $(\mathcal{D}^{\leq m}(\mathcal{A}), \mathcal{D}^{\geq m}(\mathcal{A}))$ is a t-structure in $\mathcal{D}(\mathcal{A})$ whose heart is equivalent to $\mathcal{A}$. The corresponging left and right truncation functors will be denoted by $\tau^{\leq m} : \mathcal{D}(\mathcal{A}) \to \mathcal{D}(\mathcal{A})$ and $\tau^{>m} : \mathcal{D}(\mathcal{A}) \to \mathcal{D}(\mathcal{A})$, respectively. For the case $m = 0$, the corresponding t-structure is known as the *canonical t-structure* in $\mathcal{D}(\mathcal{A})$.

(2) (Happel-Reiten-Smalø) Let $\mathcal{A}$ be an abelian category for which $\mathcal{D}(\mathcal{A})$ exists and let $\mathbf{t} = (\mathcal{T}, \mathcal{F})$ be a torsion pair in $\mathcal{A}$. The classes $\mathcal{U}_{\mathbf{t}} := \{X \in \mathcal{D}^{\leq 0}(\mathcal{A}) : H^0(X) \in \mathcal{T}\}$ and $\mathcal{W}_{\mathbf{t}} = \{X \in \mathcal{D}^{\geq -1}(\mathcal{A}) : H^{-1}(X) \in \mathcal{F}\}$ give rise a t-structure in $\mathcal{D}(\mathcal{A})$, concretely, $(\mathcal{U}_{\mathbf{t}}, \mathcal{W}_{\mathbf{t}}) = (\mathcal{U}_{\mathbf{t}}, \mathcal{U}_{\mathbf{t}}^{\perp}[1])$. It is called the *Happel-Reiten-Smalø t-structure* associated to $\mathbf{t}$. We will denote its heart by $\mathcal{H}_{\mathbf{t}}$. In next subsection we relax the hypothesis that $\mathcal{D}(\mathcal{A})$ has *Hom* sets, showing that the formation of $\mathcal{H}_{\mathbf{t}}$ from $\mathcal{A}$ and $\mathbf{t}$ is still possible sometimes.

(3) Let $\mathcal{D}$ be a triangulated category with coproducts. An object X in $\mathcal{D}$ is called *compact*, when the functor $\mathcal{D}(X, ?) : \mathcal{D} \to Ab$ preserves coproducts. If $\mathcal{S}$ is a set of compact objects in $\mathcal{D}$, then the pair $({}^{\perp}(\mathcal{S}^{\perp_{\leq 0}}), \mathcal{S}^{\perp_{<0}})$ is a t-structure in $\mathcal{D}$ (see [**KN**, Theorem 12.1]). A t-structure is said to be *compactly generated* if it is obtained in this way from a set of compact objects. We put aisle $< \mathcal{S} > := {}^{\perp}(\mathcal{S}^{\perp_{\leq 0}})$, which is the smallest full subcategory of $\mathcal{D}$ that contains $\mathcal{S}$ and is closed under coproducts, extensions and non-negative shifts.

2.3. The Happel-Reiten-Smalø (HRS) tilting process. This process stems from the seminal work in [**BBD**] and was fully developed in [**HRS**]. In our treatment here we will work in a general framework, by allowing ourselves the freedom of working for the moment with big triangulated categories. For that, let us define a *big abelian group* as a (proper) class $\mathbb{A}$ of elements together with a map between classes $+ : \mathbb{A}\times\mathbb{A} \longrightarrow \mathbb{A}$, $(a,b) \rightsquigarrow a+b$, called sum, that satisfies the axioms of usual abelian groups. When we add another map of classes $\cdot : \mathbb{A}\times\mathbb{A} \longrightarrow \mathbb{A}$, $(a,b) \rightsquigarrow ab$, called multiplication, such that $+$ and $\cdot$ satisfy the usual axioms of a(n associative unital) ring, we call $\mathbb{A}$ a *big ring*. Homomorphisms of big abelian groups and of big rings are defined as for usual groups and rings.

A *big preadditive category* $\mathcal{A}$ consists then of a class of objects $\mathrm{Ob}(\mathcal{A})$ and, for each pair $(X,Y) \in \mathrm{Ob}(\mathcal{A})\times\mathrm{Ob}(\mathcal{A})$, a big abelian group of morphisms $\mathcal{A}(X,Y)$, both satisfying the usual axioms for a preadditive category. When such a big preadditive category satisfies the usual axioms of additive, abelian or triangulated categories, we will call it a *big additive*, *big abelian* or *big triangulated* category.

Note that the use of these big categories is also implicit in [**HRS**] since the authors work with the bounded derived category $\mathcal{D}^b(\mathcal{A})$ of an abelian category, which need not have *Hom* sets as Example 2.8 below shows. This use of big additive categories allows more flexibility in the HRS tilting process and, increasing the universe if necessary, will pose no set-theoretical problems. In the rest of the paper we adopt the convention that the term *category* means a category with *Hom* sets. So the expression 'is a category' will mean 'is a category with *Hom* sets'.

The following is an expansion of an example in [**CN**].

EXAMPLE 2.8. For any big ring $\mathbb{A}$ appearing in this example, we consider the category $\mathbb{A}-\mathrm{Mod}$ of small $\mathbb{A}$-modules. Its objects are pairs (M,f) consisting of an abelian group M together with a ring homomorphism $f : \mathbb{A} \longrightarrow \mathrm{End}_{\mathbb{Z}}(M)$. The morphisms $(M,f) \longrightarrow (N,g)$ are the homomorphisms $\varphi : M \longrightarrow N$ of abelian groups such that $\varphi\circ f(a) = g(a)\circ\varphi$, for all $a \in \mathbb{A}$.

Given the class I of all ordinals and an isomorphic class of variables $\{X_\alpha\colon \alpha \in I\}$, we shall associate two big rings. The first one is the big ring of polynomials $R := \mathbb{Z}[X_\alpha\colon \alpha \in I]$, as defined in [**CN**], where we have slightly changed the notation of that paper. The second one is the big free ring $A := \mathbb{Z} < X_\alpha\colon \alpha \in I >$, i.e. its elements are finite $\mathbb{Z}$-linear combinations of words on the alphabet $\{X_\alpha : \alpha \in I\}\cup\{1\}$, and the multiplication extends by $\mathbb{Z}$-linearity the obvious juxtaposition of words with 1 as multiplicative identity. Note that giving a small A-module (M,f) amounts to giving an I-indexed class $(f_\alpha)_{\alpha\in I}$ in $\mathrm{End}_{\mathbb{Z}}(M)$, namely $f_\alpha = f(X_\alpha)$ for all $\alpha \in I$, and theses f_α are required to commutate in the case that (M,f) is a small R-module. The morphisms $\varphi : (M,f) \longrightarrow (N,g)$ (in $R-\mathrm{Mod}$ or $A-\mathrm{Mod}$) are just the homomorphisms of abelian groups $\varphi : M \longrightarrow N$ such that $\varphi\circ f_\alpha = g_\alpha\circ\varphi$, for all $\alpha \in I$.

We have an obvious fully faithful exact embedding $u : R-\mathrm{Mod} \hookrightarrow A-\mathrm{Mod}$. A *trivial small (R- or A-)module* is an (M,f) such that $f_\alpha = 0$, for all $\alpha \in I$. The forgetful functors $R-\mathrm{Mod} \longrightarrow \mathrm{Ab}$ and $A-\mathrm{Mod} \longrightarrow \mathrm{Ab}$ clearly induce an equivalence between each of the respective subcategories of trivial small modules and Ab. By [**CN**, Lemma 1.1], we know that $\mathcal{D}(R-\mathrm{Mod})(\mathbb{Z},\mathbb{Z}[1]) \cong \mathrm{Ext}^1_R(\mathbb{Z},\mathbb{Z})$ is not a set, where $\mathbb{Z}$ is the trivial small R-module associated to $\mathbb{Z}$. Since we have an obvious inclusion

$$\mathcal{D}(R-\mathrm{Mod})(\mathbb{Z},\mathbb{Z}[1]) \cong \mathrm{Ext}^1_R(\mathbb{Z},\mathbb{Z}) \hookrightarrow \mathrm{Ext}^1_A(\mathbb{Z},\mathbb{Z}) \cong \mathcal{D}(A-\mathrm{Mod})(\mathbb{Z},\mathbb{Z}[1])$$

we deduce that both $\mathcal{D}(R-\mathrm{Mod})$ and $\mathcal{D}(A-\mathrm{Mod})$, and even $\mathcal{D}^b(R-\mathrm{Mod})$ and $\mathcal{D}^b(A-\mathrm{Mod})$, are big triangulated categories, i.e. do not have Hom sets.

The following result is the version for big triangulated categories of [**Ma2**, Proposition 3.1.1 and 3.1.4]. Recall that a t-structure $\tau = (\mathcal{U},\mathcal{W})$ is *left (resp.*

right) nondegenerate when $\bigcap_{n\in\mathbb{Z}}\mathcal{U}[n]=0$ (resp. $\bigcap_{n\in\mathbb{Z}}\mathcal{W}[n]=0$), and it is called *nondegenerate* when it is left and right nondegenerate. Mattiello's proof is valid here and proves the result, except for the nondegeneracy of $\tau_{\mathbf{t}}$ and the final part of the statement, that are explicitly proved.

PROPOSITION 2.9. *Let $\mathcal{D}$ any big triangulated category, let $\tau=(\mathcal{U},\mathcal{W})$ be a nondegenerate t-structure in it and denote by $\mathcal{A}$ its heart, which is then a big abelian category, and denote by $H_\tau^0:\mathcal{D}\longrightarrow\mathcal{A}$ the associated cohomological functor. If $\mathbf{t}:=(\mathcal{T},\mathcal{F})$ is a torsion pair in $\mathcal{A}$, then the pair $\tau_{\mathbf{t}}:=(\mathcal{U}_{\mathbf{t}},\mathcal{W}_{\mathbf{t}})$ given by the following classes is again a nondegenerate t-structure in $\mathcal{D}$:*

$$\mathcal{U}_{\mathbf{t}}=\{X\in\mathcal{D}\colon H_\tau^k(X)=0,\ \text{for } k>0,\ \text{and } H_\tau^0(X)\in\mathcal{T}\}$$

$$\mathcal{W}_{\mathbf{t}}=\{Y\in\mathcal{D}\colon H_\tau^k(Y)=0,\ \text{for } k<-1,\ \text{and } H_\tau^{-1}(Y)\in\mathcal{F}\}.$$

Moreover, the pair $\bar{\mathbf{t}}:=(\mathcal{F}[1],\mathcal{T})$ is a torsion pair in the heart $\mathcal{H}_{\mathbf{t}}$ of $\tau_{\mathbf{t}}$ and, for each $M\in\mathcal{H}_{\mathbf{t}}$, the associated torsion sequence is of the form $0\to H_\tau^{-1}(M)[1]\longrightarrow M\longrightarrow H_\tau^0(M)\to 0$.

PROOF. For the nondegeneracy of $\tau_{\mathbf{t}}$, note that we clearly have that $\bigcap_{n\in\mathbb{Z}}\mathcal{U}_{\mathbf{t}}$ consists of the objects X such that $H_\tau^k(X)=0$, for all $k\in\mathbb{Z}$. The nondegeneracy of τ then implies that $X=0$ (see, e.g., [**NSZ**, Lemma 3.3], adapted to big triangulated categories). This gives the left nondegeneracy of $\tau_{\mathbf{t}}$ and the right nondegeneracy follows dually.

Recall that, due to the nondegeneracy of τ, we have that $\mathcal{U}=\{U\in\mathcal{D}\colon H_\tau^i(U)=0$, for all $i>0\}$ and $\mathcal{W}=\{W\in\mathcal{D}\colon H_\tau^i(W)=0$, for all $i<0\}$ (see [**BBD**, Proposition 1.3.7]), and hence $\mathcal{A}$ consists of the objects $A\in\mathcal{D}$ such that $H_\tau^i(A)=0$, for all integers $i\neq 0$. If now $M\in\mathcal{H}_{\mathbf{t}}=\mathcal{U}_{\mathbf{t}}\cap\mathcal{W}_{\mathbf{t}}$ is any object in the heart of $\tau_{\mathbf{t}}$, then $H_\tau^i(M[-1])=H_\tau^{i-1}(M)$ is zero, for all $i<0$, so that $M[-1]\in\mathcal{W}$ and hence $H_\tau^{-1}(M)=H_\tau^0(M[-1])\cong\tau_{\mathcal{U}}^{\leq}(M[-1])$. The associated truncation triangle with respect to τ is then

$$H_\tau^{-1}(M)\longrightarrow M[-1]\longrightarrow T[-1]\xrightarrow{+},$$

where $T\in\mathcal{W}$. By shift, we get the triangle

$$H_\tau^{-1}(M)[1]\longrightarrow M\longrightarrow T\xrightarrow{+}.$$

By taking the long exact sequence in $\mathcal{A}$ obtained by applying to the last triangle the functor $H_\tau^0:\mathcal{D}\longrightarrow\mathcal{A}$, we readily see that $H_\tau^i(T)=0$, for all $i\neq 0$, and hence $T\in\mathcal{A}$, and that the induced morphism $H_\tau^0(M)\longrightarrow H_\tau^0(T)\cong T$ is an isomorphism. We then have a triangle

$$H_\tau^{-1}(M)[1]\longrightarrow M\longrightarrow H_\tau^0(M)\xrightarrow{+}$$

in $\mathcal{D}$ with its three terms in $\mathcal{H}_{\mathbf{t}}$, and hence it gives a short exact sequence in this latter category. □

DEFINITION 2.10. The t-structure $\tau_{\mathbf{t}}$ of last proposition is said to be the *HRS-tilt of τ with respect to* $\mathbf{t}$. The *HRS process* in $\mathcal{D}$ is a 'map' $\Phi_{\mathcal{D}}$ defined on the class of pairs $(\tau,\mathbf{t})$, where τ is a nondegenerate t-structure in $\mathcal{D}$ and $\mathbf{t}$ is a torsion pair in the heart $\mathcal{H}_\tau$ of τ. It is defined by $\Phi_{\mathcal{D}}(\tau,\mathbf{t})=(\tau_{\mathbf{t}},\bar{\mathbf{t}})$.

The following is now a very natural question.

QUESTION 2.11. In the situation of Proposition 2.9, assume that $\mathcal{A}$ has *Hom* sets. When is it true that also $\mathcal{H}_{\mathbf{t}}$ has *Hom* sets?

The following is the answer:

PROPOSITION 2.12. *Let $\tau = (\mathcal{U}, \mathcal{W})$ be a nondegenerate t-structure in the big triangulated category $\mathcal{D}$ such that its heart $\mathcal{A}$ is a category. Let $\mathbf{t} = (\mathcal{T}, \mathcal{F})$ be a torsion pair in $\mathcal{A}$. The following assertions are equivalent:*

(1) *The heart $\mathcal{H}_{\mathbf{t}}$ of the tilted t-structure $\tau_{\mathbf{t}}$ is a category.*
(2) *$\mathcal{D}(T, F[1]) \cong Ext^1_{\mathcal{A}}(T, F)$ is a set, for all $T \in \mathcal{T}$ and all $F \in \mathcal{F}$.*

In such case, if one puts $\Phi^n_{\mathcal{D}}(\tau, \mathbf{t}) =: (\tau_n, \mathbf{t}_n)$, where $\Phi^n_{\mathcal{D}} = \Phi_{\mathcal{D}} \circ \overset{n}{\cdots} \circ \Phi_{\mathcal{D}}$, then the heart of τ_n has Hom sets, for each $n > 0$.

PROOF. (1) $\Longrightarrow$ (2) is clear since, by [**BBD**], we have an isomorphism $\mathcal{H}_{\mathbf{t}}(T, F[1]) \cong \mathcal{D}(T, F[1])$.

(2) $\Longrightarrow$ (1) For simplicity, put $H^k := H_\tau \circ (?[k])$ for each $k \in \mathbb{Z}$. Let M, N be objects of $\mathcal{H}_{\mathbf{t}}$. We have a triangle $H^{-1}(N)[1] \longrightarrow N \longrightarrow H^0(N) \xrightarrow{+}$ (*) in $\mathcal{D}$ (see Proposition 2.9). An application of the cohomological functor $\mathcal{D}(M, ?)$ from $\mathcal{D}$ to the category AB of big abelian groups gives an exact sequence

$$\mathcal{D}(M, H^{-1}(N)[1]) \longrightarrow \mathcal{D}(M, N) \longrightarrow \mathcal{D}(M, H^0(N)),$$

Therefore the proof is reduced to check that $\mathcal{D}(M, N)$ is a set when $N = F[1]$, for some $F \in \mathcal{F}$, or $N = T \in \mathcal{T}$, for some $T \in \mathcal{T}$.

Suppose that $N = T \in \mathcal{T}$. Taking the triangle (*) with M instead of N and applying to it the cohomological functor $\mathcal{D}(?, T)$, we obtain an exact sequence

$$\mathcal{D}(H^0(M), T) \longrightarrow \mathcal{D}(M, T) \longrightarrow \mathcal{D}(H^{-1}(M)[1], T) = 0.$$

But $\mathcal{D}(H^0(M), T) \cong \mathcal{A}(H^0(M), T)$ is a set by the hypothesis on $\mathcal{A}$. It then follows that $\mathcal{D}(M, T)$ is a set.

Suppose that $N = F[1] \in \mathcal{F}[1]$ and apply $\mathcal{D}(?, F[1])$ to the triangle of the previous paragraph. We get the exact sequence

$$0 = \mathcal{D}(H^{-1}(M)[2], F[1]) \rightarrow \mathcal{D}(H^0(M), F[1]) \rightarrow \mathcal{D}(M, F[1]) \rightarrow \mathcal{D}(H^{-1}(M)[1], F[1]).$$

But we have that $\mathcal{D}(H^{-1}(M)[1], F[1]) \cong \mathcal{D}(H^{-1}(M), F) \cong \mathcal{A}(H^{-1}(M), F)$, which is a set due to the hypothesis on $\mathcal{A}$. Then $\mathcal{D}(M, F[1])$ is a set since, by hypothesis, $\mathcal{D}(H^0(M), F[1])$ is a set.

Next we consider the tilted torsion pair $\bar{t} = (\mathcal{F}[1], \mathcal{T})$ in $\mathcal{H}_{\mathbf{t}}$. Then, by [BBD], we have that

$$\mathrm{Ext}^1_{\mathcal{H}_{\mathbf{t}}}(F[1], T) \cong \mathcal{D}(F[1], T[1]) \cong \mathcal{D}(F, T) \cong \mathcal{A}(F, T),$$

which is a set due to the hypothesis on $\mathcal{A}$. Replacing now $\mathcal{A}$ by $\mathcal{H}_{\mathbf{t}}$ in the equivalence of assertions 1 and 2, we get that the tilted t-structucture $\tau_{\bar{\mathbf{t}}}$ with respect to $\bar{\mathbf{t}}$ has a heart $\mathcal{H}_{\bar{\mathbf{t}}}$ which is a category. The last statement of the proposition is then clear. □

As a particular case, when the big triangulated category is $\mathcal{D} = \mathcal{D}(\mathcal{A})$, where $\mathcal{A}$ is an abelian category, one can consider the canonical t-structure $\tau = (\mathcal{D}^{\leq 0}(\mathcal{A}), \mathcal{D}^{\geq 0}(\mathcal{A}))$ as initial one. Its heart is $\mathcal{A}$ and the functor $H^k_\tau : \mathcal{D}(\mathcal{A}) \longrightarrow \mathcal{A}$ is the classical k-th cohomology functor. As a direct consequence of last proposition, we get:

COROLLARY 2.13. *Let $\mathcal{A}$ be an abelian category and let $\mathbf{t} = (\mathcal{T}, \mathcal{F})$ be a torsion pair in $\mathcal{A}$. The following assertions are equivalent:*

(1) *The heart $\mathcal{H}_{\mathbf{t}}$ is a category, i.e. with Hom sets.*
(2) *$Ext^1_{\mathcal{A}}(T, F)$ is a set (as opposite to a proper class), for all $T \in \mathcal{T}$ and $F \in \mathcal{F}$.*

The following shows that condition 2 of last corollary need not hold in general.

EXAMPLE 2.14. Let $A = \mathbb{Z} < X_\alpha \colon \alpha \in I >$ be as in Example 2.8 and consider the torsion pair $\mathbf{t}$ in $A - \text{Mod}$ generated by the trivial small A-module $\mathbb{Z}$, i.e. $\mathbf{t} = (\mathcal{T}, \mathcal{F}) := ({}^{\perp}(\mathbb{Z}^{\perp}), \mathbb{Z}^{\perp})$. The heart $\mathcal{H}_{\mathbf{t}}$ of the associated Happel-Reiten-Smalø t-structure in $\mathcal{D}(A - \text{Mod})$ is a big abelian category, i.e. it does not have Hom sets.

PROOF. In order to help with the intution, given a small A-module (M, f), we put $X_\alpha m := f(X_\alpha)(m)$, for all $m \in M$ and $\alpha \in I$. A trivial small A-module is then one such that $X_\alpha M = 0$, for all $\alpha \in I$. It is clear that (F, f) is in $\mathcal{F}$ if and only if it does not contain any nonzero trivial A-submodule.

We next consider two (big) ideals J and J_β of A, the second one depending on the ordinal $\beta \in I \setminus \{0\}$. The ideal J is generated by all variables X_α, with $\alpha \neq 0$, while J_β is generated by the set

$$\{X_\alpha \colon \alpha \in I \setminus \{0, \beta\}\} \cup \{X_\beta^2, X_\beta X_0\}.$$

It is clear that A/J is isomorphic to $\mathbb{Z}[X_0]$ and that it is an A-module in $\mathcal{F}$ for none of its nonzero elements is annihilated by X_0. On the other hand, the underlying abelian group of A/J_β is free with basis $\{X_0^n \colon n \in \mathbb{N}\} \cup \{X_0^n X_\beta \colon n \in \mathbb{N}\}$. We then get a morphism $\varphi : A/J \oplus A/J \cong \mathbb{Z}[X_0] \oplus \mathbb{Z}[X_0] \longrightarrow A/J_\beta$ in $A - \text{Mod}$, that takes $(P(X_0), Q(X_0)) \rightsquigarrow P(X_0)X_0 + Q(X_0)X_\beta$ and whose cokernel is the trivial small A-module $\mathbb{Z}$. Note that if $(P(X_0), Q(X_0)) \in \text{Ker}(\varphi)$, so that $P(X_0)X_0 + Q(X_0)X_\beta = 0$, we then get that $(P(X_0), Q(X_0)) = (0, 0)$ by considering the $\mathbb{Z}$-basis of A/J_β given above. Therefore φ is a monomorphism.

The last paragraph shows that, for each $\beta \in I \setminus \{0\}$, we have an exact sequence

$$0 \to A/J \oplus A/J \longrightarrow A/J_\beta \longrightarrow \mathbb{Z} \to 0$$

in $A - \text{Mod}$. It is clear that, for $\beta \neq \gamma$ in $I \setminus \{0\}$, the left A-modules A/J_β and A/J_γ cannot be isomorphic since they have different annihilators. It then follows that $\text{Ext}^1_A(\mathbb{Z}, A/J \oplus A/J)$ is not a set, which implies that $\mathcal{H}_{\mathbf{t}}$ does not have Hom sets by Corollary 2.13. □

LEMMA 2.15. *If V is an object of an abelian category $\mathcal{A}$ such that all coproducts of copies of V exist in $\mathcal{A}$ and $Ext^1_{\mathcal{A}}(V, X)$ is a set, for all $X \in \mathcal{A}$, then the canonical map $Ext^1_{\mathcal{A}}(V^{(I)}, X) \longrightarrow Ext^1_{\mathcal{A}}(V, X)^I$ is a monomorphism, and hence $Ext^1_{\mathcal{A}}(V^{(I)}, X)$ is a set, for each $X \in \mathcal{A}$ and I set.*

PROOF. Let X be an object in $\mathcal{A}$ and let I be a set. We consider $\epsilon : 0 \longrightarrow X \longrightarrow M \stackrel{f}{\longrightarrow} V^{(I)} \longrightarrow 0$ an extension in $\mathcal{A}$ such that $\phi(\epsilon) = 0$, where ϕ denote the respective canonical assignment $\text{Ext}^1_{\mathcal{A}}(V^{(I)}, X) \longrightarrow \text{Ext}^1_{\mathcal{A}}(V, X)^I$. We will show that ϵ is a split exact sequence. Indeed, for each $j \in I$, the j-th inclusion $\iota_j : V \longrightarrow V^{(I)}$ factors through $f : M \longrightarrow V^{(I)}$ since $\phi(\epsilon) = 0$. We then get a morphism $g_j : V \longrightarrow M$ such that $f \circ g_j = \iota_j$. When j varies in I, the universal property of coproducts yields a unique morphism $g : V^{(I)} \longrightarrow M$ such that $g \circ \iota_j = g_j$, for all

$j \in I$. It follows that $f \circ g \circ \iota_j = \iota_j$, for all $j \in I$, which implies that $f \circ g = 1_{V^{(I)}}$. Then ϵ is a split sequence, and so $\epsilon = 0$ in $\mathrm{Ext}^1_{\mathcal{A}}(V^{(I)}, X)$. □

In order to exhibit a very useful direct consequence of Corollary 2.13, we need to introduce a few subcategories associated to an object that will play an important role through the paper.

DEFINITION 2.16. Let $\mathcal{A}$ be an abelian category and let X and V be objects of $\mathcal{A}$, where we asume that all (set-indexed) coproducts of copies of V exist in $\mathcal{A}$. We will say that X is *V-generated* (resp. *V-presented*) when there is an epimorphism of the form $V^{(I)} \twoheadrightarrow X$ (resp. an exact sequence $V^{(J)} \longrightarrow V^{(I)} \twoheadrightarrow X$) for some set I (resp. sets I and J). We will denote by $\mathrm{Gen}(V)$ and $\mathrm{Pres}(V)$ the classes of V-generated and V-presented objects, respectively.

When $Q \in \mathrm{Ob}(\mathcal{A})$ is such that all products of copies of Q exist in $\mathcal{A}$, we get the dual notions of *Q-cogenerated* and *Q-copresented* object, and the corresponding subcategories $\mathrm{Cogen}(Q)$ and $\mathrm{Copres}(Q)$.

COROLLARY 2.17. *In the situation of last corollary, suppose that $\mathcal{T} = Pres(V)$ (resp. $\mathcal{F} = Copres(Q)$), for some object V (resp. Q) of $\mathcal{A}$ such that all coproducts (resp. products) of copies of V (resp. Q) exist in $\mathcal{A}$. The heart $\mathcal{H}_{\mathbf{t}}$ is a category if, and only if, $Ext^1_{\mathcal{A}}(V, F)$ (resp. $Ext^1_{\mathcal{A}}(T, Q)$) is a set, for all $F \in \mathcal{F}$ (resp. $T \in \mathcal{T}$).*

PROOF. The statement for Q is dual of the statement for V, so we just prove the latter one. Note that if $\mathcal{T} = \mathrm{Pres}(V)$ then the equality $\mathcal{T} = \mathrm{Gen}(V)$ also holds. Let's take any $T \in \mathcal{T}$ and consider an exact sequence $0 \to T' \longrightarrow V^{(I)} \longrightarrow T \to 0$, with $T' \in \mathcal{T}$, for some set I. If $F \in \mathcal{F}$ is any object and we apply the contravariant functor $\mathcal{A}(?, F)$, we obtain the exact sequence

$$0 = \mathcal{A}(T', F) \longrightarrow \mathrm{Ext}^1_{\mathcal{A}}(T, F) \longrightarrow \mathrm{Ext}^1_{\mathcal{A}}(V^{(I)}, F)$$

and, by Lemma 2.15, we deduce that $\mathrm{Ext}^1_{\mathcal{A}}(T, F)$ is a set, for all $T \in \mathcal{T}$. Now Corollary 2.13 applies. □

This allows us a re-interpretation of the HRS process, where the map Φ acts instead on pairs $(\mathcal{A}, \mathbf{t})$, where $\mathcal{A}$ is an abelian category and $\mathbf{t} = (\mathcal{T}, \mathcal{F})$ is torsion pair in $\mathcal{A}$ satisfying the equivalent conditions of last corollary. Concretely:

DEFINITION 2.18. A torsion pair $\mathbf{t} = (\mathcal{T}, \mathcal{F})$ in an abelian category $\mathcal{A}$ will be an *adequate torsion pair* when $\mathrm{Ext}^1_{\mathcal{A}}(T, F)$ is a set, as opposite to a proper class, for all $T \in \mathcal{T}$ and $F \in \mathcal{F}$. We will denote by $(\mathbf{AB}, \mathbf{tor})$ the class of pairs $(\mathcal{A}, \mathbf{t})$ consisting of an abelian category $\mathcal{A}$ and an adequate torsion pair $\mathbf{t}$ in it. The *Happel-Reiten-Smalø (HSR) tilting process* is the map $\Phi : (\mathbf{AB}, \mathbf{tor}) \longrightarrow (\mathbf{AB}, \mathbf{tor})$ given by $\Phi[(\mathcal{A}, \mathbf{t})] = (\mathcal{H}_{\mathbf{t}}, \bar{\mathbf{t}})$, where $\mathcal{H}_{\mathbf{t}}$ is the heart of the HRS tilt $\tau_{\mathbf{t}}$ of the canonical t-structure of $\mathcal{D}(\mathcal{A})$ with respect to $\mathbf{t}$ (see Definition 2.10).

In particular if, under the hypotheses of last corollary, we assume that $\mathrm{Ext}^1_{\mathcal{A}}(T, F)$ is a set, for all $T \in \mathcal{T}$ and all $F \in \mathcal{F}$, then, by HRS-tilting iteration, one gets gets the following diagram of abelian categories, all of them with *Hom* sets, and torsion pairs:

$$\mathcal{A} \qquad\qquad \mathcal{H}_{\mathbf{t}} \qquad\qquad \mathcal{H}_{\bar{\mathbf{t}}}$$

$$\mathbf{t} = (\mathcal{T}, \mathcal{F}) \rightsquigarrow \bar{\mathbf{t}} = (\mathcal{F}[1], \mathcal{T}[0]) \rightsquigarrow \bar{\bar{\mathbf{t}}} = (\mathcal{T}[1], \mathcal{F}[1])$$

The following is [**HRS**, Proposition 3.2].

PROPOSITION 2.19. *Let $\Phi : (\mathbf{AB}, \mathbf{tor}) \longrightarrow (\mathbf{AB}, \mathbf{tor})$ be the HRS tilting process map (see Definition 2.18). Let $(\mathcal{A}, \mathbf{t})$ be in $(\mathbf{AB}, \mathbf{tor})$ and put $(\mathcal{B}, \bar{\mathbf{t}}) := \Phi[(\mathcal{A}, \mathbf{t})]$. The torsion pair $\mathbf{t}$ is faithful (resp. co-faitful) if, and only if, $\bar{\mathbf{t}}$ is co-faithful (res. faithful). In particular we have induced maps by restriction*

$$(\mathbf{AB}, \mathbf{tor}_{faithful}) \underset{\phi}{\overset{\phi}{\rightleftarrows}} (\mathbf{AB}, \mathbf{tor}_{cofaithful}) \ ,$$

where $\mathbf{tor}_{faithful}$ (resp. $\mathbf{tor}_{cofaithful}$) denotes the class of adequate faithful (resp. co-faithful) torsion pairs.

Furthermore, we have a triangle functor $G : \mathcal{D}^b(\mathcal{H}_{\mathbf{t}}) \to \mathcal{D}^b(\mathcal{A})$ whose restriction to $\mathcal{H}_{\mathbf{t}}$ is naturally isomorphic to the inclusion functor $\mathcal{H}_{\mathbf{t}} \hookrightarrow \mathcal{D}^b(\mathcal{A})$ (see [**BBD**, Proposition 3.1.10]). The functor G is usually called the *realization functor.* The following proposition shows that in some cases the heart $\mathcal{H}_{\bar{\mathbf{t}}}$ is equivalent to $\mathcal{A}[1]$.

PROPOSITION 2.20. *[HRS, Proposition 3.4] Let $\mathcal{A}$ be an abelian category and let $\mathbf{t} = (\mathcal{T}, \mathcal{F})$ be a torsion pair in $\mathcal{A}$ such that $\mathcal{T}$ is a cogenerating class. The following assertions hold:*

1. *If $\mathcal{H}_{\mathbf{t}}$ has enough projectives, then $\mathcal{H}_{\bar{\mathbf{t}}} \cong \mathcal{A}[1]$ via the realization functor;*
2. *If $\mathcal{A}$ has enough injectives, then $\mathcal{H}_{\bar{\mathbf{t}}} \cong \mathcal{A}[1]$ via the realization functor.*

In particular, whenever $\mathcal{A}$ is a category with enough projectives or with enough injectives, $\Phi^2[(\mathcal{A}, \mathbf{t})] \cong (\mathcal{A}, \mathbf{t})$. That is, if $\Phi^2[(\mathcal{A}, \mathbf{t})] = (\mathcal{A}', \mathbf{t}')$, then there is an equivalence of categories $F : \mathcal{A} \xrightarrow{\cong} \mathcal{A}'$ which takes $\mathbf{t}$ to $\mathbf{t}'$.

3. Projective and injective objects in the heart. Quasi-(co)tilting torsion pairs

3.1. Quasi-(co)tilting objects and torsion pairs. We start with two auxiliary lemmas.

LEMMA 3.1. *Let $\mathcal{A}$ be an abelian category and let X be an object such that all coproducts (resp. products) of copies of X exist in $\mathcal{A}$. Then all coproducts (resp. products) of objects in Pres(X) (resp. Copres(X)) exist in $\mathcal{A}$.*

PROOF. The result for Copres(X) is dual of the one for Pres(X). We just do the latter one. Let $(T_\lambda)_{\lambda \in \Lambda}$ be a family in Pres(X), fix an exact sequence $X^{(J_\lambda)} \xrightarrow{f_\lambda} X^{(I_\lambda)} \longrightarrow T_\lambda \to 0$, with sets J_λ and I_λ, for each $\lambda \in \Lambda$. We then get an induced exact sequence $\coprod_{\lambda \in \Lambda} X^{(J_\lambda)} \xrightarrow{\coprod f_\lambda} \coprod_{\lambda \in \Lambda} X^{(I_\lambda)} \longrightarrow \mathrm{Coker}(\coprod f_\lambda) \to 0$. This gives the following commutative diagram of functors $\mathcal{A} \longrightarrow \mathrm{Ab}$, with exact rows:

$$\begin{array}{ccccccc}
0 \longrightarrow & \mathcal{A}(\mathrm{Coker}(\coprod f_\lambda), ?) & \longrightarrow & \mathcal{A}(\coprod_{\lambda\in\Lambda} X^{(I_\lambda)}, ?) & \longrightarrow & \mathcal{A}(\coprod_{\lambda\in\Lambda} X^{(J_\lambda)}, ?) \\
& \downarrow{\scriptstyle \alpha} & & \downarrow{\scriptstyle \wr} & & \downarrow{\scriptstyle \wr} \\
0 \longrightarrow & \prod_{\lambda\in\Lambda} \mathcal{A}(T_\lambda, ?) & \longrightarrow & \prod_{\lambda\in\Lambda} \mathcal{A}(X^{(I_\lambda)}, ?) & \longrightarrow & \prod_{\lambda\in\Lambda} \mathcal{A}(X^{(J_\lambda)}, ?)
\end{array}$$

where the left vertical arrow α exists by the universal property of kernels in Ab. By definition of coproducts, the two right vertical arrows are isomorphisms, which in turn implies that α an isomorphism. By Yoneda's lemma, for each $\mu \in \Lambda$, the composition $\mathcal{A}(\mathrm{Coker}(\coprod f_\lambda), ?) \xrightarrow{\alpha} \prod_{\lambda\in\Lambda} \mathcal{A}(T_\lambda, ?) \xrightarrow{\pi_\mu} \mathcal{A}(T_\mu, ?)$, where π_μ is the projection, is of the form $u_\mu^* = \mathcal{A}(u_\mu, ?)$, for an unique morphism $u_\mu : T_\mu \longrightarrow \mathrm{Coker}(\coprod f_\lambda)$. It immediately follows that $\mathrm{Coker}(\coprod f_\lambda)$ together with the morphisms $(u_\lambda : T_\lambda \longrightarrow \mathrm{Coker}(\coprod f_\lambda))$ is the coproduct of the T_λ in $\mathcal{A}$. □

LEMMA 3.2. *Let $\mathcal{A}$ be an abelian category and let V (resp. Q) be an object in $\mathcal{A}$ such that all coproducts (resp. products) of copies of V (resp. Q) exist in $\mathcal{A}$. If $Gen(V) \subseteq Ker(Ext^1_{\mathcal{A}}(V, ?))$ (resp. $Cogen(Q) \subseteq Ker(Ext^1_{\mathcal{A}}(?, Q))$), then the class $Gen(V)$ (resp. $Cogen(Q)$) is a torsion (resp. torsionfree) class in $\mathcal{A}$.*

PROOF. We will prove the assertion for $\mathrm{Gen}(V)$, the one for $\mathrm{Cogen}(Q)$ following by duality. We will check that the classes $\mathcal{T} := \mathrm{Gen}(V)$ and $\mathcal{F} := \mathcal{T}^\perp = V^\perp$ form a torsion pair, for which we just need to check condition 2 of Definition 2.1 since condition 1 is clearly satisfied. For any $A \in \mathrm{Ob}(\mathcal{A})$, we may consider the canonical map $\epsilon_A : V^{(\mathcal{A}(V,A))} \longrightarrow A$. This is the unique morphism such that $\epsilon_A \circ \iota_f = f$, where $\iota_f : V \longrightarrow V^{(\mathcal{A}(V,A))}$ is the f-th injection into the coproduct, for all $f \in \mathcal{A}(V, A)$. Its image is usually called the *trace of V in A* and is denoted by $\mathrm{tr}_V(A)$. We then get an exact sequence $0 \to \mathrm{tr}_V(A) \hookrightarrow A \longrightarrow A/\mathrm{tr}_V(A) \to 0$. We clearly have that $\mathrm{tr}_V(A) \in \mathrm{Gen}(V)$. Moreover, we get an induced exact sequence of abelian groups $0 \to \mathcal{A}(V, \mathrm{tr}_V(A)) \xrightarrow{\cong} \mathcal{A}(V, A) \longrightarrow \mathcal{A}(V, A/\mathrm{tr}_V(A)) \longrightarrow \mathrm{Ext}^1_{\mathcal{A}}(V, \mathrm{tr}_V(A)) = 0$. It then follows that $A/\mathrm{tr}_V(A) \in V^\perp = \mathcal{F}$, so that condition 2 of Definition 2.1 is satisfied. □

We are ready to introduce some types of objects which have special importance in the study of the heart of a t-structure. They are generalizations of corresponding notions in module categories.

DEFINITION 3.3. Let $\mathcal{A}$ be an abelian category and let V be an object such that all coproducts of copies of V exist in $\mathcal{A}$. We will say that an object X is *V-subgenerated* when it is isomorphic to a subobject of an object in $\mathrm{Gen}(V)$. The class of V-subgenerated objects will be denoted by $\overline{\mathrm{Gen}}(V)$. On the other hand, the class of objects on $\mathcal{A}$ which are isomorphic to direct summands of (resp. finite) coproducts of copies of V will be denoted by $\mathrm{Add}(V)$ (resp. $\mathrm{add}(V)$). Dually, when Q is an object such that all products of copies of Q exist in $\mathcal{A}$, we call an object *Q-subcogenerated* when it is epimorphic image of an object in $\mathrm{Cogen}(Q)$. We denote by $\underline{\mathrm{Cogen}}(Q)$ and $\mathrm{Prod}(Q)$ the subcategories consisting of Q-subcogenerated objects and objects isomorphic to direct summands of products of copies of Q, respectively.

DEFINITION 3.4. Let $\mathcal{A}$ be an abelian category. An object V (resp. Q) of $\mathcal{A}$ will be called *quasi-tilting* (resp. *quasi-cotilting*) when all coproducts (resp. products) of copies of V (resp. Q) exist in $\mathcal{A}$ and $\mathrm{Gen}(V) = \overline{\mathrm{Gen}}(V) \cap \mathrm{Ker}(\mathrm{Ext}^1_{\mathcal{A}}(V, ?))$.

(resp. $\mathrm{Cogen}(Q) = \underline{\mathrm{Cogen}}(Q) \cap \mathrm{Ker}(\mathrm{Ext}^1_{\mathcal{A}}(?, Q))$). The corresponding torsion pair $\mathbf{t} = (\mathrm{Gen}(V), V^{\perp})$ (resp. $\mathbf{t} = ({}^{\perp}Q, \mathrm{Cogen}(Q))$) (see Lemma 3.2) is called the *quasi-tilting (resp. quasi-cotilting) torsion pair* associated to V (resp. Q).

When, for such a V (resp. Q), one has $\mathrm{Gen}(V) = \mathrm{Ker}(\mathrm{Ext}^1_{\mathcal{A}}(V, ?)$ (resp. $\mathrm{Cogen}(Q) = \mathrm{Ker}(\mathrm{Ext}^1_{\mathcal{A}}(?, Q)))$ and this class is cogenerating (resp. generating) in $\mathcal{A}$, we will say that V (resp. Q) is a *1-tilting* (resp. *1-cotilting*) object. The corresponding torsion pair is called the *tilting* (resp. *cotilting*) torsion pair associated to V (resp. Q).

The proof of the following goes as in module categories (see [**CDT**, Proposition 2.1]).

COROLLARY 3.5. *If $\mathcal{A}$ is an abelian category and V (resp. Q) is a quasi-tilting (resp. quasi-cotilting) object of $\mathcal{A}$, then $Gen(V) = Pres(V)$ (resp. $Cogen(Q) = Copres(Q)$).*

The natural question of when a quasi-tilting (resp. quasi-cotilting) torsion pair has a heart that is a category, i.e. has *Hom* sets, has a clear answer:

COROLLARY 3.6. *Let V (resp. Q) be a quasi-tilting (resp. quasi-cotilting) object of the abelian category $\mathcal{A}$, and let $\mathbf{t} = (\mathcal{T}, \mathcal{F})$ the associated torsion pair in $\mathcal{A}$. The following assertions hold:*

(1) *The heart $\mathcal{H}_{\mathbf{t}}$ is a category (i.e. has Hom sets) if, and only if, $Ext^1_{\mathcal{A}}(V, F)$ (resp. $Ext^1_{\mathcal{A}}(T, Q)$) is a set, for all $F \in \mathcal{F}$ (resp. $T \in \mathcal{T}$).*
(2) *If V (resp. Q) is a 1-tilting (resp. 1-cotilting) object, then $Ext^2_{\mathcal{A}}(V, ?) = 0$ (resp. $Ext^2_{\mathcal{A}}(?, Q) = 0$). One says that the projective (resp. injective) dimension of V (resp. Q) is less or equal than 1.*
(3) *If V (resp. Q) is a 1-tilting (resp. 1-cotilting) object, then $\mathcal{H}_{\mathbf{t}}$ is a category, i.e. it has Hom sets.*

PROOF. (1) It is a direct consequence of Corollaries 2.17 and 3.5.

(2) We just do the proof for V, the one for Q being dual. Let $0 \to M \longrightarrow X \stackrel{f}{\longrightarrow} Y \longrightarrow V \to 0$ be an exact sequence in $\mathcal{A}$, representing an element $\epsilon \in \mathrm{Ext}^2_{\mathcal{A}}(V, M)$. Since $\mathcal{T}$ is a cogenerating class, we can fix a monomorphism $\mu : X \rightarrowtail T$, with $T \in \mathcal{T}$. By taking the pushout of μ and f we immediately get an exact sequence $0 \to M \longrightarrow T \stackrel{g}{\longrightarrow} T' \longrightarrow V \to 0$, where $T, T' \in \mathcal{T}$, which also represents ϵ. But then $\epsilon = 0$ since $\mathrm{Im}(g) \in \mathcal{T} = \mathrm{Ker}(\mathrm{Ext}^1_{\mathcal{A}}(V, ?))$.

(3) Let $F \in \mathcal{F}$ be any object. Using the cogenerating condition of $\mathcal{T}$, we take an exact sequence $0 \to F \longrightarrow T_0 \longrightarrow T_1 \to 0$, where $T_0, T_1 \in \mathcal{T}$. We then get an exact sequence of (in principle big) abelian groups

$$\mathcal{A}(V, T_1) \longrightarrow \mathrm{Ext}^1_{\mathcal{A}}(V, F) \longrightarrow \mathrm{Ext}^1_{\mathcal{A}}(V, T_0) = 0.$$

It then follows that $\mathrm{Ext}^1_{\mathcal{A}}(V, F)$ is a set, which, by Corollary 2.13, implies that $\mathcal{H}_{\mathbf{t}}$ has *Hom* sets. □

3.2. When does the heart of a co-faithful (resp. faithful) torsion pair have a projective generator (resp. injective cogenerator)? To answer the question of the title of this subsection we need a few preliminary results.

LEMMA 3.7. *Let $\mathcal{D}$ be a big triangulated category and $\tau = (\mathcal{U}, \mathcal{W})$ be a nondegenerate t-structure in $\mathcal{D}$ whose heart $\mathcal{A} := \mathcal{U} \cap \mathcal{W}$ is a category, i.e. it has Hom sets. Let $\mathbf{t} = (\mathcal{T}, \mathcal{F})$ be a torsion pair in $\mathcal{A}$ such that $Ext^1_{\mathcal{A}}(T, F) \cong \mathcal{D}(T, F[1])$ is* *a set (as opposite to a proper class), for all $T \in \mathcal{T}$ and all $F \in \mathcal{F}$ (see Proposition 2.12). The following assertions hold, where $\mathcal{H}_{\mathbf{t}}$ denotes the heart of the tilted t-structure $\tau_{\mathbf{t}}$:*

(1) *The functor $(H^0_\tau)_{|\mathcal{H}_{\mathbf{t}}} : \mathcal{H}_{\mathbf{t}} \longrightarrow \mathcal{A}$ is left adjoint of the functor $\mathcal{A} \longrightarrow \mathcal{H}_{\mathbf{t}}$ taking $A \rightsquigarrow t(A)$, where $t : \mathcal{A} \longrightarrow \mathcal{T}$ is the torsion radical associated to $\mathbf{t}$. In particular $(H^0_\tau)_{|\mathcal{H}_{\mathbf{t}}} : \mathcal{H}_{\mathbf{t}} \longrightarrow \mathcal{A}$ preserves all colimits that exist in $\mathcal{H}_{\mathbf{t}}$.*
(2) *The functor $(H^{-1}_\tau)_{|\mathcal{H}_{\mathbf{t}}} : \mathcal{H}_{\mathbf{t}} \longrightarrow \mathcal{A}$ is right adjoint of the functor $\mathcal{A} \longrightarrow \mathcal{H}_{\mathbf{t}}$ taking $A \rightsquigarrow (1 : t)(A)[1]$. In particular $(H^{-1}_\tau)_{|\mathcal{H}_{\mathbf{t}}} : \mathcal{H}_{\mathbf{t}} \longrightarrow \mathcal{A}$ preserves all limits that exist in $\mathcal{H}_{\mathbf{t}}$.*

PROOF. We just prove assertion 1 since assertion 2 follows by duality. By Proposition 2.9, given $M \in \mathcal{H}_{\mathbf{t}}$, we have an exact sequence in $\mathcal{H}_{\mathbf{t}}$

$$0 \to H^{-1}_\tau(M)[1] \longrightarrow M \longrightarrow H^0_\tau(M) \to 0.$$

This sequence is precisely the one associated to the torsion pair $\bar{\mathbf{t}} = (\mathcal{F}[1], \mathcal{T})$. Then the associated torsion radical $\bar{t}$ and coradical $(1 : \bar{t})$ with respect to this torsion pair act on objects as $M \rightsquigarrow \bar{t}(M) = H^{-1}_\tau(M)[1]$ and $M \rightsquigarrow (1 : \bar{t})(M) = H^0_\tau(M)$, respectively. We can then decompose $(H^0_\tau)_{|\mathcal{H}_{\mathbf{t}}} : \mathcal{H}_{\mathbf{t}} \longrightarrow \mathcal{A}$ as the composition $\mathcal{H}_{\mathbf{t}} \overset{(1:\bar{t})}{\longrightarrow} \mathcal{T} \overset{\iota}{\hookrightarrow} \mathcal{A}$, where the right arrow is the inclusion functor. Each of the two functors in this composition has a right adjoint, which implies that $(H^0_\tau)_{|\mathcal{H}_{\mathbf{t}}} : \mathcal{H}_{\mathbf{t}} \longrightarrow \mathcal{A}$ has a right adjoint which is the composition $\mathcal{A} \overset{t}{\longrightarrow} \mathcal{T} \hookrightarrow \mathcal{H}_{\mathbf{t}}$. □

The importance of quasi-(co)tilting objects in the study of hearts of HRS t-structures stems from the following fact:

PROPOSITION 3.8. *Let $\mathcal{A}$ be an abelian category and let $\mathbf{t} = (\mathcal{T}, \mathcal{F})$ be a torsion pair in $\mathcal{A}$. If $\mathcal{H}_{\mathbf{t}}$ is an AB3 (resp. AB3*) abelian category with a projective generator (resp. injective cogenerator) P (resp. E), then $H^0(P)$ (resp. $H^{-1}(E)$) is a quasi-tilting (resp. quasi-cotilting) object and $\mathbf{t}$ is the associated quasi-tilting (resp. quasi-cotilting) torsion pair.*

PROOF. The statement for the injective cogenerator is dual to the one for projective generator. We just do the last one. Let P be as above and put $V := H^0(P)$, and let $P^{(I)}$ denote the coproduct of I copies of it in $\mathcal{H}_{\mathbf{t}}$. We warn that it might not coincide with the corresponding coproduct in $\mathcal{D}(\mathcal{A})$, if this one exists. By applying Lemma 3.7 with $\tau = (\mathcal{D}^{\leq 0}(\mathcal{A}), \mathcal{D}^{\geq 0}(\mathcal{A}))$ the canonical t-structure, we have an isomorphism $H^0(P^{(I)}) \cong H^0(P)^{(I)} = V^{(I)}$ in $\mathcal{A}$, so that all coproducts of copies of V exist in $\mathcal{A}$.

If $T \in \mathcal{T}$ is any object, then, due to the fact that P is a projective generator of $\mathcal{H}_{\mathbf{t}}$, we have an exact sequence $P^{(I)} \longrightarrow P^{(J)} \longrightarrow T[0] \to 0$ in $\mathcal{H}_{\mathbf{t}}$. By last paragraph, we get an exact sequence $H^0(P)^{(I)} \longrightarrow H^0(P)^{(J)} \longrightarrow H^0(T[0]) = T \to 0$ in $\mathcal{A}$. We then get that $\mathcal{T} \subseteq \text{Pres}(V)$, the converse inclusion being obvious. So we have that $\mathcal{T} = \text{Gen}(V) = \text{Pres}(V)$.

Moreover, if we consider the short exact sequence $0 \to H^{-1}(P)[1] \longrightarrow P \longrightarrow V[0] \to 0$ in $\mathcal{H}_{\mathbf{t}}$ and apply to it the long exact sequence of $\text{Ext}^*_{\mathcal{H}_{\mathbf{t}}}(?, T[0])$, we get an exact sequence

$$0 = \mathcal{H}_{\mathbf{t}}(H^{-1}(P)[1], T[0]) \longrightarrow \text{Ext}^1_{\mathcal{H}_{\mathbf{t}}}(V[0], T[0]) \longrightarrow \text{Ext}^1_{\mathcal{H}_{\mathbf{t}}}(P, T[0]) = 0,$$

from which we get that $\mathrm{Ext}^1_{\mathcal{A}}(V,T) \cong \mathrm{Ext}^1_{\mathcal{H}_{\mathbf{t}}}(V[0],T[0]) = 0$, for all $T \in \mathcal{T}$. It then follows that $\mathcal{T} \subseteq \mathrm{Ker}(\mathrm{Ext}^1_{\mathcal{A}}(V,?)) =: V^{\perp_1}$, and so $\mathcal{T} = \mathrm{Gen}(V) \subseteq \overline{\mathrm{Gen}}(V) \cap \mathrm{Ker}(\mathrm{Ext}^1_{\mathcal{A}}(V,?))$.

For the reverse inclusion, given $M \in \overline{\mathrm{Gen}}(V) \cap V^{\perp_1}$, there exist T_1, $T_2 \in \mathcal{T}$ and an exact sequence in $\mathcal{A}$ as follows:

$$0 \longrightarrow M \longrightarrow T_1 \longrightarrow T_2 \longrightarrow 0.$$

Since $\mathrm{Pres}(V) = \mathrm{Gen}(V) = \mathcal{T}$, we can take an epimorphism $q\colon V^{(\alpha)} \to T_2$ whose kernel belongs to $\mathcal{T}$. Consider the following pullback diagram

$$\begin{array}{ccccccccc} 0 & \longrightarrow & M & \longrightarrow & Z & \longrightarrow & V^{(\alpha)} & \longrightarrow & 0 \\ & & \| & & \downarrow & \text{P.B.} & \downarrow q & & \\ 0 & \longrightarrow & M & \longrightarrow & T_1 & \longrightarrow & T_2 & \longrightarrow & 0 \end{array}$$

Notice that Z is an extension of T_1 and the kernel of q, so that $Z \in \mathcal{T}$. Taking into account that $M \in V^{\perp_1} = \mathrm{Ker}(\mathrm{Ext}^1_{\mathcal{A}}(V,?)) = \mathrm{Ker}(\mathrm{Ext}^1_{\mathcal{A}}(V^{(I)},?))$, for each set $I \neq \emptyset$, we get that the first row in the diagram splits, so that $M \in \mathcal{T}$. □

A first lesson of last proposition is that, in order to identify torsion pairs whose associated heart is a Grothendieck category, one can restrict to the quasi-cotilting ones. The proposition also helps in the following answer to the title of the subsection:

THEOREM 3.9. *Let $\mathcal{A}$ be an abelian category and let $\mathbf{t} = (\mathcal{T},\mathcal{F})$ be a torsion pair in $\mathcal{A}$. The following assertions are equivalent:*

(1) $\mathbf{t}$ *is a tilting torsion pair.*
(2) $\mathbf{t}$ *is a co-faithful torsion pair whose heart $\mathcal{H}_{\mathbf{t}}$ is an AB3 abelian category with a projective generator.*
(3) $\mathcal{H}_{\mathbf{t}}$ *is an AB3 abelian category with a projective generator and $\bar{\mathbf{t}} = (\mathcal{F}[1],\mathcal{T}[0])$ is a faithful torsion pair in $\mathcal{H}_{\mathbf{t}}$.*

In this case, V is a 1-tilting object such that $\mathcal{T} = Gen(V)$ if, and only if, $V[0]$ is a projective generator of $\mathcal{H}_{\mathbf{t}}$. Moreover, an object P of $\mathcal{H}_{\mathbf{t}}$ is a projective generator of this latter category if, and only if, it is isomorphic to $V[0]$ for some 1-tilting object V of $\mathcal{A}$ such that $\mathcal{T} = Gen(V)$.

PROOF. Note that in any of assertions (1)-(3) the class $\mathcal{T}$ is cogenerating in $\mathcal{A}$. This is clear in assertions (1) and (2), and for assertion (3) it follows from Proposition 2.19.

(2) $\iff$ (3) is a consequence of this last mentioned proposition (= [**HRS**, Proposition 3.2]).

(1) $\Longrightarrow$ (2) Let V be a 1-tilting object of $\mathcal{A}$ such that $\mathbf{t} = (\mathrm{Gen}(V),V^{\perp})$. We start by proving that $V[0]$ is a projective object of $\mathcal{H}_{\mathbf{t}}$, i.e. that $\mathrm{Ext}^1_{\mathcal{H}_{\mathbf{t}}}(V[0],M) = 0$, for all $M \in \mathcal{H}_{\mathbf{t}}$. But, taking into account the associated exact sequence $0 \to H^{-1}(M)[1] \longrightarrow M \longrightarrow H^0(M)[0] \to 0$, the task reduces to the case when $M \in \mathcal{T}[0] \cup \mathcal{F}[1]$. If $M = T[0]$, with $T \in \mathcal{T} = \mathrm{Ker}(\mathrm{Ext}^1_{\mathcal{A}}(V,?))$, then we have $\mathrm{Ext}^1_{\mathcal{H}_{\mathbf{t}}}(V[0],T[0]) \cong \mathrm{Ext}^1_{\mathcal{A}}(V,T) = 0$. On the other hand, if $F \in \mathcal{F}$ we have $\mathrm{Ext}^1_{\mathcal{H}_{\mathbf{t}}}(V[0],F[1]) \cong \mathrm{Ext}^2_{\mathcal{A}}(V,F) = 0$ (see [**BBD**, Remarque 3.1.17] and Corollary 3.6). Note that what we have done with V can be done with $V^{(I)}$, for any set $I \neq \emptyset$.

That is, the argument also proves that $V^{(I)}[0]$ is a projective object of $\mathcal{H}_{\mathbf{t}}$, for for all sets I.

Lemma 3.10 below says now that the stalk complex $V^{(I)}[0]$ is the coproduct in $\mathcal{H}_{\mathbf{t}}$ of I copies of $V[0]$. Moreover $\mathcal{T}[0]$ is a generating class in $\mathcal{H}_{\mathbf{t}}$ since $\bar{\mathbf{t}} = (\mathcal{F}[1], \mathcal{T}[0])$ is a faithful torsion pair due to Proposition 2.19. By the equality $\mathcal{T} = \mathrm{Pres}_{\mathcal{A}}(V)$, we then get that $\mathcal{T}[0] \subseteq \mathrm{Gen}_{\mathcal{H}_{\mathbf{t}}}(V[0])$, from which one immediately gets that $\mathcal{H}_{\mathbf{t}} = \mathrm{Gen}_{\mathcal{H}_{\mathbf{t}}}(V[0]) = \mathrm{Pres}_{\mathcal{H}_{\mathbf{t}}}(V[0])$. Applying now Lemma 3.1, we conclude that arbitrary coproducts exist in $\mathcal{H}_{\mathbf{t}}$, so that this is an AB3 abelian category, with $V[0]$ as a projective generator.

(2) $\Longleftrightarrow$ (3) $\Longrightarrow$ (1) By Proposition 3.8 we know that $\mathbf{t}$ is a quasi-tilting torsion pair. Let V be a quasi-tilting object such that $\mathcal{T} = \mathrm{Gen}(V)$. Since $\mathbf{t}$ is co-faithful, i.e. $\mathcal{T}$ is a cogenerating class in $\mathcal{A}$ we get that $\overline{\mathrm{Gen}}(V) = \mathcal{A}$, which then implies that $\overline{\mathrm{Gen}}(V) \cap \mathrm{Ker}(\mathrm{Ext}^1_{\mathcal{A}}(V, ?)) = \mathrm{Ker}(\mathrm{Ext}^1_{\mathcal{A}}(V, ?))$, so that $\mathrm{Gen}(V) = \mathrm{Ker}(\mathrm{Ext}^1_{\mathcal{A}}(V, ?))$ and, hence, V is a 1-tilting object.

For the final statement, the proof of implication (1) $\Longrightarrow$ (2) shows that if V is a 1-tilting object of $\mathcal{A}$ defining $\mathbf{t}$, then $V[0]$ is a projective generator of $\mathcal{H}_{\mathbf{t}}$. It remains to prove that if P is projective generator of $\mathcal{H}_{\mathbf{t}}$ then $P \cong V[0]$ for such a 1-tilting object. By Proposition 3.8 we know that $V := H^0(P)$ is a quasi-tilting object associated to $\mathbf{t}$, and by the argument in (2) $\Longrightarrow$ (1), it is even a 1-tilting object of $\mathcal{A}$. Then, by implication (1) $\Longrightarrow$ (2), we also know that $V[0]$ is a projective generator of $\mathcal{H}_{\mathbf{t}}$. It then follows that P is a direct summand of the coproduct in $\mathcal{H}_{\mathbf{t}}$ of I copies of $V[0]$, for some set I. By Lemma 3.10 below, we then get that P is a direct summand of $V^{(I)}[0]$, which implies that $H^{-1}(P) = 0$ and hence that $P \cong V[0]$. □

LEMMA 3.10. *Let $\mathcal{A}$ be an abelian category, let V be a 1-tilting object, let $\mathbf{t} = (\mathrm{Gen}(V), V^{\perp})$ be the associated torsion pair in $\mathcal{A}$ and let $\mathcal{H}_{\mathbf{t}}$ be the heart of the associated HRS t-structure in $\mathcal{D}(\mathcal{A})$. For each set I, the coproduct of I copies of $V[0]$ exists in $\mathcal{H}_{\mathbf{t}}$ and it is precisely the stalk complex $V^{(I)}[0]$.*

PROOF. Let $\iota_j : V \longrightarrow V^{(I)}$ denote the j-th injection into the coproduct in $\mathcal{A}$, for each $j \in I$. For each $N \in \mathcal{H}_{\mathbf{t}}$ we have an induced morphism $\gamma_N : \mathcal{H}_{\mathbf{t}}(V^{(I)}[0], N) \longrightarrow \mathcal{H}_{\mathbf{t}}(V[0], N)^I$, which is the unique morphism of abelian groups such that the j-th projection $\pi_j : \mathcal{H}_{\mathbf{t}}(V[0], N)^I \longrightarrow \mathcal{H}_{\mathbf{t}}(V[0], N)$ satisfies $\pi_j \circ \gamma_N = (\iota_j[0])^*(N) = \mathcal{H}_{\mathbf{t}}(\iota_j[0], N) : \mathcal{H}_{\mathbf{t}}(V^{(I)}[0], N) \longrightarrow \mathcal{H}_{\mathbf{t}}(V[0], N)$, for all $j \in I$.

Our task reduces to prove that γ_N is an isomorphism, for all $N \in \mathcal{H}_{\mathbf{t}}$. To do that we consider the exact sequence $0 \to H^{-1}(N)[1] \longrightarrow N \longrightarrow H^0(N)[0] \to 0$ in $\mathcal{H}_{\mathbf{t}}$. Note that, by the first paragraph of the proof of implication (1) $\Longrightarrow$ (2) of last theorem, we know that $V^{(I)}[0]$ is projective in $\mathcal{H}_{\mathbf{t}}$, for all sets I. This gives the following commutative diagram with exact rows:

$$\begin{array}{ccccccccc}
0 & \longrightarrow & \mathcal{H}_{\mathbf{t}}(V^{(I)}[0], H^{-1}(N)[1]) & \longrightarrow & \mathcal{H}_{\mathbf{t}}(V^{(I)}[0], N) & \longrightarrow & \mathcal{H}_{\mathbf{t}}(V^{(I)}[0], H^0(N)[0]) & \longrightarrow & 0 \\
 & & \downarrow{\scriptstyle \gamma_{H^{-1}(N)[1]}} & & \downarrow{\scriptstyle \gamma_N} & & \downarrow{\scriptstyle \gamma_{H^0(N)[0]}} & & \\
0 & \longrightarrow & \mathcal{H}_{\mathbf{t}}(V[0], H^{-1}(N)[1])^I & \longrightarrow & \mathcal{H}_{\mathbf{t}}(V[0], N)^I & \longrightarrow & \mathcal{H}_{\mathbf{t}}(V[0], H^0(N)[0])^I & \longrightarrow & 0
\end{array}$$

$\gamma_{H^0(N)[0]}$ is clearly an isomorphism since it can be identified with the canonical map $\mathcal{A}(V^{(I)}, H^0(N)) \longrightarrow \mathcal{A}(V, H^0(N))^I$, which is an isomorphism by definition of the coproduct $V^{(I)}$ in $\mathcal{A}$. The task is further reduced to prove that $\gamma_{H^{-1}(N)[1]}$ is an isomorphism. But this latter map gets identified with the canonical morphism γ'_F :

$Ext^1_{\mathcal{A}}(V^{(I)},F) \longrightarrow \text{Ext}^1_{\mathcal{A}}(V,F)^I$, where $F := H^{-1}(M)$. We just need to prove that γ'_F is an isomorphism, for all $F \in \mathcal{F}$. For this we use the cogenerating condition of $\mathcal{T} = \text{Gen}(V)$ and, given $F \in \mathcal{F}$, we fix an exact sequence $0 \to F \longrightarrow T \longrightarrow T' \to 0$, with $T, T' \in \mathcal{T}$. Bearing in mind that $\text{Ext}^1_{\mathcal{A}}(V^{(J)}, ?)_{|\mathcal{T}} = 0$, for all sets J, we get the following commutative diagram with exact rows, where the two left vertical arrows are the canonical isomorphisms induced by definition of the coproduct $V^{(I)}$ in $\mathcal{A}$:

$$\begin{array}{ccccccc} \mathcal{A}(V^{(I)},T) & \longrightarrow & \mathcal{A}(V^{(I)},T') & \longrightarrow & \text{Ext}^1_{\mathcal{A}}(V^{(I)},F) & \longrightarrow & 0 \\ \downarrow & & \downarrow & & \downarrow{\scriptstyle \gamma'_F} & & \\ \mathcal{A}(V,T)^I & \longrightarrow & \mathcal{A}(V,T')^I & \longrightarrow & \text{Ext}^1_{\mathcal{A}}(V,F)^I & \longrightarrow & 0 \end{array}$$

It follows that γ'_F is also an isomorphism as desired. □

Due to its importance, it is worth stating explicitly the dual of Theorem 3.9:

THEOREM 3.11. *Let $\mathcal{A}$ be an abelian category and let $\mathbf{t} = (\mathcal{T}, \mathcal{F})$ be a torsion pair in $\mathcal{A}$. The following assertions are equivalent:*

(1) *$\mathbf{t}$ is a cotilting torsion pair.*
(2) *$\mathbf{t}$ is a faithful torsion pair whose heart $\mathcal{H}_{\mathbf{t}}$ is an AB3* abelian category with an injective cogenerator.*
(3) *$\mathcal{H}_{\mathbf{t}}$ is an AB3* abelian category with an injective cogenerator and $\bar{\mathbf{t}} = (\mathcal{F}[1], \mathcal{T}[0])$ is a co-faithful torsion pair in $\mathcal{H}_{\mathbf{t}}$.*

In this case Q is a 1-cotilting object such that $\mathcal{F} = Cogen(Q)$ if, and only if, $Q[1]$ is an injective cogenerator of $\mathcal{H}_{\mathbf{t}}$. Moreover, an object E of $\mathcal{H}_{\mathbf{t}}$ is an injective cogenertor of this category if, and only if, $E \cong Q[1]$ for some 1-cotilting object of $\mathcal{A}$ defining $\mathbf{t}$.

We have now the following sort of reverse consequence:

COROLLARY 3.12. *Let $\mathcal{A}$ be an abelian category and let $\mathbf{t} = (\mathcal{T}, \mathcal{F})$ be a torsion pair in $\mathcal{A}$. The following assertions hold:*

(1) *$\mathcal{A}$ is AB3 with a projective generator and $\mathbf{t}$ is a faithful torsion pair in $\mathcal{A}$ if, and only if, $\bar{\mathbf{t}} = (\mathcal{F}[1], \mathcal{T}[0])$ is a tilting torsion pair in $\mathcal{H}_{\mathbf{t}}$. In such case, P is a projective generator of $\mathcal{A}$ if and only if $P[1]$ is a 1-tilting object of $\mathcal{H}_{\mathbf{t}}$ such that $\mathcal{F}[1] = Gen_{\mathcal{H}_{\mathbf{t}}}(P[1])$.*
(2) *$\mathcal{A}$ is AB3* with an injective cogenerator and $\mathbf{t}$ is a co-faithful torsion pair in $\mathcal{A}$ if, and only if, $\bar{\mathbf{t}} = (\mathcal{F}[1], \mathcal{T}[0])$ is a cotilting torsion pair in $\mathcal{H}_{\mathbf{t}}$. In such case, E is an injective cogenerator of $\mathcal{A}$ if and only if $E[0]$ is a 1-cotilting object of $\mathcal{H}_{\mathbf{t}}$ such that $\mathcal{T}[0] = Cogen_{\mathcal{H}_{\mathbf{t}}}(E[0])$.*

PROOF. Obviously, each assertion is obtained from the other one by duality. We just prove assertion 2. By [**HRS**, Proposition 3.2] we know that $\mathcal{T}[0]$ is a generating class in $\mathcal{H}_{\mathbf{t}}$, and, by [**HRS**, Proposition I.3.4] and using the terminology of that article, we have that $\Phi[(\mathcal{H}_{\mathbf{t}}, \bar{\mathbf{t}})]$ is equivalent to $(\mathcal{A}, \mathbf{t})$, in fact it is equal to $(\mathcal{A}[1], \mathbf{t}[1])$. Moreover, by [**HRS**, Theorem 3.3] we even know that $\mathcal{D}^b(\mathcal{A})$ and $\mathcal{D}^b(\mathcal{H}_{\mathbf{t}})$ are equivalent triangulated categories. This allows us to apply Theorem 3.11, replacing $\mathcal{A}$ by $\mathcal{H}_{\mathbf{t}}$ and $\mathbf{t}$ by $\bar{\mathbf{t}}$ in that theorem, to conclude that $\bar{\mathbf{t}}$ is a cotilting torsion pair in $\mathcal{H}_{\mathbf{t}}$. The last statement is also a consequence of Theorem 3.11. □

3.3. Hearts that are module categories. In order to study those hearts which are module categories, we need the following concepts:

DEFINITION 3.13. Let $\mathcal{A}$ be an abelian category and $\mathcal{T}_0$ be a set of objects such that arbitrary coproducts of objects of $\mathcal{T}_0$ exist in $\mathcal{A}$. We shall say that $\mathcal{T}_0$ is:

(1) a *1-tilting set* when $\coprod_{T\in\mathcal{T}_0} T$ is a 1-tilting object;
(2) a *self-small set* when, for each $T \in \mathcal{T}_0$ and each family $(T_\lambda)_{\lambda\in\Lambda}$ in $\mathcal{T}_0$, the canonical map $\coprod_{\lambda\in\Lambda} \mathcal{A}(T, T_\lambda) \longrightarrow \mathcal{A}(T, \coprod_{\lambda\in\Lambda} T_\lambda)$ is an isomorphism.
(3) a *classical 1-tilting set* when it is 1-tilting and self-small.

When $\mathcal{T}_0 = \{T\}$ we say that T is, respectively, a 1-tilting, a self-small and a classical 1-tilting object.

The following is the version that we will need of a theorem of Gabriel and Mitchell (see [**Po**, Corollary 3.6.4]):

PROPOSITION 3.14. *Let $\mathcal{A}$ be any category. The following assertions are equivalent:*

(1) *$\mathcal{A}$ is equivalent to $Mod - \mathcal{B}$ (resp. $Mod - R$), for some small pre-additive category $\mathcal{B}$ (resp. some ring R);*
(2) *$\mathcal{A}$ is an AB3 abelian category that admits a self-small set of projective generators (resp. a self-small projective generator).*

PROOF. The equivalence for $\text{Mod}-R$ is a particular case of the one for $\text{Mod}-\mathcal{B}$, for $\mathcal{B}$ a small pre-additive category, since a ring is the same as a pre-additive category with just one object. The classical version of Gabriel-Mitchell theorem states that assertion 1 holds if, and only if, $\mathcal{A}$ is AB3 and has a set of small(=compact) projective generators (see, e.g., [**Po**, Corollary 3.6.4]). We just need to check that in any AB3 abelian category, if $\mathcal{P}_0$ is a self-small set of projective generators, then $\mathcal{P}_0$ consists of small objects. Indeed, let $(A_\lambda)_{\lambda\in\Lambda}$ be any family of objects in $\mathcal{A}$. For each $\lambda \in \Lambda$, we then have an exact sequence $\coprod_{P\in\mathcal{P}_0} P^{(I_{P,\lambda})} \xrightarrow{f_\lambda} \coprod_{P\in\mathcal{P}_0} P^{(J_{P,\lambda})} \xrightarrow{p_\lambda} A_\lambda \to 0$ in $\mathcal{A}$, for some sets $I_{P,\lambda}$ and $J_{P,\lambda}$. Due to right exactness of coproducts, we then get an exact sequence

$$\coprod_{\lambda\in\Lambda} \coprod_{P\in\mathcal{P}_0} P^{(I_{P,\lambda})} \xrightarrow{\coprod f_\lambda} \coprod_{\lambda\in\Lambda} \coprod_{P\in\mathcal{P}_0} P^{(J_{P,\lambda})} \xrightarrow{\coprod p_\lambda} \coprod_{\lambda\in\Lambda} A_\lambda \to 0.$$

If now $P' \in \mathcal{P}_0$ is arbitrary and we apply $\mathcal{A}(P', ?)$ to the last exact sequence, using the projectivity of P' and the self-smallness of $\mathcal{P}_0$ we readily get that the canonical map $\coprod_{\lambda\in\Lambda} \mathcal{A}(P', A_\lambda) \longrightarrow \mathcal{A}(P', \coprod_{\lambda\in\Lambda} A_\lambda)$ is an isomorphism, so that P' is small (=compact) in $\mathcal{A}$. □

COROLLARY 3.15. *Let $\mathcal{A}$ be an abelian category and let $\mathbf{t} = (\mathcal{T}, \mathcal{F})$ be a torsion pair in $\mathcal{A}$. The following assertions are equivalent:*

(1) *There is a classical 1-tilting set $\mathcal{T}_0$ (resp. a classical 1-tilting object V) such that $\mathcal{T} = Gen(\mathcal{T}_0)$ (resp. $\mathcal{T} = Gen(V)$).*
(2) *$\mathbf{t}$ is a co-faithful torsion pair whose heart $\mathcal{H}_\mathbf{t}$ is equivalent to the module category over a small pre-additive category (resp. over a ring).*
(3) *$\mathcal{H}_\mathbf{t}$ is equivalent to the module category over a small pre-additive category (resp. over a ring) and $\bar{\mathbf{t}} = (\mathcal{F}[1], \mathcal{T}[0])$ is a faithful torsion pair in $\mathcal{H}_\mathbf{t}$.*

PROOF. (2) $\Longleftrightarrow$ (3) is a consequence of Proposition 2.19.

(1) $\Longrightarrow$ (2) Since $V := \coprod_{T\in\mathcal{T}_0} T$ is a 1-tilting object it follows from Theorem 3.9 that $V[0]$ is a projective generator of $\mathcal{H}_{\mathbf{t}}$, which in turns implies that $\mathcal{T}_0[0] = \{T[0]\colon T \in \mathcal{T}_0\}$ is a set of projective generators of $\mathcal{H}_{\mathbf{t}}$. An easy adaptation of the proof of Lemma 3.10 shows that if $(T_\lambda)_{\lambda\in\Lambda}$ is a family in $\mathcal{T}_0$, then the coproduct of the $T_\lambda[0]$ in $\mathcal{H}_{\mathbf{t}}$ exists and is the stalk complex $(\coprod_{\lambda\in\Lambda} T_\lambda)[0]$. Here the μ-th injection into the coproduct, for each $\mu \in \Lambda$, is the map $\iota_\mu[0] : T_\mu[0] \longrightarrow (\coprod_{\lambda\in\Lambda} T_\lambda)[0]$, where $\iota_\mu : T_\mu \longrightarrow \coprod_{\lambda\in\Lambda} T_\lambda$ is the μ-th injection into the coproduct in $\mathcal{A}$. Given now $T \in \mathcal{T}_0$ arbitrary, we have a sequence of isomorphisms:

$$\coprod_{\lambda\in\Lambda} \mathcal{H}_{\mathbf{t}}(T[0], T_\lambda[0]) \cong \coprod_{\lambda\in\Lambda} \mathcal{A}(T, T_\lambda) \overset{canonical}{\longrightarrow} \mathcal{A}(T, \coprod_{\lambda\in\Lambda} T_\lambda) \cong \mathcal{H}_{\mathbf{t}}(T[0], (\coprod_{\lambda\in\Lambda} T_\lambda)[0]).$$

We claim that the composition of these isomorphisms, denoted by ψ in the sequel, is precisely the canonical morphism

$$\coprod_{\lambda\in\Lambda} \mathcal{H}_{\mathbf{t}}(T[0], T_\lambda[0]) \longrightarrow \mathcal{H}_{\mathbf{t}}(T[0], \coprod_{\lambda\in\Lambda}^{\mathcal{H}_{\mathbf{t}}} (T_\lambda[0])) = \mathcal{H}_{\mathbf{t}}(T[0], (\coprod_{\lambda\in\Lambda} T_\lambda)[0]). \qquad (\star)$$

To see this, for each $\mu \in \Lambda$ we let $u_\mu : \mathcal{H}_{\mathbf{t}}(T[0], T_\mu[0]) \longrightarrow \coprod_{\lambda\in\Lambda} \mathcal{H}_{\mathbf{t}}(T[0], T_\lambda[0])$ the μ-th injection into the coproduct in Ab. We need to check that $\psi \circ u_\mu = (\iota_\mu[0])_* = \mathcal{H}_{\mathbf{t}}(T[0], \iota_\mu[0])$, for all $\mu \in \Lambda$. But this follows immediately from the equivalence of categories $\mathcal{T} \cong \mathcal{T}[0]$ and the fact that the composition

$$\mathcal{A}(T, T_\mu) \overset{\tilde{u}_\mu}{\longrightarrow} \coprod_{\lambda\in\Lambda} \mathcal{A}(T, T_\lambda) \overset{canonical}{\longrightarrow} \mathcal{A}(T, \coprod_{\lambda\in\Lambda} T_\lambda),$$

where $\tilde{u}_\mu$ is the canonical injection into the coproduct, is precisely the morphism $(\iota_\mu)_* = \mathcal{A}(T, \iota_\mu)$.

Therefore $\mathcal{T}_0[0]$ is a self-small set of projective generators of $\mathcal{H}_{\mathbf{t}}$ and, by Proposition 3.14, we conclude that $\mathcal{H}_{\mathbf{t}} \cong \text{Mod} - \mathcal{B}$, for some small pre-additive category $\mathcal{B}$.

(2) $\Longleftrightarrow$ (3) $\Longrightarrow$ (1) Due to the co-faithful condition on $\mathbf{t}$ and Proposition 2.19, the class $\mathcal{T}[0]$ is generating in $\mathcal{H}_{\mathbf{t}}$. Hence any projective object of $\mathcal{H}_{\mathbf{t}}$ is in $\mathcal{T}[0]$. Then any self-small set of projective generators of $\mathcal{H}_{\mathbf{t}}$ is of the form $\mathcal{T}_0[0]$, for some set $\mathcal{T}_0 \subset \mathcal{T}$. By Theorem 3.9 and its proof, we get that $\mathbf{t}$ is the tilting torsion pair defined by the 1-tilting object $\hat{T} := \coprod_{T\in\mathcal{T}_0} T$. It just remains to check that $\mathcal{T}_0$ is a self-small set. But this is a direct consequence of the self-smallness of $\mathcal{T}_0[0]$ since we have an equivalence of categories $\mathcal{T} \cong \mathcal{T}[0]$ and coproducts in $\mathcal{T}$ and $\mathcal{T}[0]$ are calculated as in $\mathcal{A}$ and $\mathcal{H}_{\mathbf{t}}$, respectively. □

3.4. Bijections induced by the HRS tilting process. The previous results and the HRS tilting process give rise to a nice series of bijections that we gather in our next corollary. We continue with the terminology of Definition 2.18 and Proposition 2.19. They give induced maps $(\mathbf{AB}, \mathbf{tor}_{faithful}) \rightleftarrows (\mathbf{AB}, \mathbf{tor}_{cofaithful})$. By Theorems 3.9 and 3.11 and Corollary 3.12, we get the bijections in 1 and 2 of next corollary, and its bijections 3 and 4 follow from Corollary 3.15.

COROLLARY 3.16. *Let $\Phi : (\mathbf{AB}, \mathbf{tor}) \longrightarrow (\mathbf{AB}, \mathbf{tor})$ be the map induced by the HRS tilting process (see Definition 2.18). By restriction, Φ defines bijections, which are inverse of themselves (i.e. $(\Phi \circ \Phi)_{|\mathcal{C}} = 1_{\mathcal{C}}$, for $\mathcal{C}$ any subclass in the list):*

(1) *Between* $(\mathbf{AB}, \mathbf{tor}_{tilt})$ *and* $(\mathbf{AB3}_{proj}, \mathbf{tor}_{faithful})$, *where* $\mathbf{tor}_{tilt}$ *and* $\mathbf{tor}_{faithful}$ *denote the subclasses of* $\mathbf{tor}$ *consisting of the tilting and the faithful torsion pairs, respectively, and* $\mathbf{AB3}_{proj}$ *denotes the class of AB3 abelian categories with a projective generator;*

(2) *Between* $(\mathbf{AB}, \mathbf{tor}_{cotilt})$ *and* $(\mathbf{AB3}^*_{inj}, \mathbf{tor}_{cofaithful})$, *where* $\mathbf{tor}_{cotilt}$ *and* $\mathbf{tor}_{cofaithful}$ *denote the subclasses of* $\mathbf{tor}$ *consisting of the cotilting and the co-faithful torsion pairs, respectively, and* $\mathbf{AB3}^*_{inj}$ *denotes the class of AB3* abelian categories with an injective cogenerator;*

(3) *Between* $(\mathbf{AB}, \mathbf{tor}_{stilt-class})$ *and* $(\mathbf{Mod}_{paddt}, \mathbf{tor}_{faithful})$, *where* $\mathbf{tor}_{stilt-class}$ *denotes the subclass of* $\mathbf{tor}_{tilt}$ *consisting of those torsion pairs associated to a classical tilting set of objects and* $\mathbf{Mod}_{paddt}$ *is the class of abelian categories which are equivalent to module categories over small pre-additive categories.*

(4) *Between* $(\mathbf{AB}, \mathbf{tor}_{class-tilt})$ *and* $(\mathbf{Mod}_{ring}, \mathbf{tor}_{faithful})$, *where* $\mathbf{tor}_{class-tilt}$ *denotes the subclass of* $\mathbf{tor}_{tilt}$ *consisting of the torsion pairs associated to classical tilting objects and* $\mathbf{Mod}_{ring}$ *denotes the class of categories equivalent to module categories over rings.*

Positselski and Stovicek have recently shown that complete and cocomplete abelian categories with an injective cogenerator and an n-tilting object correspond bijectively to complete and cocomplete abelian categories with a projective generator and an n-cotilting object [**Po-St2**, Corollary 4.12]. One can now recover the case $n = 1$ in their result.

COROLLARY 3.17 (Positselski-Stovicek). *The HRS tilting process gives a one-to-one correspondence between:*

(1) *The pairs* $(\mathcal{A}, \mathbf{t})$ *consisting of an AB3* abelian category* $\mathcal{A}$ *with an injective cogenerator and a tilting torsion pair* $\mathbf{t}$ *in* $\mathcal{A}$*;*

(2) *The pairs* $(\mathcal{B}, \bar{\mathbf{t}})$ *consisting of an AB3 abelian category* $\mathcal{B}$ *with a projective generator and a cotilting torsion pair* $\bar{\mathbf{t}}$ *in* $\mathcal{B}$

Moreover the categories of assertion 1 are also AB4 and those of assertion 2 are also AB4.*

PROOF. We start by proving that any AB3* abelian category $\mathcal{A}$ with an injective cogenerator is AB4, that, together with its dual, will prove the last sentence of the corollary. Note that it is enough to prove that $\mathcal{A}$ is AB3 for it is well-known that any AB3 abelian category with an injective cogenerator is AB4 (see [**Po**, Corollary 3.2.9]). But proving the AB3 condition amounts to prove that if I is any set, viewed as a small category, then the constant diagram functor $\kappa : \mathcal{A} \longrightarrow \mathcal{A}^I$ has a left adjoint. This follows from Freyd's special adjoint theorem and its consequences (see [**Freyd**, Chapter 3, Exercises M, N]).

By Corollary 3.16 we have induced bijections

$$\Phi : (\mathbf{AB}, \mathbf{tor}_{tilt}) \xrightarrow{\cong} (\mathbf{AB3}_{proj}, \mathbf{tor}_{faithful})$$

and

$$\Phi : (\mathbf{AB3}^*_{inj}, \mathbf{tor}_{cofaithful}) \xrightarrow{\cong} (\mathbf{AB}, \mathbf{tor}_{cotilt}).$$

By restriction, we then get a bijection between the intersection of the domains and the intersection of the codomains. The intersection of the domains is precisely the class of pairs in 1 (note that the fact that $\mathcal{T}$ is cogenerating, equivalently that

$\mathbf{t} \in \mathbf{tor}_{cofaithful}$, is automatic). Similarly, the intersection of codomains is precisely the class of pairs in 2. □

4. When is the heart of a torsion pair a Grothendieck category?

4.1. Initial results. The work on the problem started with a series of papers [**CG**], [**CGM**], [**CMT**], [**MT**], which we now review in the terminology of this manuscript. Suppose that $\mathcal{A}$ is an abelian category with a classical tilting torsion $\mathbf{t}$ given by a 1-tilting object V. According to [**HRS**, Theorem 3.3] the realization functor gives an equivalence of triangulated categories $G : \mathcal{D}^b(\mathcal{H}_\mathbf{t}) \xrightarrow{\cong} \mathcal{D}^b(\mathcal{A})$. On the other hand, by Corollary 3.15, we know that $\mathcal{H}_\mathbf{t}$ is a module category, actually via the equivalence of categories $\mathcal{H}_\mathbf{t}(V[0], ?) : \mathcal{H}_\mathbf{t} \xrightarrow{\cong} \text{Mod} - R$, where $R = \text{End}_{\mathcal{H}_\mathbf{t}}(V[0]) \cong \text{End}_\mathcal{A}(V)$. Then we also have an equivalence of triangulated categories $\mathcal{D}^b(\text{Mod}-R) \xrightarrow{\cong} \mathcal{D}^b(\mathcal{H}_\mathbf{t})$, and taking the composition, we get an induced equivalence of triangulated categories $\mathcal{D}^b(\text{Mod} - R) \xrightarrow{\cong} \mathcal{D}^b(\mathcal{A})$, taking R to V. We can think of the inverse of this functor as a sort of right derived functor $\mathbb{R}H_V : \mathcal{D}(\mathcal{A}) \longrightarrow \mathcal{D}(\text{Mod} - R)$ of the canonical functor $H_V := \mathcal{A}(V, ?) : \mathcal{A} \longrightarrow \text{Mod} - R$. This last functor turns out to have a left adjoint $T_V : \text{Mod} - R \longrightarrow \mathcal{A}$ of which we can think as a sort of 'tensor product by V'. We can then think of the equivalence $\mathbb{L}T_V : \mathcal{D}(\text{Mod} - R) \xrightarrow{\cong} \mathcal{D}(\mathcal{A})$ as a left derived functor of T_V. Note that for $A \in \mathcal{A}$ (resp. for $M \in \text{Mod} - R$), $\mathbb{R}H_V(A)$ (resp. $\mathbb{L}T_V(M)$) is a complex and not just an R-module (resp. not just an object of $\mathcal{A}$). Concretely, due to the fact that $\text{Ext}^2_\mathcal{A}(V, ?) = 0$, one actually has that $\mathbb{R}H_V(A)$ has cohomology concentrated in degrees $0, 1$, with $H^0(\mathbb{R}H_V(A)) = \mathcal{A}(V, A)$ and $H^1(\mathbb{R}H_V(A)) = \text{Ext}^1_\mathcal{A}(V, A)$, for all $A \in \mathcal{A}$. Dually, $\mathbb{L}T_V(M)$ has cohomology concentrated in degrees $-1, 0$, with $H^0(\mathbb{L}T_V(M)) = T_V(M)$ and $H^{-1}(\mathbb{L}T_V(M)) = T'_V(M)$, where $T'_V : \text{Mod} - R \longrightarrow \mathcal{A}$ is the first left derived functor of T_V in the classical sense, for all R-modules M.

Due to the definition of the torsion pair $\mathbf{t}$, one then has $\mathbb{R}H_V(T) = \mathcal{A}(V, T)[0]$ and $\mathbb{R}H_V(F) = \text{Ext}^1_\mathcal{A}(V, F)[-1]$. This implies that the equivalence $\mathbb{R}H_V : \mathcal{D}(A) \longrightarrow \mathcal{D}(\text{Mod} - R)$ induces equivalences of categories

$$\mathcal{F}[1] \xrightarrow{\cong} \mathcal{X} := \{X \in \text{Mod} - R : \; X \cong \text{Ext}^1_\mathcal{A}(V, F), \text{ with } F \in \mathcal{F}\}$$

$$\mathcal{T}[0] \xrightarrow{\cong} \mathcal{Y} := \{Y \in \text{Mod} - R : \; Y \cong \mathcal{A}(V, T), \text{ with } T \in \mathcal{T}\}.$$

Note that we then get induced equivalences of categories $H_V = \mathcal{A}(V, ?) : \mathcal{T} \xrightarrow{\cong} \mathcal{Y}$ and $H'_V = \text{Ext}^1_\mathcal{A}(V, ?) : \mathcal{F} \xrightarrow{\cong} X$ whose quasi-inverses are necessarily $T_V : \mathcal{Y} \xrightarrow{\cong} \mathcal{T}$ and $T'_V : \mathcal{X} \xrightarrow{\cong} \mathcal{F}$. This essentially gives the proof of the following generalization of Brenner-Butler's theorem (see [**BB**]), due to Colpi and Fuller (see [**CF**, Theorem 3.2]):

THEOREM 4.1. *Let $\mathcal{A}$ be an abelian category, let V be a classical 1-tilting object in $\mathcal{A}$ and let $R = End_\mathcal{A}(V)$ the ring of endomorphisms of V. With the notation above, we have an equality of pairs $(\mathcal{X}, \mathcal{Y}) = (Ker(T_V), Ker(T'_V))$, and this is a faitful torsion pair $\mathbf{t}'$ in $Mod - R$. Moreover, we have induced equivalences of categories* $\mathcal{T} \underset{T_V}{\overset{H_V}{\rightleftarrows}} \mathcal{Y}$ *and* $\mathcal{F} \underset{T'_V}{\overset{H'_V}{\rightleftarrows}} \mathcal{X}$

In addition, by the paragraphs above, the torsion pair $\mathbf{t}'$ is sent to $\bar{\mathbf{t}} = (\mathcal{F}[1], \mathcal{T}[0])$ by the equivalence of categories $\text{Mod} - R \xrightarrow{\cong} \mathcal{H}_{\mathbf{t}}$. Then, using Proposition 2.20, one gets the following initial result:

PROPOSITION 4.2. *[CGM, Corollary 2.4] $\mathcal{A}$ is equivalent to $\mathcal{H}_{\mathbf{t}'}$, where $\mathbf{t}'$ is as above.*

On the other hand, we have a dual situation, starting with (Mod-$R, \mathbf{s}$) a pair in $(\mathbf{Mod}_{ring}, \mathbf{tor}_{faithful})$. It then follows that $R[1]$ is a classical 1-tilting object of $\mathcal{H}_{\mathbf{s}}$ (see Corollary 3.12) so that $\Phi[(\text{Mod-}R, \mathbf{s})] = (\mathcal{H}_{\mathbf{s}}, \bar{\mathbf{s}}) \in (\mathbf{AB}, \mathbf{tor}_{tilt})$. For this reason, the last result indicates that Question 1.3 for faithful torsion pairs in modules categories is equivalent to the question of when an abelian category $\mathcal{A}$ with a classical 1-tilting object is a Grothendieck category. This fact was exploited by Colpi, Gregorio and Mantese, who obtained the first partial answer to Question 1.3.

THEOREM 4.3. *[CGM, Theorem 3.7] Let $(\mathcal{A}, \mathbf{t}) \in (\mathbf{AB}, \mathbf{tor}_{tilt})$ and we consider $\mathbf{t}'$ as above. Then, the following assertions are equivalent:*

1. *$\mathcal{A} \cong \mathcal{H}_{\mathbf{t}'}$ is a Grothendieck category;*
2. *for any direct system $(X_\lambda)_\lambda$ in $\mathcal{H}_{\mathbf{t}'}$ the canonical morphism $\varinjlim H_V(X_\lambda) \to H_V(\varinjlim_{\mathcal{H}_{\mathbf{t}'}} X_\lambda)$ is a monomorphism;*
3. *the functor H_V preserves direct limits.*

If $\mathbf{t}'$ is of finite type, then the previous conditions are equivalent to the condition that the functor $T_V \circ H_V$ preserve direct limits.

Already in [**CGM**] the authors gave necessary conditions for a faithful torsion pair in a module category to have a heart which is a Grothendieck category, a condition that was shown to be also sufficient in an unpublished paper by Colpi and Gregorio [**CG**] (see [**Ma**, Theorem 6.2]).

THEOREM 4.4. [**CGM**, Proposition 3.8] *and* [**CG**, Theorem 1.3] *Let R be a ring, let $\mathbf{t} = (\mathcal{T}, \mathcal{F})$ be a faithful torsion pair in Mod-R and let $\mathcal{H}_{\mathbf{t}}$ be the heart of the associated Happel-Reiten-Smalø t-structure in $\mathcal{D}(\text{Mod} - R)$. Then $\mathcal{H}_{\mathbf{t}}$ is a Grothendieck category if, and only if, $\mathbf{t} = (\mathcal{T}, \mathcal{F})$ is a cotilting torsion pair.*

4.2. The solution of the problem. The solution to the problem was given by the authors in [**PS1**] and [**PS2**]. We realized that the hard part of the problem was to deal with the AB5 condition on $\mathcal{H}_{\mathbf{t}}$. This naturally led to a detailed study of direct limits in the heart. And, in order to understand those direct limits, it was a preliminary step to understand the behavior of the stalk complexes in the heart with respect to direct limits.

PROPOSITION 4.5. *[PS1, Lemma 4.1 and Proposition 4.2] Let $\mathbf{t} = (\mathcal{T}, \mathcal{F})$ be a torsion pair in the Grothendieck category $\mathcal{G}$. The following assertions hold:*

1. *The functor $H^0 : \mathcal{H}_{\mathbf{t}} \to \mathcal{G}$ is right exact and preserve coproducts;*
2. *The functor $H^{-1} : \mathcal{H}_{\mathbf{t}} \to \mathcal{G}$ is left exact and preserve coproducts;*
3. *For every (M_λ) direct system in $\mathcal{H}_{\mathbf{t}}$, the induced morphism $\varinjlim H^k(M_\lambda) \to H^k(\varinjlim_{\mathcal{H}_{\mathbf{t}}} M_\lambda)$ is an epimorphism, for $k = -1$, and an isomorphism, for $k \neq -1$.*
4. *the pair $\bar{t} = (\mathcal{F}[1], \mathcal{T}[0])$ in $\mathcal{H}_{\mathbf{t}}$ is of finite type.*

(5) *For each direct system* $(F_\lambda)_\lambda$ *in* $\mathcal{F}$, *we have a canonical isomorphism* $\varinjlim_{\mathcal{H}_\mathbf{t}} F_\lambda[1] \cong (1:t)(\varinjlim F_\lambda)[1]$.

For instance, using the previous result and Proposition 2.20, we immediately get a necessary condition for a positive answer in the case of a co-faithful torsion pair.

LEMMA 4.6. *Let* $\mathcal{G}$ *be a Grothendieck category and let* $\mathbf{t} = (\mathcal{T}, \mathcal{F})$ *be a co-faithful torsion pair. If* $\mathcal{H}_\mathbf{t}$ *is a Grothendieck category, then* $\mathbf{t}$ *is of finite type.*

PROOF. Suppose that $\mathcal{H}_\mathbf{t}$ is a Grothendieck category. Since $\mathcal{G}$ has enough injectives, from Proposition 2.20 we get that $\mathcal{H}_{\overline{\mathbf{t}}}$ is equivalent to $\mathcal{G}[1]$, via realization functor, where $\overline{\mathbf{t}} = (\mathcal{F}[1], \mathcal{T}[0])$ is the corresponding torsion pair in $\mathcal{H}_\mathbf{t}$. Using assertion 4 of the previous proposition, we deduce that $\overline{\overline{\mathbf{t}}} = (\mathcal{T}[1], \mathcal{F}[1])$ is a torsion pair of finite type in $\mathcal{H}_{\overline{\mathbf{t}}} \cong \mathcal{G}[1]$. That is, given a family $(F_\lambda)_\lambda$ in $\mathcal{F}$, we have that $\varinjlim_{\mathcal{G}[1]} F_\lambda[1] \in \mathcal{F}[1]$, and this implies that $\varinjlim F_\lambda \in \mathcal{F}$ due to the canonical equivalence $\mathcal{G} \cong \mathcal{G}[1]$, which restricts to $\mathcal{F} \cong \mathcal{F}[1]$. □

Another point of the strategy of the authors was to use the canonical cohomology functors $H^k : \mathcal{D}(\mathcal{G}) \longrightarrow \mathcal{G}$ to approach the problem. In that way one gets sufficient conditions:

PROPOSITION 4.7. *[PS1, Proposition 3.4] Let* $\mathcal{G}$ *be a Grothendieck category and let* $\sigma = (\mathcal{U}, \mathcal{U}^\perp[1])$ *be a t-structure on* $\mathcal{D}(\mathcal{G})$. *We denote its heart by* $\mathcal{H}_\sigma$. *If the classical cohomological functors* $H^k : \mathcal{H}_\sigma \to \mathcal{G}$ *preserve direct limits, for all integer* k, *then* $\mathcal{H}_\sigma$ *is an AB5 abelian category.*

The following is now a natural question that remains open.

QUESTION 4.8. Given a Grothendieck category $\mathcal{G}$ and a t-structure $\sigma = (\mathcal{U}, \mathcal{U}^\perp[1])$ in $\mathcal{D}(\mathcal{G})$ such that its heart $\mathcal{H}_\sigma$ is an AB5 abelian category, do the classical cohomological functors $H^m : \mathcal{H}_\sigma \to \mathcal{G}$ preserve direct limits, for all $m \in \mathbb{Z}$?

Recently, Chen, Han and Zhou have given necessary and sufficient conditions to have an equivalence $\mathcal{H}_{\overline{\mathbf{t}}}$ is equivalent to $\mathcal{G}[1]$.

THEOREM 4.9. *[CHZ, Theorem A] Let* $\mathcal{G}$ *be a Grothendieck category and let* $\mathbf{t} = (\mathcal{T}, \mathcal{F})$ *be a torsion pair in* $\mathcal{G}$. *The following assertions are equivalent:*

(1) $\mathcal{H}_{\overline{\mathbf{t}}}$ *is equivalent to* $\mathcal{G}[1]$, *via the realization functor;*
(2) *Each object* X *in* $\mathcal{G}$ *fits into an exact sequence*

$$0 \longrightarrow F^0 \longrightarrow F^1 \longrightarrow X \longrightarrow T^0 \longrightarrow T^1 \longrightarrow 0$$

with $F^i \in \mathcal{F}$ *and* $T^i \in \mathcal{T}$, *for* $i = 0, 1$, *and* $Ext^3_{\mathcal{G}}(T^1, F^0) = 0$.

The key point in the proof of Lemma 4.6 to guarantee that the class $\mathcal{F}$ is closed under direct limits is the fact that $\mathcal{H}_{\overline{\mathbf{t}}}$ is in that situation equivalent to $\mathcal{G}[1]$ via the realization functor. Then, keeping the same proof of that lemma, one immediately deduces from Theorem 4.9:

COROLLARY 4.10. *Let* $\mathcal{G}$ *be a Grothendieck category and let* $\mathbf{t} = (\mathcal{T}, \mathcal{F})$ *be a torsion pair in* $\mathcal{G}$ *that satisfies condition 2 of last theorem (e.g. any (co)faithful torsion pair). If* $\mathcal{H}_\mathbf{t}$ *is a Grothendieck category, then* $\mathbf{t}$ *is of finite type.*

The general answer to Question 1.3 was given by the authors.

THEOREM 4.11. *[PS2, Theorem 1.2] Let $\mathcal{G}$ be a Grothendendieck and let $\mathbf{t} = (\mathcal{T}, \mathcal{F})$ be a torsion pair in $\mathcal{G}$. Then, $\mathcal{H}_{\mathbf{t}}$ is a Grothendieck category if, and only if, $\mathbf{t}$ is of finite type.*

The main and harder part was to prove that the finite type of $\mathbf{t}$ is a necessary condition. That is done in [**PS1**, Theorem 4.8] and the preliminary results leading to it. We will give a new proof in Section 4.4. Conversely, for a torsion pair of finite type $\mathbf{t} = (\mathcal{T}, \mathcal{F})$, the authors proved that $\mathcal{T} = \text{Pres}(V)$, for some object V, and then, for a fixed generator G of $\mathcal{G}$, they showed in [**PS1**, Proposition 4.7] that the skeletally small subclass $\mathcal{N}$ of $\mathcal{H}_{\mathbf{t}}$ consisting of those complexes N such that $H^{-1}(N)$ is a subquotient of G^m and $H^0(N) \cong V^n$, for some $m, n \in \mathbb{N}$, is a class of generators of $\mathcal{H}_{\mathbf{t}}$. Therefore, in order to answer Question 1.3, the only thing remaining was to prove that if $\mathbf{t}$ is of finite type then $\mathcal{H}_{\mathbf{t}}$ is AB5. This was done in [**PS2**, Theorem 1.2].

4.3. A side problem: When is the heart of a tilting torsion pair a Grothendieck category? Since tilting torsion pairs have hearts with a projective generator, it is good to know when that heart is a Grothendieck category, because such a heart would be very close to a module category. The question of the title of this subsection has been recently answered [**BHPST**, Corollary 2.5] for the case when the ambient Grothendieck category $\mathcal{G}$ is the module category over a ring. Recall that a *pure exact sequence* in Mod $-$ R is a short exact sequence $0 \to L \longrightarrow M \longrightarrow N \to 0$ that remains exact after applying the functor $? \otimes_R X$, for all left R-modules X. A module $P \in \text{Mod} - R$ is *pure-projective* when the functor $\text{Hom}_R(P, ?) : \text{Mod-}R \longrightarrow \text{Ab}$ preserves exactness of pure exact sequences.

THEOREM 4.12. *[BHPST, Corollary 2.5] Let R be a ring, let V be a 1-tilting (right) R-module and let $\mathbf{t} = (Gen(V), V^{\perp})$ be the associated torsion pair in Mod$-R$. The following assertions are equivalent:*

(1) *V is pure-projective.*
(2) *$\mathbf{t}$ is of finite type (equivalently, the heart $\mathcal{H}_{\mathbf{t}}$ is a Grothendieck category)*

It is well-known that a module is pure-projective if, and only if, it is a direct summand of a coproduct (=direct sum) of finitely presented modules. So when the heart of a tilting torsion pair $\mathbf{t} = (\text{Gen}(V), V^{\perp})$ is a Grothendieck category, the projective generator of the heart $V[0]$ is determined by a set of 'small objects'. Therefore the following question, first risen in [**PS1**, Question 5.5], is apropos. We will call two modules M and N *Add*-equivalent when $\text{Add}(M) = \text{Add}(N)$.

QUESTION 4.13. Let V be a 1-tilting R-module whose associated torsion pair is of finite type (equivalently, such that the heart $\mathcal{H}_{\mathbf{t}}$ is a Grothendieck category). Is V *Add*-equivalent to a classical 1-tilting module?. Equivalently, is the heart $\mathcal{H}_{\mathbf{t}}$ equivalent to the module category over a ring?

It turns out that the answer to this question is negative in general, with counterexamples already existing when R is a noetherian ring (see [**BHPST**, Section 4]). However, the following is true:

THEOREM 4.14. *Let the ring R satisfy one of the following conditions:*

(1) *R is a commutative ring;*
(2) *R is a Krull-Schmidt ring, i.e. every finitely presented (right) R-module is a direct sum of modules with local endomorphism ring (e.g. R is right Artinian);*

(3) *every pure-projective (right) R-module is a coproduct of finitely presented modules.*

Then, a 1-tilting R-module is pure-projective if, and only if, it is Add-equivalent to a classical 1-tilting R-modules. Said in equivalent words, the heart of a tilting torsion pair in $Mod - R$ is a Grothendieck category if, and only if, it is equivalent to a module category over a ring.

PROOF. See [**BHPST**, Corollary 2.8 and Theorem 3.7]. □

4.4. A new approach using purity. Using now a recent result of Positselski and Stovicek [**Po-St**] we can actually identify the cotilting torsion pairs in an abelian category for which the heart is an AB5 abelian category. We need the following definition.

DEFINITION 4.15. Let $\mathcal{A}$ be any additive category. We shall say that an object Y of $\mathcal{A}$ is *pure-injective* if the following two conditions hold:

(1) The product of Y^I exists in $\mathcal{A}$, for all sets I;
(2) For each nonempty set I, there is a map $\phi : Y^I \longrightarrow Y$ such that $\phi \circ \iota_j = 1_Y$, for all $j \in I$. Here $\iota_j : Y \longrightarrow Y^I$ is the unique morphism such that $\pi_i \circ \iota_j = \delta_{ij} 1_Y$, with δ_{ij} the Kronecker symbol and $\pi_i : Y^I \longrightarrow Y$ the i-th projection.

We call the morphism ι_j in *j-th injection into the product.*

Note that, if for $\mathcal{A}$ and Y as in last definition, also the coproduct $Y^{(I)}$ exists for all sets I, then there is a canonical morphism $\kappa_Y : Y^{(I)} \longrightarrow Y^I$, uniquely determined by the fact that the composition $Y \xrightarrow{\lambda_j} Y^{(I)} \xrightarrow{\kappa_Y} Y^I \xrightarrow{\pi_k} Y$ equals $\delta_{jk} 1_Y$, where λ_j and π_k are the j-th injection into the coproduct and π_k is the k-th projection from the product, for all $j, k \in I$. We also have a *summation map* $s_Y : Y^{(I)} \longrightarrow Y$, which is the only morphism such that $s_Y \circ \lambda_j = 1_Y$, for all $j \in I$. We leave as an easy exercise for the reader to check that in this situation Y is pure-injective if, and only if, this summation map s_Y factors through κ_Y, for all sets I. This completes the proof of the following result, which is crucial for us:

LEMMA 4.16. *([**Po-St**, Theorem 3.3 (dual)]) Let $\mathcal{A}$ be an AB3* abelian category with an injective cogenerator E (whence $\mathcal{A}$ is also AB3 by the first paragraph of the proof of Corollary 3.17). The following assertions are equivalent:*

(1) *Direct limits are exact in $\mathcal{A}$, i.e. $\mathcal{A}$ is AB5.*
(2) *The summation map $s_E : E^{(I)} \longrightarrow E$ factors through $\kappa_E : E^{(I)} \longrightarrow E^I$, for all sets I.*
(3) *E is a pure-injective object of $\mathcal{A}$.*

COROLLARY 4.17. *Let $\mathcal{A}$ be an abelian category and $\mathbf{t} = (\mathcal{T}, \mathcal{F})$ be a torsion pair in $\mathcal{A}$. The following assertions are equivalent:*

(1) *There is a pure-injective 1-cotilting object Q of $\mathcal{A}$ such that $\mathcal{F} = Cogen(Q)$;*
(2) *$\mathbf{t}$ is a faithful torsion pair in $\mathcal{A}$ whose heart is an AB5 abelian category with an injective cogenerator;*
(3) *The heart $\mathcal{H}_{\mathbf{t}}$ is an AB5 abelian category with an injective cogenerator and $\bar{\mathbf{t}}$ is a co-faithful torsion pair in $\mathcal{H}_{\mathbf{t}}$.*

In particular, the HRS process gives a bijection 'inverse of itself' $(\mathbf{AB}, \mathbf{tor}_{cotilt-pinj}) \xrightarrow{\cong} (\mathbf{AB5}_{inj}, \mathbf{tor}_{cofaithful})$, where:

(a) $(\mathbf{AB}, \mathbf{tor}_{cotilt-pinj})$ *consists of the pairs* $(\mathcal{A}, \mathbf{t})$, *where* $\mathcal{A}$ *is an abelian category and* $\mathbf{t} = (\mathcal{T}, \mathcal{F})$ *is a torsion pair, with* $\mathcal{F} = Cogen(Q)$ *for* Q *a 1-coltilting pure-injective object.*
(b) $(\mathbf{AB5}_{inj}, \mathbf{tor}_{cofaithful})$ *consists of the pairs* $(\mathcal{B}, \bar{\mathbf{t}})$, *where* $\mathcal{B}$ *is an AB5 abelian category with an injective cogenerator and* $\bar{\mathbf{t}} = (\mathcal{X}, \mathcal{Y})$ *is a co-faithful torsion pair in* $\mathcal{B}$.

PROOF. (2) $\Longleftrightarrow$ (3) It is a consequence of Proposition 2.19.

(1) $\Longrightarrow$ (2) By dualizing the proof of Theorem 3.9 and Lemma 3.10, we know that $Q[1]$ is an injective cogenerator of $\mathcal{H}_\mathbf{t}$ and that the stalk complex $Q^I[1]$ is the product of I copies of $Q[1]$ in $\mathcal{H}_\mathbf{t}$, for all sets I. If $\phi : Q^I \longrightarrow Q$ is a map such that $\phi \circ \iota_j = 1_Y$, for all $j \in I$, with the notation of Definition 4.15, we then get that $\phi[1] : Q^I[1] \longrightarrow Q[1]$ satisfies that $\phi[1] \circ \iota_j[1] = 1_{Y[1]}$, for all $j \in I$. But $\iota_j[1] : Y[1] \longrightarrow Y^I[1]$ is cleary in j-th injection into the product in $\mathcal{H}_\mathbf{t}$. Therefore $Q[1]$ is pure-injective in $\mathcal{H}_\mathbf{t}$. Since, by Theorem 3.11, we know that $\mathcal{H}_\mathbf{t}$ is AB3*, we can apply Lemma 4.16 to conclude that $\mathcal{H}_\mathbf{t}$ is AB5.

(2) $\Longrightarrow$ (1) By Lemma 4.16 again, we know that $\mathcal{H}_\mathbf{t}$ admits an injective cogenerator E which is pure-injective. But since $\mathcal{F}[1]$ is a cogenerating class in $\mathcal{H}_\mathbf{t}$ we necessarily have that $E = Q[1]$, for some $Q \in \mathcal{F}$. Now the dual of the proof of (2) $\Longleftrightarrow$ (3) $\Longrightarrow$ (1) in Theorem 3.9 shows that Q is is a 1-cotilting object of $\mathcal{A}$ such that $\mathcal{F} = \text{Cogen}(Q)$ and $\mathcal{F}$ is a generating class in $\mathcal{A}$. The argument in the proof of (1) $\Longrightarrow$ (2) proves that Q is pure-injective in $\mathcal{A}$ if and only if $Q[1] = E$ is pure-injective in $\mathcal{H}_\mathbf{t}$, something that we know by hypothesis. □

Recall that if $\mathcal{A}$ is an abelian category, then an *abelian exact subcategory* is a full subcategory $\mathcal{B}$ that is abelian and such that the inclusion functor $\mathcal{B} \hookrightarrow \mathcal{A}$ is exact. This is equivalent to say that $\mathcal{B}$ is closed under taking finite coproducts, kernels and cokernels in $\mathcal{A}$. We are now in a position to re-prove the hard part of [**PS2**, Theorem 1.2], that is the proof of [**PS1**, Theorem 4.8], by using recent results in the literature.

THEOREM 4.18. *Let* $\mathcal{G}$ *be a Grothendieck category and let* $\mathbf{t} = (\mathcal{T}, \mathcal{F})$ *be a torsion pair in* $\mathcal{G}$. *If the heart* $\mathcal{H}_\mathbf{t}$ *of the associated Happel-Reiten-Smalo t-structure in* $\mathcal{D}(\mathcal{G})$ *is a Grothendieck category, then* $\mathbf{t}$ *is of finite type, i.e.* $\mathcal{F}$ *is closed under taking direct limits in* $\mathcal{G}$.

PROOF. By Proposition 3.8 we know that $\mathcal{F} = \text{Cogen}(Q) = \text{Copres}(Q)$, for some quasi-cotilting object Q. Consider now the subcategory $\underline{\mathcal{F}} = \underline{\text{Cogen}}(Q)$ of $\mathcal{G}$ (see Definition 3.3). This subcategory is clearly closed under taking subobjects, quotients and coproducts, so that it is an abelian exact subcategory where colimits are calculated as in $\mathcal{G}$. In particular $\underline{\mathcal{F}}$ is an AB5 abelian category. Moreover, if X is a generator of $\mathcal{G}$ one readily gets that $(1 : t)(X)$ is a generator of $\underline{\mathcal{F}}$, so that this subcategory is actually a Grothendieck category.

Note also that the inclusion functor $\iota : \underline{\mathcal{F}} \hookrightarrow \mathcal{G}$ has a right adjoint $\rho : \mathcal{G} \longrightarrow \underline{\mathcal{F}}$. The action on objects is given by $\rho(M) = \text{tr}_\mathcal{F}(M)$, where $\text{tr}_\mathcal{F}(M)$ is the trace of $\mathcal{F}$ in M, i.e. the subobject sum of all subobjects of M which are in $\underline{\mathcal{F}}$. We leave to the reader the easy verification that (ι, ρ) is an adjoint pair. We can then derive these functors. Due to the exactness of ι, the left derived of ι, $\mathbb{L}\iota = \iota$ 'is' ι itself, i.e. it just takes a complex $X^\bullet \in \mathcal{D}(\underline{\mathcal{F}})$ to the same complex viewed as an object of $\mathcal{D}(\mathcal{G})$. The right derived $\mathbb{R}\rho : \mathcal{D}(\mathcal{G}) \longrightarrow \mathcal{D}(\underline{\mathcal{F}})$ is defined in the usual way, namely, it is the composition $\mathcal{D}(\mathcal{G}) \xrightarrow{\mathbf{i}} \mathcal{K}(\mathcal{G}) \xrightarrow{\rho} \mathcal{K}(\underline{\mathcal{F}}) \xrightarrow{q} \mathcal{D}(\underline{\mathcal{F}})$, where $\mathbf{i}$ is the homotopically injective resolution functor, (abusing of notation) ρ is the obvious functor induced at the level of homotopy categories, and q is the canonical localization functor. Then, by classical properties of derived functors and derived categories, we get that $(\iota, \mathbb{R}\rho)$ is an adjoint pair of triangulated functors.

Consider now the restricted torsion pair $\mathbf{t}' = (\mathcal{T} \cap \underline{\mathcal{F}}, \mathcal{F})$ in $\underline{\mathcal{F}}$. By [**PS1**, Proposition 3.2], we know that its heart $\mathcal{H}_{\mathbf{t}'}$ is an AB3 abelian category. Moreover, the triangulated functor $\iota : \mathcal{D}(\underline{\mathcal{F}}) \longrightarrow \mathcal{D}(\mathcal{G})$ clearly satisfies that $\iota(\mathcal{H}_{\mathbf{t}'}) \subseteq \mathcal{H}_{\mathbf{t}}$. We therefore get and induced functor $\iota : \mathcal{H}_{\mathbf{t}'} \longrightarrow \mathcal{H}_{\mathbf{t}}$, which is necessarily exact since short exact sequences in hearts are the triangles in the ambient triangulated category with their three vertices in that heart. We claim that the composition of functors $\rho' : \mathcal{H}_{\mathbf{t}} \hookrightarrow \mathcal{D}(\mathcal{G}) \xrightarrow{\mathbf{R}\rho} \mathcal{D}(\underline{\mathcal{F}}) \xrightarrow{H^0_{\mathbf{t}'}} \mathcal{H}_{\mathbf{t}'}$ is right adjoint of $\iota : \mathcal{H}_{\mathbf{t}'} \longrightarrow \mathcal{H}_{\mathbf{t}}$. Let $X \in \mathcal{H}_{\mathbf{t}'}$ and $M \in \mathcal{H}_{\mathbf{t}}$ be arbitrary objects. Note that we can identify M with a complex $\cdots \to 0 \to E^{-1} \to E^0 \to E^1 \to \cdots$ of injective objects of $\mathcal{G}$ concentrated in degrees ≥ -1. Then $\mathbb{R}\rho(M)$ is the complex $\cdots \to 0 \to \rho(E^{-1}) \to \rho(E^0) \to \rho(E^1) \to \cdots$. As a right adjoint, the functor $\rho : \mathcal{G} \longrightarrow \underline{\mathcal{F}}$ is left exact, and this implies that $H^{-1}(\mathbb{R}\rho(M)) \cong \rho(H^{-1}(M)) \cong H^{-1}(M)$ since $H^{-1}(M) \in \mathcal{F}$. This implies that $\mathbb{R}\rho(M) \in \mathcal{W}_{\mathbf{t}'}$, where $\mathcal{W}_{\mathbf{t}'}$ is the coaisle of the HRS t-structure in $\mathcal{D}(\underline{\mathcal{F}})$ associated to $\mathbf{t}'$. Remember that the restriction of $H^0_{\mathbf{t}'}$ to $\mathcal{W}_{\mathbf{t}'}$ is right adjoint of the inclusion functor $\mathcal{H}_{\mathbf{t}'} \hookrightarrow \mathcal{W}_{\mathbf{t}'}$ (see [**PS1**, Lemma 3.1(2)]). We then have a sequence of isomorphisms, natural on both variables:

$$\begin{aligned} \mathcal{H}_{\mathbf{t}'}(X, \rho'(M)) &= \mathcal{H}_{\mathbf{t}'}(X, (H^0_{\mathbf{t}'} \circ \mathbb{R}\rho)(M)) \cong \mathcal{W}_{\mathbf{t}'}(X, \mathbb{R}\rho(M)) \\ &= \mathcal{D}(\underline{\mathcal{F}})(X, \mathbb{R}\rho(M)) \\ &\cong \mathcal{D}(\mathcal{G})(\iota(X), M) \\ &\cong \mathcal{H}_{\mathbf{t}}(\iota(X), M), \end{aligned}$$

which implies that (ι, ρ') is an adjoint pair.

We then get that the exact functor $\iota : \mathcal{H}_{\mathbf{t}'} \longrightarrow \mathcal{H}_{\mathbf{t}}$ preserves direct limits. Moreover, it reflects zero objects since $\iota(X) = 0$ means that X is acyclic, viewed as a complex of objects of $\mathcal{G}$, which is the same as being acyclic when viewed as a complex of objects in $\underline{\mathcal{F}}$. Consider now a direct system $(0 \to L_i \overset{u_i}{\rightarrowtail} M_i)_{i \in I}$ of monomorphisms in $\mathcal{H}_{\mathbf{t}'}$ and, putting $u := \varinjlim(u_i)$, consider the exact sequence in $\mathcal{H}_{\mathbf{t}'}$

$$0 \to \mathrm{Ker}_{\mathcal{H}_{\mathbf{t}'}}(u) \longrightarrow \varinjlim_{\mathcal{H}_{\mathbf{t}'}} L_i \xrightarrow{u} \varinjlim_{\mathcal{H}_{\mathbf{t}'}} M_i.$$

By exactness and preservation of direct limits by $\iota : \mathcal{H}_{\mathbf{t}'} \longrightarrow \mathcal{H}_{\mathbf{t}}$, we get an exact sequence in $\mathcal{H}_{\mathbf{t}}$

$$0 \to \iota(\mathrm{Ker}(u)) \longrightarrow \varinjlim_{\mathcal{H}_{\mathbf{t}}} \iota(L_i) \xrightarrow{\varinjlim \iota(u_i)} \varinjlim_{\mathcal{H}_{\mathbf{t}}} \iota(M_i).$$

Since the $\iota(u_i)$ are monomorphism and $\mathcal{H}_{\mathbf{t}}$ is AB5 we get that $\varinjlim \iota(u_i)$ is a monomorphism, so that $\iota(\mathrm{Ker}(u)) = 0$. The fact that ι reflects zero objects then implies that u is a monomorphism. Therefore $\mathcal{H}_{\mathbf{t}'}$ is also AB5.

On the other hand, we claim that Q is a 1-cotilting object of $\underline{\mathcal{F}}$ and that $\mathbf{t}'$ is its associated torsion pair in $\underline{\mathcal{F}}$. Indeed, since we know that $\mathcal{F} = \mathrm{Cogen}(Q) = \underline{\mathcal{F}} \cap \mathrm{Ker}(\mathrm{Ext}^1_{\mathcal{G}}(?, Q))$, it is enough to check that $\underline{\mathcal{F}} \cap \mathrm{Ker}(\mathrm{Ext}^1_{\mathcal{G}}(?, Q)) = \mathrm{Ker}(\mathrm{Ext}^1_{\underline{\mathcal{F}}}(?, Q))$.

The inclusion "$\subseteq$" is clear. For the converse, let $M \in \mathrm{Ker}(\mathrm{Ext}^1_{\underline{\mathcal{F}}}(?, Q))$ and fix two exact sequences

$$0 \to F' \xrightarrow{u} F \longrightarrow M \to 0 \qquad \text{and} \qquad 0 \to F' \xrightarrow{v} Q^I \longrightarrow F'' \to 0$$

with F, F', $F'' \in \mathcal{F}$ and I some set (where for the second exact sequence we used that $\mathcal{F} = \mathrm{Copres}(Q)$). Taking the pushout of u and v, we obtain the following commutative diagram with exact rows and columns:

$$\begin{array}{ccccccccc}
 & & 0 & & 0 & & & & \\
 & & \downarrow & & \downarrow & & & & \\
0 & \longrightarrow & F' & \xrightarrow{u} & F & \longrightarrow & M & \longrightarrow & 0 \\
 & & {\scriptstyle v}\downarrow & \text{P.O.} & \downarrow & & \| & & \\
0 & \longrightarrow & Q^I & \longrightarrow & X & \longrightarrow & M & \longrightarrow & 0 \\
 & & \downarrow & & \downarrow & & & & \\
 & & F'' & = & F'' & & & & \\
 & & \downarrow & & \downarrow & & & & \\
 & & 0 & & 0 & & & &
\end{array}$$

We then obtain that $X \in \mathcal{F}$ (as it is an extension of F and $F'' \in \mathcal{F}$), so that Q^I, X, and M all belong in $\underline{\mathcal{F}}$. By the choice of M, the second row of the diagram splits, so that $M \in \mathcal{F}$ since it is isomorphic to a direct summand of X.

It now follows from Theorem 3.11, Corollary 4.17 and the proof of the latter that Q is pure-injective in $\underline{\mathcal{F}}$. But then $\mathcal{F}$ is closed under taking direct limits in the Grothendieck category $\underline{\mathcal{F}}$ (see [**Cou-St**, Theorem 3.9]), which is equivalent to say that is is closed under taking direct limits in $\mathcal{G}$. That is, $\mathbf{t} = (\mathcal{T}, \mathcal{F})$ is a torsion pair of finite type, as desired. □

5. Beyond the HRS case: Some recent results

After Question 1.3 was solved, as said in the introduction, it is Question 1.2 the one that has deserved more attention. So far, the work was mainly concentrated on the case when the t-structure $(\mathcal{U}, \mathcal{W})$ is compactly generated. Then one can even assume that the ambient triangulated category $\mathcal{D}$ is compactly generated. This is due to the fact that $\mathcal{L} := \mathrm{Loc}_{\mathcal{D}}(\mathcal{U})$, the smallest triangulated subcategory of $\mathcal{D}$ containing $\mathcal{U}$ and closed under taking arbitrary coproducts, is compactly generated and the restricted t-structure $\tau' = (\mathcal{U}, \mathcal{U} \cap \mathcal{L})$ has the same heart as τ.

In the compactly generated case, partial answers to the question were obtained by using different techniques, such as functor categories ([**AMV**], [**Bo**]), stable ∞-categories [**Lurie**] and the theory of derivators [**SSV**], see also [**PS3**] and [**Bazz**] for particular cases. These investigations suggest that for all compactly generated t-structures appearing in nature the heart is a Grothendieck category. The concluding result in this vein has been recently obtained independently in [**Bo2**] and [**SS**]:

THEOREM 5.1 ([**Bo2**] and [**SS**]). *Let $\mathcal{D}$ be a triangulated category with coproducts and $\tau = (\mathcal{U}, \mathcal{W})$ be a compactly generated t-structure in $\mathcal{D}$. Then the heart $\mathcal{H}_\tau = \mathcal{U} \cap \mathcal{W}$ is a Grothendieck category.*

In the development via derivators of [**SSV**], the new concept of *homotopically smashing t-structure* (with respect to a strong stable derivator) was introduced. We refer to that reference for the definition and to [**Groth**] for all the terminology concerning derivators. All compactly generated t-structures that appear as the base of a strong stable derivator are homotopically smashing. The latter t-structures are always smashing, but the converse is not true. For instance the HRS t-structure is always smashing, but it is homotopically smashing exactly when the torsion pair is of finite type (see [**SSV**, Proposition 6.1]). The following is a combination of [**SSV**, Theorems B and C], and we refer to that reference for all unexplained terminology appearing in the statement:

THEOREM 5.2. *Let* $\mathbb{D} : Cat^{op} \longrightarrow CAT$ *be a strong stable derivator, with base* $\mathcal{D} := \mathbb{D}(\mathbf{1})$*, and let* $\tau = (\mathcal{U}, \mathcal{W})$ *be a t-structure in* $\mathcal{D}$ *that is homotopically smashing with respect to* $\mathbb{D}$*, then the heart* $\mathcal{H}_\tau$ *is an AB5 abelian category. When, in addition,* $\mathbb{D}$ *is the derivator associated to the homotopy category of a stable combinatorial model structure and* τ *is generated by a set, that heart is a Grothendieck category.*

Soon after [**SSV**] appeared, Rosanna Laking [**L**] proved the following result:

THEOREM 5.3. *Let* $\mathcal{D}$ *be a compactly generated triangulated category that is the base of a strong stable derivator* $\mathbb{D}$*, and let* $\tau = (\mathcal{U}, \mathcal{W})$ *be a left nondegenerate t-structure in* $\mathcal{D}$*. The following assertions are equivalent:*

(1) τ *is homotopically smashing with respect to* $\mathbb{D}$*.*
(2) τ *is smashing and the heart* $\mathcal{H}_\tau$ *of* τ *is a Grothendieck category.*

The last two results suggest the following open question:

QUESTION 5.4. Let $\mathcal{D}$ be a well-generated triangulated category (e.g. a compactly generated one) that is the base of a strong stable derivator $\mathbb{D}$. Are the following two conditions equivalent for a t-structure $\tau = (\mathcal{U}, \mathcal{W})$ in $\mathcal{D}$?

(1) τ is homotopically smashing with respect to $\mathbb{D}$.
(2) τ is smashing and the heart $\mathcal{H}_\tau$ of τ is a Grothendieck category.

In order to get (a partial version of) this question in a derivator-free way, a hint comes from [**L**, Theorem 4.6] (see also [**LV**, Theorem 4.7]), where the author proves that 'homotopically smashing' and 'definable' are synonymous terms for the co-aisle of left nondegenerated t-structures, when the ambient triangulated category is the compactly generated base of a strong stable derivator (see [**L**] for the definition of definable subcategory of a compactly generated triangulated category). This suggests the following question:

QUESTION 5.5. Let $\mathcal{D}$ be a compactly generated triangulated category. Are the following two conditions equivalent for a t-structure $\tau = (\mathcal{U}, \mathcal{W})$ in $\mathcal{D}$?:

(1) $\mathcal{W}$ is definable.
(2) τ is smashing and its heart $\mathcal{H}_\tau$ is a Grothendieck category.

Acknowledgments

Both authors warmly thank Simone Virili for his comments, suggestions and help during the preparation of the paper.

References

[AMV] Lidia Angeleri Hügel, Frederik Marks, and Jorge Vitória, *Torsion pairs in silting theory*, Pacific J. Math. **291** (2017), no. 2, 257–278, DOI 10.2140/pjm.2017.291.257. MR3703444

[Bazz] Silvana Bazzoni, *The t-structure induced by an n-tilting module*, Trans. Amer. Math. Soc. **371** (2019), no. 9, 6309–6340, DOI 10.1090/tran/7488. MR3937326

[BHPST] Silvana Bazzoni, Ivo Herzog, Pavel Příhoda, Jan Šaroch, and Jan Trlifaj, *Pure projective tilting modules*, Doc. Math. **25** (2020), 401–424. MR4106898

[BBD] A. A. Beĭlinson, J. Bernstein, and P. Deligne, *Faisceaux pervers* (French), Analysis and topology on singular spaces, I (Luminy, 1981), Astérisque, vol. 100, Soc. Math. France, Paris, 1982, pp. 5–171. MR751966

[Bo] M. Bondarko, *On torsion pairs, (well-generated) weight structures, adjacent t-structures, and related (co)homological functors.* Preprint available at `arXiv:1611.00754`

[Bo2] M. Bondarko, *On perfectly generated weight structures and adjacent t-structures.* Preprint available at `arXiv:1909.12819`

[BB] Sheila Brenner and M. C. R. Butler, *Generalizations of the Bernstein-Gelfand-Ponomarev reflection functors*, Representation theory, II (Proc. Second Internat. Conf., Carleton Univ., Ottawa, Ont., 1979), Lecture Notes in Math., vol. 832, Springer, Berlin-New York, 1980, pp. 103–169. MR607151

[CN] Carles Casacuberta and Amnon Neeman, *Brown representability does not come for free*, Math. Res. Lett. **16** (2009), no. 1, 1–5, DOI 10.4310/MRL.2009.v16.n1.a1. MR2480555

[CHZ] X. Chen, Z. Han, Y. Zhou, *Derived equivalences via HRS-tilting*, preprint (2018), Preprint available at: `arXiv:1804.05629`

[CDT] Riccardo Colpi, Gabriella D'Este, and Alberto Tonolo, *Quasi-tilting modules and counter equivalences*, J. Algebra **191** (1997), no. 2, 461–494, DOI 10.1006/jabr.1997.6873. MR1448804

[CF] Riccardo Colpi and Kent R. Fuller, *Tilting objects in abelian categories and quasitilted rings*, Trans. Amer. Math. Soc. **359** (2007), no. 2, 741–765, DOI 10.1090/S0002-9947-06-03909-2. MR2255195

[CG] R. Colpi, E. Gregorio, *The Heart of cotilting theory pair is a Grothendieck category*, Preprint.

[CGM] Riccardo Colpi, Enrico Gregorio, and Francesca Mantese, *On the heart of a faithful torsion theory*, J. Algebra **307** (2007), no. 2, 841–863, DOI 10.1016/j.jalgebra.2006.01.020. MR2275375

[CMT] Riccardo Colpi, Francesca Mantese, and Alberto Tonolo, *When the heart of a faithful torsion pair is a module category*, J. Pure Appl. Algebra **215** (2011), no. 12, 2923–2936, DOI 10.1016/j.jpaa.2011.04.013. MR2811575

[Cou-St] Pavel Čoupek and Jan Šťovíček, *Cotilting sheaves on Noetherian schemes*, Math. Z. **296** (2020), no. 1-2, 275–312, DOI 10.1007/s00209-019-02404-8. MR4140742

[Freyd] P. Freyd, *Abelian categories*, Harper and Row (1964).

[GP] Nicolae Popesco and Pierre Gabriel, *Caractérisation des catégories abéliennes avec générateurs et limites inductives exactes* (French), C. R. Acad. Sci. Paris **258** (1964), 4188–4190. MR166241

[Groth] Moritz Groth, *Derivators, pointed derivators and stable derivators*, Algebr. Geom. Topol. **13** (2013), no. 1, 313–374, DOI 10.2140/agt.2013.13.313. MR3031644

[G] Alexander Grothendieck, *Sur quelques points d'algèbre homologique* (French), Tohoku Math. J. (2) **9** (1957), 119–221, DOI 10.2748/tmj/1178244839. MR102537

[HRS] Dieter Happel, Idun Reiten, and Sverre O. Smalø, *Tilting in abelian categories and quasitilted algebras*, Mem. Amer. Math. Soc. **120** (1996), no. 575, viii+ 88, DOI 10.1090/memo/0575. MR1327209

[KN] Bernhard Keller and Pedro Nicolás, *Weight structures and simple dg modules for positive dg algebras*, Int. Math. Res. Not. IMRN **5** (2013), 1028–1078, DOI 10.1093/imrn/rns009. MR3031826

[L] Rosanna Laking, *Purity in compactly generated derivators and t-structures with Grothendieck hearts*, Math. Z. **295** (2020), no. 3-4, 1615–1641, DOI 10.1007/s00209-019-02411-9. MR4125704

[LV] R. Laking, J. Vitoria, *Definability and approximations in triangulated categories.* To appear in Pacific J. Math. Preprint available at `arXiv:1811.00340`

[Lurie] *Higher Algebra* (2017). Available at `http://www.math.harvard.edu/~lurie/papers/HA.pdf`

[MT] Francesca Mantese and Alberto Tonolo, *On the heart associated with a torsion pair*, Topology Appl. **159** (2012), no. 9, 2483–2489, DOI 10.1016/j.topol.2011.08.032. MR2921836

[Ma] F. Mattiello, *On the heart associated to a faithful torsion pair*, Available on `www.algant.eu/documents/theses/mttiello.pdf`, (2011).

[Ma2] F. Mattiello, *New trends in Tilting Theory.* PhD Thesis. University of Padova. 2015. Available at `http://paduaresearch.cab.unipd.it/7499/`

[N] Amnon Neeman, *Triangulated categories*, Annals of Mathematics Studies, vol. 148, Princeton University Press, Princeton, NJ, 2001. MR1812507

[NSZ] Pedro Nicolás, Manuel Saorín, and Alexandra Zvonareva, *Silting theory in triangulated categories with coproducts*, J. Pure Appl. Algebra **223** (2019), no. 6, 2273–2319, DOI 10.1016/j.jpaa.2018.07.016. MR3906550

[PS1] Carlos E. Parra and Manuel Saorín, *Direct limits in the heart of a t-structure: the case of a torsion pair*, J. Pure Appl. Algebra **219** (2015), no. 9, 4117–4143, DOI 10.1016/j.jpaa.2015.02.011. MR3336001

[PS2] Carlos E. Parra and Manuel Saorín, *Addendum to "Direct limits in the heart of a t-structure: the case of a torsion pair" [J. Pure Appl. Algebra 219 (9) (2015) 4117–4143] [MR3336001]*, J. Pure Appl. Algebra **220** (2016), no. 6, 2467–2469, DOI 10.1016/j.jpaa.2015.11.013. MR3448805

[PS3] Carlos E. Parra and Manuel Saorín, *Hearts of t-structures in the derived category of a commutative Noetherian ring*, Trans. Amer. Math. Soc. **369** (2017), no. 11, 7789–7827, DOI 10.1090/tran/6875. MR3695845

[Po] N. Popescu, *Abelian categories with applications to rings and modules*, Academic Press, London-New York, 1973. London Mathematical Society Monographs, No. 3. MR0340375

[Po-St] Leonid Positselski and Jan Šťovíček, *Exactness of direct limits for abelian categories with an injective cogenerator*, J. Pure Appl. Algebra **223** (2019), no. 8, 3330–3340, DOI 10.1016/j.jpaa.2018.11.004. MR3926214

[Po-St2] L. Positselski, J. Šťovíček, *The tilting-cotilting correspondence*, Int. Math. Res. Notices, rnz116 (2019), htpps://doi.org/10.1093/imrn/rnz116.

[SS] M. Saorín, J. Stovicek, *t-structures with Grothendieck hearts via functor categories.* Preprint available at `arXiv:2003.01401`

[SSV] M. Saorín, J. Stovicek, S. Virili, *t-Structures on stable derivators and Grothendieck hearts.* Preprint available at `arXiv:1708.07540`.

[S] Bo Stenström, *Rings of quotients*, Springer-Verlag, New York-Heidelberg, 1975. Die Grundlehren der Mathematischen Wissenschaften, Band 217; An introduction to methods of ring theory. MR0389953

Instituto de Ciencias Físicas y Matemáticas, Edificio Emilio Pugin, Campus Isla Teja, Universidad Austral de Chile, 5090000 Valdivia, Chile
Email address: `carlos.parra@uach.cl`

Departamento de Matemáticas, Universidad de Murcia, Aptdo. 4021, 30100 Espinardo, Murcia, Spain
Email address: `msaorinc@um.es`

PUBLISHED TITLES IN THIS SERIES

769 **Alex Martsinkovsky, Kiyoshi Igusa, and Gordana Todorov, Editors,** Representations of Algebras, Geometry and Physics, 2021

765 **Michael Aschbacher,** Quaternion Fusion Packets, 2021

764 **Gabriel Cunningham, Mark Mixer, and Egon Schulte, Editors,** Polytopes and Discrete Geometry, 2021

763 **Tyler J. Jarvis and Nathan Priddis, Editors,** Singularities, Mirror Symmetry, and the Gauged Linear Sigma Model, 2021

762 **Atsushi Ichino and Kartik Prasanna,** Periods of Quaternionic Shimura Varieties. I., 2021

761 **Ibrahim Assem, Christof Geiß, and Sonia Trepode, Editors,** Advances in Representation Theory of Algebras, 2021

760 **Olivier Collin, Stefan Friedl, Cameron Gordon, Stephan Tillmann, and Liam Watson, Editors,** Characters in Low-Dimensional Topology, 2020

759 **Omayra Ortega, Emille Davie Lawrence, and Edray Herber Goins, Editors,** The Golden Anniversary Celebration of the National Association of Mathematicians, 2020

758 **Jan Šťovíček and Jan Trlifaj, Editors,** Representation Theory and Beyond, 2020

757 **Kaïs Ammari and Stéphane Gerbi, Editors,** Identification and Control: Some New Challenges, 2020

756 **Joeri Van der Veken, Alfonso Carriazo, Ivko Dimitrić, Yun Myung Oh, Bogdan D. Suceavă, and Luc Vrancken, Editors,** Geometry of Submanifolds, 2020

755 **Marion Scheepers and Ondřej Zindulka, Editors,** Centenary of the Borel Conjecture, 2020

754 **Susanne C. Brenner, Igor Shparlinski, Chi-Wang Shu, and Daniel B. Szyld, Editors,** 75 Years of Mathematics of Computation, 2020

753 **Matthew Krauel, Michael Tuite, and Gaywalee Yamskulna, Editors,** Vertex Operator Algebras, Number Theory and Related Topics, 2020

752 **Samuel Coskey and Grigor Sargsyan, Editors,** Trends in Set Theory, 2020

751 **Ashish K. Srivastava, André Leroy, Ivo Herzog, and Pedro A. Guil Asensio, Editors,** Categorical, Homological and Combinatorial Methods in Algebra, 2020

750 **A. Bourhim, J. Mashreghi, L. Oubbi, and Z. Abdelali, Editors,** Linear and Multilinear Algebra and Function Spaces, 2020

749 **Guillermo Cortiñas and Charles A. Weibel, Editors,** K-theory in Algebra, Analysis and Topology, 2020

748 **Donatella Danielli and Irina Mitrea, Editors,** Advances in Harmonic Analysis and Partial Differential Equations, 2020

747 **Paul Bruillard, Carlos Ortiz Marrero, and Julia Plavnik, Editors,** Topological Phases of Matter and Quantum Computation, 2020

746 **Erica Flapan and Helen Wong, Editors,** Topology and Geometry of Biopolymers, 2020

745 **Federico Binda, Marc Levine, Manh Toan Nguyen, and Oliver Röndigs, Editors,** Motivic Homotopy Theory and Refined Enumerative Geometry, 2020

744 **Pieter Moree, Anke Pohl, L'ubomír Snoha, and Tom Ward, Editors,** Dynamics: Topology and Numbers, 2020

743 **H. Garth Dales, Dmitry Khavinson, and Javad Mashreghi, Editors,** Complex Analysis and Spectral Theory, 2020

742 **Francisco-Jesús Castro-Jiménez, David Bradley Massey, Bernard Teissier, and Meral Tosun, Editors,** A Panorama of Singularities, 2020

741 **Houssam Abdul-Rahman, Robert Sims, and Amanda Young, Editors,** Analytic Trends in Mathematical Physics, 2020

740 **Alina Bucur and David Zureick-Brown, Editors,** Analytic Methods in Arithmetic Geometry, 2019

739 **Yaiza Canzani, Linan Chen, and Dmitry Jakobson, Editors,** Probabilistic Methods in Geometry, Topology and Spectral Theory, 2019